Cybersecurity for Medical Devices

Cybersecurity for Medical Devices

Rebecca M. Basta
Waqar Anwar

WILEY

Published by John Wiley & Sons, Inc., Hoboken, New Jersey.

Library of Congress Cataloging-in-Publication Data applied for:

Hardback ISBN: 9781394433681

Cover Design: Wiley
Cover Image: vectorfusionart/Shutterstock

Set in 9.5/12.5pt STIXTwoText by Straive, Chennai, India

To my father, Dr. Alfred Basta

Dad, every page of this book carries the echo of your voice encouraging me to reach higher, dream bigger, and never settle for less than my best. Your unwavering belief in me, even when I doubted myself, has been the foundation upon which I've built not just this work, but my entire career. You didn't just teach me to think critically; you taught me to live authentically. The late-night conversations, the gentle corrections, the proud smiles—they're all woven into these words.

To my beloved mother, whose physical presence I miss with every breath

Mom, though you're no longer here to hold this book in your hands, I feel you in every sentence I've written. Your voice still guides me through difficult decisions, your laughter still brightens my darkest moments, and your unwavering love continues to be the compass that points me toward what matters most. This book exists because you taught me that knowledge without compassion is hollow, and that success without integrity is meaningless.

To Stavros Basta, my constant companion in this journey

Stavros, you've been my research partner, my sounding board, my technical support, and most importantly, my reminder that no great work is ever accomplished in isolation. Your patience during my countless revisions, your insights that sharpened my thinking, and your quiet confidence in this project sustained me through every challenge. You didn't just help me write this book; you helped me become the person capable of writing it.

This book is more than words on paper; it's a love letter to the family that made me who I am.

Contents

About the Author

Rebecca M. Basta is a distinguished bioinformatics and cybersecurity education specialist with deep expertise at the critical intersection of healthcare technology, data science, and cybersecurity. Holding a Master of Science in Bioinformatics and a Bachelor of Science in Biochemistry with a Minor in Biotechnology, Basta combines advanced academic credentials with hands-on laboratory and educational expertise.

With an impressive portfolio of 30+ professional certifications spanning healthcare data analytics, AI governance, cybersecurity, risk management, and emerging security technologies, Basta demonstrates exceptional breadth across intersecting disciplines. Her credentials include Certified Health Data Analyst (CHDA), Certified Information Privacy Professional/United States (CIPP/US), Artificial Intelligence Governance Professional (AIGP), ISO 27001 and ISO 42001 Certified Lead Auditor, Certified Analytics Professional Expert (CAP), Certified Lean Six Sigma Master Black Belt, Certified Red Team Operations Management (CRTOM), and Certified Artificial Intelligence Security & Risk (CAISR). This diverse certification landscape reflects her commitment to understanding security from multiple angles: defensive, offensive, and governance perspectives.

Basta's career demonstrates a strategic progression from curriculum development to research and academia. As a Grant Writer, she developed age-appropriate cybersecurity awareness programs for K-12 students and secured funding for initiatives targeting minors and families. Her experience bridging technical concepts with diverse audiences has established her as a skilled communicator of complex scientific and cybersecurity topics.

Basta's research and publication record is exceptionally prolific, focusing on the convergence of healthcare security and advanced data analytics. She is the author of "**Introduction to Data Science in Medical Research**," "**Cybersecurity for Medical Devices**," "**When Frameworks Fail: Why Traditional Cybersecurity Standards Leave Industrial Control Systems Vulnerable**," "**Cryptography: A Math-Free Introduction**," "**Legal, Ethical, and Social Issues in Information Security**," "**Governance, Risk and Compliance**," and "**Cybersecurity Leadership**." Her peer-reviewed research includes groundbreaking work on "Adversarial Predictive Analytics for Resilient Supply Chains: An Integrated Framework for AI-Driven Threat Detection and Data Integrity Assurance" and "Poisoning-Resistant Anomaly Detection for Industrial Control Systems: A Byzantine-Robust Data Analytics Framework," demonstrating her ability to make complex topics accessible to nontechnical audiences.

Her technical expertise spans Python, C++, data analysis with R, AI/machine learning, log analysis, supply chain security, threat intelligence, and healthcare-specific security challenges. Basta's interdisciplinary background uniquely positions her to bridge the critical gap between cybersecurity professionals and healthcare practitioners; two communities that often struggle to communicate effectively about risk and compliance requirements. Active in professional organizations, she is a member of the American Health Information Management Association (AHIMA), where her expertise contributes to advancing healthcare data security practices. Her active involvement in governance, AI strategy, and compliance frameworks (ISO/IEC standards, Cybersecurity Maturity Model Certification [CMMC] limitations, healthcare regulations) makes her a trusted voice in helping organizations understand both the technical and organizational dimensions of healthcare cybersecurity and data protection.

Waqar Anwar is a Cybersecurity Curriculum Specialist with over 10 years of experience in the field. He also develops and delivers training to faculty and staff on cybersecurity topics and conducts research on cybersecurity topics. Mr. Anwar is a frequent speaker at industry conferences. He is also a member of several cybersecurity organizations including SysAdmin, Audit, Network and Security (SANS), CYBRARY, and Information Systems Security Association International (ISSA).

Foreword

Throughout my decades-long career in cybersecurity, from conducting penetration tests on critical infrastructure to authoring textbooks that have shaped the education of thousands of security professionals, I have witnessed the profound transformation of our digital landscape. I have held numerous advanced certifications, from Offensive Security Certified Professional (OSCP) to Certified Chief Information Security Officer (CCISO), from Certified Red Team Operator (CRTO) to Certified HIPAA Privacy Security Expert (CHPSE), each representing a different facet of the complex security challenges we face. Yet among all the domains I have studied, assessed, and defended, none carries the weight of consequence quite like medical device cybersecurity.

The stakes in healthcare security are fundamentally different from those in any other sector. When we assess the security posture of a financial institution or a retail network, the potential losses are measured in dollars, customer trust, and regulatory penalties. These are serious consequences, to be sure. Nevertheless, when we examine the security of medical devices such as pacemakers, insulin pumps, ventilators, and infusion systems, the potential losses are measured in human lives. A vulnerability in a hospital's networked devices is not merely a technical flaw; it is a direct pathway to patient harm. This reality demands a different level of rigor, urgency, and ethical commitment.

As someone who has conducted penetration testing engagements across diverse industries and documented methodologies in works such as "Penetration Testing from Contract to Report" and "Computer Security and Penetration Testing," I have seen firsthand how organizations approach security, or fail to. In too many cases, security is treated as an afterthought, a checkbox exercise performed to satisfy compliance requirements rather than a fundamental design principle. This approach is dangerous in any context, but in healthcare, it is unconscionable.

The healthcare sector faces unique challenges that complicate achieving robust security. Legacy medical devices, some deployed decades ago when cybersecurity was barely considered, continue to operate in modern networked environments. These devices were designed in an era when isolation provided protection; today, that same isolation has been replaced by connectivity that not only enables remote monitoring, data sharing, and coordinated care, but also creates expansive attack surfaces. Hospital networks must balance the clinical imperative of accessibility and interoperability with the security imperative of controlled access and segmentation. Resource constraints limit many healthcare organizations' ability to invest in security infrastructure and expertise at levels required to keep pace with evolving threats.

Throughout my work as a Certified Lead CMMC Assessor (LCCA) and my experience with ISO 27001 auditing, I have evaluated countless security programs. The most effective programs share common characteristics: they embed security throughout the entire life cycle, from initial design through deployment and maintenance; they treat security as everyone's responsibility, not just the

IT department's; they implement defense-in-depth strategies that assume breach rather than relying on perimeter protection alone; they continuously monitor, assess, and improve their security posture based on emerging threats and incident lessons learned.

These principles are not theoretical abstractions. They represent hard-won lessons from real-world breaches, from penetration tests that revealed critical vulnerabilities, and from incident response engagements where we worked to contain damage and restore operations. In my role, conducting security assessments and red team operations, I have exploited vulnerabilities in systems that their operators believed were secure. I have demonstrated how an attacker with determination and skill can bypass controls, escalate privileges, move laterally through networks, and ultimately compromise critical assets. The techniques I have employed in authorized testing engagements are the same techniques that malicious actors use in actual attacks.

This brings me to *Cybersecurity for Medical Devices* and why this work represents such an essential contribution to the field. What distinguishes this book is its comprehensive, life cycle-oriented approach to medical device security. Rather than focusing narrowly on individual technical controls or specific device categories, it addresses the full spectrum of challenges facing healthcare organizations and medical device manufacturers.

For security practitioners and penetration testers like me, the book provides detailed technical guidance on assessing and securing medical devices, covering embedded systems security, real-time operating systems, wireless protocols, and secure update mechanisms. The treatment of threat modeling, vulnerability assessment, and penetration testing methodologies is grounded in practical reality rather than academic theory.

For healthcare IT professionals managing complex device ecosystems, the book offers actionable strategies for network segmentation, endpoint security, access control, and incident response specifically tailored to the clinical environment. The emphasis on balancing security with clinical workflow requirements reflects a deep understanding of the operational realities facing healthcare organizations.

For regulatory and compliance professionals navigating the evolving landscape of the U.S. Food and Drug Administration (FDA) guidance, EU Medical Device Regulation/in vitro Diagnostic Regulation (MDR/IVDR) requirements, and international standards, the book provides comprehensive coverage of regulatory expectations and compliance frameworks. As someone who has worked extensively with compliance frameworks, from the Health Insurance Portability and Accountability Act (HIPAA) to Cybersecurity Maturity Model Certification (CMMC) to ISO 27001, I appreciate the clarity with which regulatory obligations are explained and contextualized.

For medical device manufacturers and engineers, the book provides essential guidance on security-by-design principles, secure development life cycles, supply chain security, and integrating security testing throughout the development process. The connection between security decisions and patient safety outcomes is made explicit and unavoidable.

Perhaps most importantly, the book establishes the ethical foundation that must guide all work in medical device security. Through my various certifications in ethical hacking (CEH), incident handling (ECIH), and security analysis (ECSA), I have internalized the ethical obligations inherent in security work. Those obligations are magnified exponentially when the systems we are protecting directly sustain human life. Every design decision, every resource allocation, and every security control implementation must be evaluated through the lens of patient safety.

The inclusion of real-world case studies provides invaluable context. The WannaCry outbreak, insulin pump vulnerabilities, and pacemaker exploits have shaped regulatory policy and industry practice. By examining both the technical root causes and organizational responses, readers gain insight into what can go wrong and how to respond effectively when it does. As someone who has conducted forensic investigations following security incidents (leveraging my Computer Hacking Forensic Investigator [CHFI] certification and experience), I can attest to the value of learning from documented failures.

The book's treatment of emerging technologies, artificial intelligence and machine learning integration, advanced wireless protocols, and wearable and mobile health platforms positions readers to anticipate future challenges rather than merely react to current ones. In my work with cloud security (as an AWS Certified Solutions Architect Professional and Google Cloud Certified Professional Cloud Architect), I have seen how rapidly technology evolves and how security must evolve in parallel. Medical devices are becoming increasingly autonomous, interconnected, and data-driven. The security challenges will only intensify, making foundational knowledge and strategic frameworks essential.

Throughout my career, whether developing penetration testing methodologies that became the basis for copyrighted frameworks, implementing security operations centers, or writing technical books that have educated the next generation of security professionals, I have been driven by a commitment to practical, actionable guidance. This book exemplifies that same commitment. It does not merely describe problems; it provides solutions. It does not simply catalog threats; it offers concrete mitigation strategies. It does not stop at technical controls; it addresses the organizational, regulatory, and human factors that ultimately determine security outcomes.

The multidisciplinary nature of medical device security demands collaboration between security professionals, biomedical engineers, clinical staff, regulatory specialists, and organizational leadership. Too often, these groups operate in silos, speaking different languages and pursuing different priorities. This book serves as a common reference point, a shared foundation of knowledge that enables productive collaboration. Having worked in environments ranging from Department of Defense contractors (where my CMMC expertise was critical) to healthcare organizations (where I leveraged my CHPSE certification), I understand the importance of bridging these professional divides.

To the readers preparing to engage with this material, I offer the following perspective based on decades of security work: cybersecurity is not a destination but a continuous journey. There is no final state of "secure," only varying degrees of risk managed through layered defenses, continuous monitoring, and adaptive response. The threat landscape evolves constantly, driven by technological change, geopolitical dynamics, and economic incentives. Our defenses must evolve in parallel.

Approach this book with the seriousness it demands. Whether you are designing medical devices, managing hospital networks, writing regulatory policy, conducting security assessments, or studying to enter this critical field, the knowledge contained in these pages will serve you well. But knowledge alone is insufficient. It must be paired with commitment: commitment to patient safety, to continuous improvement, to collaboration across disciplines, and to the ethical imperative that must guide all work in healthcare technology.

The healthcare system is increasingly dependent on connected medical devices. That dependence will only grow as we embrace wearable sensors, implantable monitors, AI-driven diagnostics, and telemedicine platforms. The tremendous promise of these technologies can only be realized if we build security into their foundation. We cannot sacrifice security for innovation, nor can we allow security concerns to stifle beneficial technological advancement. We must do both: innovate and secure, because patients deserve nothing less.

I have spent my career breaking into systems, identifying vulnerabilities, and helping organizations strengthen their defenses. I have written books, developed methodologies, and trained professionals across the security spectrum. However, I am humbled by the complexity and criticality of medical device security. The challenges are formidable, but they are not insurmountable.

With the right knowledge, tools, and commitment, we can build medical device ecosystems that are both innovative and secure.

Cybersecurity for Medical Devices provides the roadmap for achieving that balance. It is comprehensive without being overwhelming, technical without being inaccessible, and realistic without being defeatist. It challenges us to do better while providing the tools to make that improvement possible.

I commend this work to you with the conviction that it represents an essential contribution to a field where technical excellence and ethical commitment must be inseparable. May the knowledge you gain inspire not only learning but action; action that will ultimately save lives and preserve the trust upon which our healthcare system depends.

Dr. Alfred Basta

Preface

In the spring of 2017, I sat in a hospital room watching monitors flicker with vital signs, their soft beeps marking each heartbeat of a loved one. As I observed the intricate web of devices—infusion pumps delivering medication, cardiac monitors tracking every rhythm, ventilators supporting each breath—I was struck by a profound realization: these machines, miraculous in their life-sustaining capabilities, were also potentially vulnerable gateways to harm. At that moment, my dual identity as a concerned family member and a cybersecurity professional crystallized into an urgent question: How secure are the devices we place our absolute trust in?

That question became the seed from which this book grew.

The journey of writing *Cybersecurity for Medical Devices* has been one of discovery, frustration, inspiration, and, ultimately, purpose. What began as personal curiosity evolved into years of research, countless interviews with medical device engineers, clinical staff, cybersecurity experts, and regulatory specialists. Each conversation revealed the same troubling pattern: despite the exponential growth in connected medical technology, cybersecurity remained an afterthought, a box to be checked rather than a foundational principle embedded from conception through deployment.

I witnessed the dangerous myth of "security by obscurity" perpetuated across boardrooms and engineering labs alike. I encountered brilliant clinicians who understood human anatomy with breathtaking precision yet remained wholly unaware of the digital vulnerabilities lurking in their examination rooms. I met dedicated engineers constrained by budget limitations, accelerated timelines, and a regulatory landscape struggling to keep pace with innovation. Most troubling of all, I observed a pervasive complacency born from the false belief that "no one would ever attack a hospital" or "our devices are too specialized for hackers to understand."

These observations were shattered by reality. The WannaCry ransomware attack of 2017 crippled hospitals across continents, forcing the cancellation of surgeries and the diversion of critical patients. Insulin pumps were demonstrated to be remotely hackable. Pacemakers required urgent firmware updates to close vulnerabilities that could prove fatal. Each incident underscored an uncomfortable truth: in our rush to connect and innovate, we had created life-critical systems with catastrophic security flaws.

This book is my response to that crisis.

Cybersecurity for Medical Devices is not merely a technical manual, though it provides comprehensive technical guidance. It is not simply a regulatory roadmap, though it navigates the complex compliance landscape. Rather, it is a call to action: a demand that every stakeholder in the healthcare technology ecosystem recognize cybersecurity as inseparable from patient safety, as fundamental as sterile technique or proper dosing protocols.

My purpose in writing this book is threefold:

First, to demystify medical device cybersecurity for those who feel overwhelmed by its complexity. Whether you are a biomedical engineer designing next-generation implantables, a hospital IT manager defending sprawling networks, a regulatory specialist interpreting evolving standards, or a clinical engineer maintaining deployed devices, you will find practical, actionable guidance tailored to your role.

Second, to challenge the dangerous misconceptions that perpetuate vulnerability. Security by obscurity is not security. Compliance is not equivalent to safety. Innovation without security is recklessness, not progress. Throughout these chapters, I confront these myths directly, replacing them with evidence-based strategies grounded in real-world case studies and validated frameworks.

Third, to establish an ethical imperative that transcends technical requirements. The devices we discuss are not abstract systems; they are the technological extensions of human care. A compromised insulin pump doesn't just mean a data breach; it means a child is at risk. A ransomware-frozen ventilator doesn't merely signal network failure; it signals interrupted breathing, potential brain damage, and possible death. We must approach medical device security with the gravity it demands.

The structure of this book reflects a deliberate progression. Early chapters establish the critical foundations: why medical device cybersecurity cannot be ignored, the unique landscape of healthcare threats, and the architectural building blocks of secure devices. Middle chapters dive deep into technical implementation: secure software development, network security, data protection, and risk assessment methodologies. Later chapters expand into specialized domains: artificial intelligence integration, wireless device security, incident response, and future directions. Each chapter builds on the last, creating a comprehensive framework applicable throughout the device life cycle.

Throughout the writing process, I have been guided by several core principles:

Accessibility without oversimplification: Complex technical concepts are explained clearly without sacrificing accuracy or depth.

Multidisciplinary integration: Medical device security requires collaboration between engineering, clinical, regulatory, and security domains. This book speaks to all these audiences, fostering the cross-functional understanding essential for success.

Evidence-based recommendations: Every strategy, framework, and best practice is grounded in published research, regulatory guidance, industry standards, or documented incident analysis.

Ethical grounding: Technical competence must be matched by moral responsibility. Patient safety is not negotiable.

Writing this book has challenged me in ways I did not anticipate. Translating between the languages of cybersecurity, biomedical engineering, clinical practice, and regulatory compliance required constant learning and humility. Balancing technical rigor with readable prose demanded countless revisions. Confronting the enormity of the problem—the thousands of vulnerable devices already deployed, the millions of patients potentially at risk—was, at times, overwhelming.

Yet, I have also been profoundly inspired. I have encountered manufacturers who, upon discovering vulnerabilities, immediately prioritized patient safety over profit. I have worked alongside clinical engineers who champion security despite resource constraints. I have witnessed regulatory bodies evolving rapidly to establish higher standards. These dedicated professionals prove that change is possible when we unite behind a shared mission.

To the readers who will engage with this material, I ask not for passive consumption but for active commitment. Read these chapters with the understanding that the knowledge you gain carries responsibility. Challenge your organizations to prioritize security. Advocate for resources, training, and cultural transformation. Share this knowledge widely, with colleagues, leadership, students, and the broader healthcare community. Most importantly, never forget that behind every device specification, every line of code, and every network architecture diagram are human lives depending on your diligence.

The future of healthcare is undeniably connected. Wearable devices, implantable sensors, AI-driven diagnostics, telemedicine platforms—these innovations promise to extend care beyond hospital walls, to personalize treatment with unprecedented precision, and to save lives in ways we are only beginning to imagine. Nevertheless, this future can only be realized safely if we build security into its foundation.

This book represents my contribution to that essential work. It is offered with deep respect for the complexity of the challenges ahead, profound appreciation for those already laboring in this field, and unwavering optimism that together, we can secure the devices upon which our lives depend.

The monitors in that hospital room years ago continued their steady rhythm, and my loved one recovered. But the question that arose in that moment has never left me. Now, through these pages, I invite you to join me in answering it, not with complacency or fear, but with knowledge, skill, and resolute determination to protect every patient, every time.

Let us begin.

List of Acronyms

AES	Advanced Encryption Standard
AI	Artificial Intelligence
API	Application Programming Interface
APT	Advanced Persistent Threat
BLE	Bluetooth Low Energy
BYOD	Bring Your Own Device
CASP+	CompTIA Advanced Security Practitioner
CCISO	Certified Chief Information Security Officer
CCP	Certified CMMC Professional
CCSE	Certified Cloud Security Engineer
CEH	Certified Ethical Hacker
CHFI	Computer Hacking Forensic Investigator
CHPSE	Certified HIPAA Privacy Security Expert
CISA	Certified Information Systems Auditor
CISM	Certified Information Security Manager
CMMC	Cybersecurity Maturity Model Certification
CPU	Central Processing Unit
CRISC	Certified in Risk and Information Systems Control
CRM	Customer Relationship Management
DDoS	Distributed Denial of Service
DHS	Department of Homeland Security
DNS	Domain Name System
DoS	Denial of Service
ECG	Electrocardiogram
ECIH	Certified Incident Handler
ECSA	Certified Security Analyst
EDR	Endpoint Detection and Response
EHR	Electronic Health Record
EMR	Electronic Medical Record
EU	European Union
FDA	Food and Drug Administration
FIPS	Federal Information Processing Standards
FOTA	Firmware Over-the-Air
GDPR	General Data Protection Regulation
GSLC	GIAC Security Leadership Certification

HHS	Department of Health and Human Services
HIPAA	Health Insurance Portability and Accountability Act
HTTPS	Hypertext Transfer Protocol Secure
ICS	Industrial Control Systems
ICU	Intensive Care Unit
IDS	Intrusion Detection System
IEC	International Electrotechnical Commission
IoMT	Internet of Medical Things
IoT	Internet of Things
IP	Internet Protocol
IPS	Intrusion Prevention System
ISO	International Organization for Standardization
IT	Information Technology
ITIL	Information Technology Infrastructure Library
IVDR	in Vitro Diagnostic Regulation
LCCA	Lead CMMC Assessor
MAC	Media Access Control
MDR	Medical Device Regulation
MFA	Multi-Factor Authentication
ML	Machine Learning
MRI	Magnetic Resonance Imaging
NFC	Near Field Communication
NIST	National Institute of Standards and Technology
OS	Operating System
OSCP	Offensive Security Certified Professional
OT	Operational Technology
PHI	Protected Health Information
PII	Personally Identifiable Information
PKI	Public Key Infrastructure
PMP	Project Management Professional
RAM	Random Access Memory
RBAC	Role-Based Access Control
RF	Radio Frequency
RFID	Radio-Frequency Identification
ROM	Read-Only Memory
RSA	Rivest-Shamir-Adleman (encryption algorithm)
RTOS	Real-Time Operating System
SBOM	Software Bill of Materials
SCADA	Supervisory Control and Data Acquisition
SDLC	Secure Development Lifecycle / Software Development Lifecycle
SHA	Secure Hash Algorithm
SIEM	Security Information and Event Management
SOC	Security Operations Center
SQL	Structured Query Language
SSH	Secure Shell
SSL	Secure Sockets Layer
TCP	Transmission Control Protocol

TLS	Transport Layer Security
TPM	Trusted Platform Module
UDP	User Datagram Protocol
UI	User Interface
URL	Uniform Resource Locator
US	United States
USB	Universal Serial Bus
VLAN	Virtual Local Area Network
VPN	Virtual Private Network
WAN	Wide Area Network
Wi-Fi	Wireless Fidelity (IEEE 802.11)
WPA	Wi-Fi Protected Access
XML	Extensible Markup Language
XSS	Cross-Site Scripting
ZTA	Zero Trust Architecture

Cybersecurity for Medical Devices

Cybersecurity for Medical Devices addresses one of the most critical challenges in modern healthcare technology. As medical devices become increasingly connected and digitized (from pacemakers and insulin pumps to hospital networks and diagnostic equipment), they present unique cybersecurity vulnerabilities that could directly impact patient safety and privacy. This technical book bridges the gap between cybersecurity expertise and medical device engineering, providing essential knowledge for professionals working at the intersection of healthcare and information security.

Objectives

- Provide a comprehensive understanding of the cybersecurity risks specific to medical devices, detailing how vulnerabilities in connected health technologies can affect patient safety and privacy.
- Equip healthcare professionals, engineers, and cybersecurity experts with practical strategies and best practices for securing medical devices through design, implementation, and ongoing management.
- Explore regulatory requirements and standards governing medical device cybersecurity, helping readers navigate compliance and foster multidisciplinary collaboration between security and medical device sectors.

This book delivers in-depth technical guidance on protecting connected medical devices from cybersecurity threats, focusing on safeguarding patient safety and data privacy.

Target Audience

- Healthcare cybersecurity professionals: Individuals responsible for securing hospital networks, medical devices, and patient data who need specialized knowledge of healthcare-specific risks and controls.
- Medical device engineers and designers: Professionals involved in the development and manufacturing of medical devices, looking to integrate cybersecurity principles into their processes.
- Regulatory and compliance officers: Specialists overseeing adherence to healthcare cybersecurity regulations, standards, and quality assurance in medical technology.

- Healthcare IT managers and clinical engineers: Those managing healthcare infrastructure who require awareness of medical device vulnerabilities and risk management strategies.
- Researchers and advanced students in cybersecurity and biomedical engineering seeking technical insights into the intersection of these fields.

Chapter Summaries

Chapter 1: Why Medical Device Cybersecurity Cannot Be Ignored

This chapter highlights the critical importance of medical device cybersecurity by exploring the human risks of device exploits, the evolution of medical technology into interconnected systems, the variety of attackers targeting these devices, the fallacy of security through obscurity, and the moral and professional imperatives demanding serious attention and accountability.

Chapter 2: Understanding the Medical Device Cyber Landscape

This chapter provides an essential foundation for comprehending the complex medical device ecosystem, the common threat models targeting connected medical devices, the global regulatory environment, challenges posed by legacy devices, and the critical distinctions and interactions between data privacy and data security in healthcare. Together, these elements reveal the multifaceted landscape of medical device cybersecurity and frame the challenges and considerations for securing healthcare technologies.

Chapter 3: Architectures and Technologies: The Building Blocks of Secure Devices

This chapter provides an in-depth technical primer on the key architectures and technologies that form the foundation of secure medical devices. It covers hardware root of trust mechanisms, software security fundamentals, networking protocols in healthcare, the role of embedded operating systems with real-time constraints, and cloud-based remote access methods. Together, these building blocks are essential to ensure the confidentiality, integrity, and availability of life-critical medical device systems.

Chapter 4: Cyber Threats Targeting Medical Devices

This chapter exposes the risks posed by various cyber threats targeting medical devices, highlighting how malware, unauthorized access, denial of service attacks, supply-chain compromises, and wireless interface vulnerabilities can undermine device integrity, availability, and patient safety. It underscores the importance of understanding these threats and implementing effective protections in clinical environments.

Chapter 5: Regulatory and Legal Landscape: Raising the Bar

This chapter explores the tightening regulatory and legal environment surrounding medical device cybersecurity. It traces the evolution of regulatory demands, examines key Food and Drug Administration (FDA) guidance and international compliance challenges, highlights the intersection of privacy laws with device security, and discusses the serious legal consequences of negligence in cybersecurity practices. Together, these elements illustrate how the policy landscape is elevating standards and accountability in protecting medical technology.

Chapter 6: Risk Assessment: The Critical Starting Point

This chapter emphasizes the foundational importance of risk assessment in medical device cybersecurity. It guides readers through identifying and disclosing device vulnerabilities, applying threat modeling tailored to healthcare technology, and quantifying risks using established frameworks. The chapter highlights the relevance of cyber risks to clinical workflows and patient outcomes while advocating a continuous life cycle approach to monitoring and managing risks for improved device safety.

Chapter 7: Designing Security into the Medical Device Life Cycle

This chapter emphasizes an integrated approach to embedding security throughout the medical device life cycle, covering foundational design principles, practical development processes, human factors, secure supply-chain management, and device hardening techniques. It guides readers on how to incorporate robust security measures from initial design to deployment, ensuring medical devices resist evolving threats and operate safely in clinical environments.

Chapter 8: Secure Software Development for Embedded Medical Devices

This chapter provides hands-on guidance on developing secure software for embedded medical devices, focusing on secure coding standards, memory safety, update mechanisms, secure APIs, and advanced testing techniques to ensure device reliability, safety, and regulatory compliance in clinical environments.

Chapter 9: Network Security in Healthcare Environments

This chapter explains the critical aspects of securing healthcare networks, focusing on design and operational strategies to protect sensitive devices and data. It covers network segmentation to limit breach impact, comprehensive endpoint security, wireless network protections, rapid incident isolation protocols, and secure methods for remote access and telemedicine. Together, these elements form a holistic approach to maintaining network integrity and patient safety in complex healthcare environments.

Chapter 10: Data Security and Patient Privacy Enhancement

This chapter explores comprehensive strategies and technologies to enhance data security and protect patient privacy in healthcare. It covers safeguarding stored health data, securing data transmission, advanced anonymization techniques, digital consent management, and effective detection and response to privacy breaches. Together, these topics provide a framework to maintain confidentiality, ensure compliance, and foster patient trust in the digital era.

Chapter 11: Clinical Workflow Security: Bridging IT and Patient Care

This chapter explores the essential integration of clinical workflow processes with cybersecurity measures, emphasizing how routine hospital procedures can both introduce vulnerabilities and serve as opportunities to enhance device and data security. It covers the implementation of role-based access controls tailored to real healthcare environments, best practices for device usage policies, the importance of tailored cybersecurity training for healthcare staff, and strategies for managing mobile and Bring Your Own Device (BYOD) use securely within clinical settings.

Chapter 12: The Human Firewall: Addressing Insider Threats

This chapter explores the unique risks insider threats pose to healthcare settings, highlighting their complexity and potential damage. It covers profiling insider threat types, detection methods, cultivating a security-conscious culture, managing access controls and offboarding processes, and effectively responding to insider incidents to strengthen organizational resilience.

Chapter 13: Incident Detection, Response, and Recovery

This chapter covers the critical phases of managing cybersecurity incidents involving medical devices, including detection using advanced technologies, immediate response actions after a breach, forensic investigation methodologies, safe recovery and restoration protocols, and postincident reviews to improve security measures. It emphasizes maintaining patient safety, regulatory compliance, and continuous learning to strengthen defenses against future incidents.

Chapter 14: Third-Party Vendor and Supply-Chain Security

This chapter explores the critical role of third-party vendors and supply chains in maintaining medical device security. It covers the systematic evaluation of vendor security practices, secure management of updates and patches, safeguarding logistics and device transit, fostering collaborative supplier relationships, and responding effectively to third-party security breaches. Together, these strategies help ensure the integrity and safety of medical devices throughout their life cycle.

Chapter 15: Penetration Testing and Red Team Exercises

This chapter explores the critical role of offensive security testing in healthcare environments, highlighting how penetration testing and red team exercises verify the resilience of medical devices and hospital systems. It covers practical guidance on structuring tests, managing the unique challenges of clinical settings, analyzing outcomes, and planning remediation. Ethical and legal considerations are addressed to ensure these activities are conducted safely and compliantly, ultimately enhancing patient safety and organizational readiness.

Chapter 16: Medical Device Cybersecurity in the Era of AI and Machine Learning

This chapter explores how artificial intelligence (AI) and machine learning (ML) are transforming medical devices by enhancing capabilities while introducing new cybersecurity challenges. It examines the integration of AI in healthcare technology, unique vulnerabilities arising from autonomous systems, strategies to secure AI pipelines, regulatory and ethical considerations, and future-proofing security against evolving threats.

Chapter 17: Wireless, Mobile, and Wearable Device Security

This chapter explores the unique security challenges and risks associated with wireless, mobile, and wearable medical devices. It addresses the ecosystem of wearables, vulnerabilities in short-range communication protocols, protection strategies for mobile health applications and data, incident response for lost or stolen devices, and best practices for integrating personal devices in clinical environments.

Chapter 18: Securing Medical Device Updates and Maintenance

This chapter focuses on the comprehensive strategies and best practices for securing medical device updates and maintenance throughout their life cycle. It covers secure delivery mechanisms, planning for safe updates within clinical environments, proper decommissioning and disposal procedures, ongoing vulnerability management, and regulatory certification requirements to ensure device safety, reliability, and compliance.

Chapter 19: Medical Device Security Incident Case Studies

This chapter explores significant medical device security incidents, revealing the vulnerabilities exploited, the lessons learned from both failures and ethical research, effective response strategies, the consequences of flawed policies, and how these incidents have driven lasting improvements in security practices across healthcare organizations.

Chapter 20: Future Directions and Ultimate Mandate: Securing the Connected Patient

This chapter presents a visionary call to action emphasizing the urgent need to achieve uncompromising cybersecurity in connected medical devices to protect patient safety. It highlights the ethical imperative for perfection in device security, the integration of cybersecurity education within healthcare and engineering training, the importance of proactive policy and advocacy, the necessity of continuous innovation in security practices, and the critical personal and organizational commitments required to safeguard the future of connected patient care.

Chapter 21: Conclusion

This chapter synthesizes the overarching mission of medical device cybersecurity, affirming that the protection of patient safety stands at the core of healthcare technology. The evolution from isolated devices to complex, interconnected systems has intensified the urgency of cybersecurity, demanding ethical accountability and collaborative resilience. The chapter reviews key threat actors, regulatory frameworks, and the necessity for ongoing risk assessment, security-by-design, and incident response protocols. It highlights the ethical imperative for transparency, lifelong learning, and professional growth, encouraging readers to actively contribute to sector-wide knowledge sharing, advocacy, and training. Ultimately, it calls for a united, multidisciplinary commitment—across technical, regulatory, and clinical domains—to ensure that innovation in healthcare is always matched by steadfast dedication to security and the preservation of public trust.

About the Companion Website

This book is accompanied by a companion website:

www.wiley.com/go/bastacmd

To access the website materials, click on the link under Related Resources on the book's page on wiley.com. You can also find this by searching www.wiley.com for the ISBN 9781394433681.

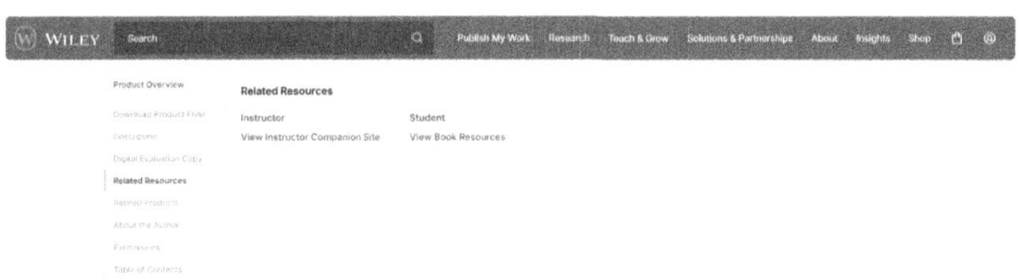

The website includes:
- Test bank
- PPT

Introduction

In the dimly lit control room of a bustling hospital, an urgent alert flashes on a monitor: a connected insulin pump has been accessed without authorization. Behind the scenes, commands have been altered, risking a patient's life in real time. Minutes earlier, clinicians were unaware that a silent attack had begun, exploiting vulnerabilities hidden deep within the device's software. This moment is not an anomaly but a glimpse into the fragile reality of modern healthcare technology; a world where digital safety and human life are inseparably linked. The challenge to secure medical devices is complex, layered with technical intricacies, regulatory demands, and ethical responsibility. Navigating this evolving landscape requires clear understanding, practical tools, and unwavering commitment. As you move forward, this introduction invites you to engage with those realities thoughtfully, recognizing both the difficulties involved and the critical importance of your role in protecting patients and preserving trust in healthcare systems.

Purpose

In an intensive care unit humming with life-sustaining technology, a patient lies tethered to an insulin pump, their blood glucose levels under the careful watch of digital sensors. The pump's stream of medicine is controlled by remote commands—a marvel of modern engineering and connectivity. Yet, it only takes a few keystrokes from a distant attacker to override that flow. One notable incident involved the successful hacking of an insulin pump, changing its operation so it delivered a lethal dosage, all without the patient or attending physician noticing the danger until it was almost too late (Williams & Woodward, 2015). In another headline-grabbing case, researchers demonstrated the possibility of remotely disabling a pacemaker or rewriting its instructions, threatening to stop a beating heart at the touch of a button (Williams & Woodward, 2015). These are not hypothetical risks. Patients' lives have depended on engineers racing against time to protect devices from unseen invasions, while families watched helplessly as everyday hospital equipment became a battlefield.

These moments bring into acute relief what is truly at stake. Vulnerabilities in medical device cybersecurity translate instantly from lines of code to matters of life and death. A malfunctioning infusion pump, altered by a malicious actor, can introduce a toxic drug dose instead of a therapeutic one. Imaging systems, hijacked in the midst of surgery or critical intervention, can relay false results or freeze at the worst possible instant. Heart monitors and neurostimulators—devices trusted to buffer life's most fragile moments—can be silenced or manipulated from afar. The fallout stretches beyond physical harm. Breaches in connected medical device environments have allowed attackers to siphon off massive volumes of personal health data, exposing patients to identity theft, blackmail, and the lifelong erosion of privacy (Aldosari, 2025). For clinicians, every corrupted record risks an erroneous treatment decision; for organizations, every attack draws legal scrutiny, financial loss, and irreparable blows to reputation (Aldosari, 2025).

These vulnerabilities have escalated as healthcare has embraced the benefits of wireless devices, remote monitoring, and interconnected clinical environments. Implantable devices once isolated from outside networks are now subject to reprogramming in clinics and real-time telemetry, increasing patient autonomy but at the expense of a much broader digital attack surface (Williams & Woodward, 2015). Medical devices that were never designed with robust cybersecurity as a priority now transmit vital information across hospital networks and even directly to the cloud. The result is a swift shift in the risk landscape: from stand-alone safety to tangled, systemic vulnerability. The number and diversity of threats are climbing: from exploitation of default administrative passwords and unsecured device web interfaces to sophisticated ransomware campaigns that can hold entire treatment units hostage (Aldosari, 2025; Zaid & Garai, 2024).

It is against this backdrop of escalating and deeply personal risk that this book offers a practical, technical roadmap to securing medical devices. Each page is formulated to respond to the sense of urgency felt by professionals on the frontlines: engineers designing next-generation implants, IT leaders defending intricate health networks, regulatory specialists charged with oversight, and advanced students preparing to shape the field. The book goes beyond surface-level advice. It guides readers through the core technologies powering modern devices, demystifies evolving cyber threats, and addresses the particularities that make healthcare environments vulnerable yet critical. It draws from validated industry frameworks and real-world breach analyses, empowering readers with the strategic depth needed to assess risk and implement layered protections (Aldosari, 2025; Williams & Woodward, 2015).

Readers are supported through actionable, stepwise instruction in risk assessment, threat modeling, device architecture review, regulatory compliance, and incident response. Content is tailored to provide clarity, whether you are negotiating FDA regulatory pathways, architecting a device with embedded connectivity, or evaluating third-party software risks. Emphasis is placed on practical strategies: segmentation of device networks, implementation of robust encryption, thorough access controls, and industry-tested response tactics that anticipate—not simply react to—emerging threats.

This book arms you with the skills to master complex device ecosystems, foresee and block evolving cyber threats, design devices from the ground up with security in mind, and confidently navigate a shifting landscape of legal and technical requirements. These are not abstract competencies but tangible skills that result in better patient safety and unshakable organizational integrity. Investing in this knowledge means shifting from a posture of reaction to one of anticipation and mastery in a domain where the cost of delay is measured in more than dollars; it is measured in human lives.

The urgency of these challenges demands more than technical proficiency. Protecting interconnected devices is an ethical imperative, one grounded in the trust patients place in their caregivers and technology. The responsibility stretches from boardrooms to bedside monitors; everyone engaged in the ecosystem of modern healthcare must recognize cybersecurity as synonymous with patient safety. By internalizing and applying the principles found here, readers pledge themselves not only to innovation but to empathy, safeguarding life at every digital touchpoint (Aldosari, 2025; Williams & Woodward, 2015).

Challenges and Industry Misconceptions

In the current landscape of healthcare technology, the urgency to address medical device cybersecurity grows sharper with each passing year. Daily headlines and industry reports alike have cultivated a pervasive awareness among stakeholders: vulnerabilities in connected medical devices can not only disrupt clinical operations but also directly endanger patient safety. Whether the concern is a networked infusion pump in constant use at a major hospital or an implantable cardiac device that a patient relies on at home, the notion that a single weak link might lead to catastrophic consequences weighs heavily on everyone involved in device development, administration, and security oversight (Yaacoub et al., 2020). A palpable sense of anxiety permeates the field, fueled by the rapid expansion of equipment connectivity, the exponential growth of sensitive health data, and the ever-evolving sophistication of cyber threats.

For a medical device engineer working on next-generation software updates, the task of meeting emerging encryption standards often feels overwhelming. Each technical advance in cryptography or secure networking brings both a measure of relief and a new set of compliance headaches. The balancing act here is not just technical; it is also strategic. Protection against zero-day vulnerabilities, for example, requires relentless vigilance in updating firmware and operating systems, but these safeguards can introduce unacceptable network latencies or operational delays; the unspoken trade-offs that haunt implementation teams (Yaacoub et al., 2020). On the other hand, IT managers on the hospital side find themselves trapped between demand for high-grade protection and significant resource constraints. Budgets, personnel, and time remain limited, even as the expectations for security and compliance continue to rise, now further complicated by newly available demographic data in postmarket device reports that must be analyzed to prevent disproportionate impacts across patient populations (Epstein, 2025).

A frequent source of frustration arises from the relentless pace of technological innovation, which seems to outstrip the capacity of regulatory frameworks to adapt. While agencies such as the FDA issue timely new guidance—sometimes introducing expanded surveillance requirements or clarifying postmarket obligations—manufacturers can feel as though the bar for compliance is perpetually in motion, creating a landscape where competitive pressures and regulatory risks continuously collide (Epstein, 2025). Navigating this unexplored terrain, device makers worry about being left behind, plagued by uncertainty around how and when to implement required changes.

Among the most persistent and dangerous misconceptions confronting the industry is the myth of "security by obscurity": the belief that the mere complexity, proprietary design, or "hidden-ness" of a system can serve as a sufficient defense against cyberattack. This allure of obscurity is understandable, but history shows it to be deeply flawed. Consider the case of legacy infusion pumps, where manufacturers, wrongly assuming that proprietary communication protocols shielded devices from adversaries, delayed critical firmware updates. The result: vulnerabilities remained open, exposing both organizational and patient safety to direct risk once researchers or malicious actors discovered undocumented interfaces or exploited default credentials (Yaacoub et al., 2020). When security is not systematically addressed—and when obscurity becomes a substitute for best practice—organizations may fall into a pattern of complacency that gives attackers exactly what they seek: easy, poorly monitored access points.

Security through obscurity also invites a false sense of confidence that can permeate vendor-client relationships, yielding cultures of risk aversion and delayed disclosure. Just as products circulated with unaddressed operating system flaws have been shown to compromise entire hospital networks, the underlying assumption that "nobody will ever try" grows ever more dangerous as adversaries become increasingly sophisticated and well-resourced (Yaacoub et al., 2020).

Overlaying these technical and cultural barriers are the daunting realities of industry-specific jargon and standards. For many, the lexicon of healthcare cybersecurity—ranging from cryptographic primitives to FDA-specific 510(k) postmarket reporting, from network segmentation mandates to supply chain risk management—feels inaccessible or even paralyzing. Decision-makers may hesitate to act for fear of "doing it wrong"; engineers and administrators alike may struggle to determine where to begin or which standard applies to their product class (Epstein, 2025). This leads to inertia, further compounding exposure to threats.

Despite these formidable barriers, it is crucial to validate the reader's concerns. These anxieties are not a sign of weakness or lack of preparation; rather, they reflect the fundamental tensions inherent in an industry striving to innovate while safeguarding both patients and organizations. By naming these challenges directly and describing lived experiences—overworked engineers, under-resourced compliance teams, myths perpetuated through word-of-mouth—the hope is to cultivate empathy and set a tone of validation. This approach not only acknowledges the complexity of the terrain but also reinforces the fact that support, solutions, and stepwise guidance do exist.

The following chapters confront these obstacles head-on, offering myth-busting clarifications and concrete tools crafted to demystify even the most technical security, clinical, and compliance considerations. As readers move forward, they should feel not only seen but increasingly confident that the knowledge and strategies provided will empower them to close gaps, enhance safety, and champion resilience throughout the life cycle of connected care (Epstein, 2025; Yaacoub et al., 2020).

Why Do We Need to Read this Book?

Mastering medical device cybersecurity is not only about understanding intricate technical systems. It is about protecting the lives of those who depend on safe and functional devices every day in clinical settings and at home. When a pacemaker is at risk because of an unpatched vulnerability, or an insulin pump can be manipulated remotely, the consequences extend far beyond technical inconvenience; they threaten individual patient lives, shatter institutional trust, and put broader public health at risk (Aldosari, 2025). Actual incidents have shown that system failures or cyber attacks can delay care, cause harmful interventions, or even result in patient death. Cybersecurity, therefore, must be recognized as an ethical imperative as much as a technical discipline.

The urgency at the heart of this field is driven by concrete risks. Consider a scenario in which a hospital network is hit with ransomware. Diagnostic machines grind to a halt, vital records become inaccessible, and patient care slows or stops altogether. Those seeking treatment suddenly find that the very systems designed to ensure their safety have been compromised. In another case, researchers have demonstrated how vulnerabilities in implantable cardiac devices could allow a hacker to alter device settings or drain its battery, endangering the user without their knowledge (Aldosari, 2025). These frightening examples illustrate what can happen when complacency, neglect, or a false sense of security takes root. No healthcare environment—be it a busy urban hospital or a small outpatient clinic—is immune.

The risks are magnified by the increasing complexity and interconnectedness of healthcare delivery. Today's medical devices often rely on networked communication, mobile apps, cloud computing, and real-time data sharing. Each link in this chain, from the secure design of a device to the transmission of its data, represents a potential entry point for attackers (Alaeifar et al., 2024;

Cory, 2024; Mensah et al., 2024). Even technologies like digital health records, praised for improving safety and efficiency, open new avenues for data breaches, unauthorized access, and human error when not properly safeguarded. The stakes could not be higher: a single breach does not simply put data at risk but can lead to identity theft, financial loss, and, most dangerously, compromised patient safety (Mensah et al., 2024).

At the core of this challenge is a pressing ethical responsibility. Healthcare professionals, engineers, administrators, and policymakers work with a dual obligation: to embrace technological progress for its potential to heal, and to guard fiercely against any threat that might turn these life-saving advances into sources of harm (Aldosari, 2025; Goktas and Grzybowski, 2025). Regulatory frameworks such as the Health Insurance Portability and Accountability Act (HIPAA) and the General Data Protection Regulation (GDPR) exist to provide structure, but true security relies on the vigilance, integrity, and shared commitment of everyone responsible for deploying, managing, or using medical devices (Aldosari, 2025; Mensah et al., 2024). This book approaches the topic by breaking down sophisticated concepts and offering clear, step-by-step guidance. Whether you come from a technical or a clinical background, you will find accessible explanations, practical tools, and real-world examples that make each key idea tangible.

A common misconception is the belief that "obscure" or lesser-known technology is immune from attack. In reality, relying on security through obscurity leads to dangerous blind spots. Many high-profile data breaches and device vulnerabilities have occurred precisely because of overlooked assumptions or failures to follow established best practices. Complacency can lead to permanent consequences, especially when human error remains a significant vulnerability (Mensah et al., 2024). Cyber incidents often start with something as simple as a weak password, an unpatched system, or an employee unaware of phishing risks. Overlooking these basics can result in disastrous breaches, putting patients and institutional reputations at risk.

The challenge goes further: rapid innovation in healthcare exerts pressure to bring new devices to market quickly, while regulatory and ethical demands press for robust, thorough security. Balancing these pressures is not a one-time achievement but a continuous journey that requires thoughtful engagement and ongoing learning (Aldosari, 2025). This book exists to empower you with the confidence and skills necessary to confront these tensions.

Every person involved in medical device security holds a powerfully important role. Whether designing new devices, managing compliance, or overseeing clinical integration, your actions directly affect patient safety. The requirement to adopt cybersecurity best practices is more than just technical; it reflects a profound moral commitment.

Imagine a device designed to save lives but, if left unprotected, could be turned into a weapon by a remote attacker. Would you feel comfortable knowing that your device or institution could become that scenario? Each reader has the agency to drive change and prevent such outcomes. The mission is clear and urgent: let the knowledge you gain here be your guide in building resilient, trustworthy healthcare systems that protect every life placed in your care (Aldosari, 2025; Mensah et al., 2024).

Final Thoughts

Now that we understand the critical stakes involved in medical device cybersecurity—where technical vulnerabilities translate directly into risks to patient safety and institutional trust—we are better equipped to face these challenges with clarity and purpose. This book provides a comprehensive framework grounded in real-world experience, rigorous standards, and practical tools designed to empower professionals across disciplines to anticipate threats, implement effective defenses, and navigate complex regulatory landscapes. By embracing this knowledge, readers can transform uncertainty and anxiety into confident action, fostering resilient healthcare environments where innovation thrives without compromising security. The path ahead demands ongoing vigilance, collaboration, and ethical commitment; armed with the guidance offered here, you are prepared to lead the way in securing the devices that safeguard human lives every day.

References

Alaeifar, P., Pal, S., Jadidi, Z., Hussain, M., Foo, E. (2024, June 1). *Current Approaches and Future Directions for Cyber Threat Intelligence Sharing: A Survey*. Journal of Information Security and Applications; Elsevier BV. https://doi.org/10.1016/j.jisa.2024.103786.

Aldosari, B. (2025, May 6). *Cybersecurity in Healthcare: New Threat to Patient Safety*. Cureus; Springer Science and Business Media LLC. https://doi.org/10.7759/cureus.83614.

Cory, N. (2024, April 22). *Technical and Legal Criteria for Assessing Cloud Trustworthiness*. Itif.org; Information Technology and Innovation Foundation | ITIF. https://itif.org/publications/2024/04/22/technical-legal-criteria-for-assessing-cloud-trustworthiness/.

Epstein. (2025, July 31). *Health Law Advisor*. @Ebglaw. https://www.healthlawadvisor.com/category/fda.

Goktas, P., Grzybowski, A. (2025, February 27). *Shaping the Future of Healthcare: Ethical Clinical Challenges and Pathways to Trustworthy AI*. Journal of Clinical Medicine; Multidisciplinary Digital Publishing Institute. https://doi.org/10.3390/jcm14051605.

Mensah, N. K., Adzakpah, G., Kissi, J., Taylor-Abdulai, H., Johnson, S. B., Agbeshie, P. A., Opoku, C., Abakah, J., Osei, E., Agyekum, A. Y., Boadu, R. O. (2024, January 1). *Health Professionals' Ethical, Security, and Patient Safety Concerns Using Digital Health Technologies: A Mixed Method Research Study*. Health Services Insights; SAGE Publishing. https://doi.org/10.1177/11786329241303379.

Williams, P., Woodward, A. (2015, July). *Cybersecurity Vulnerabilities in Medical Devices: A Complex Environment and Multifaceted Problem*. Medical Devices: Evidence and Research. https://doi.org/10.2147/mder.s50048.

Yaacoub, J.-P. A., Salman, O., Noura, H. N., Kaaniche, N., Chehab, A., Malli, M. (2020, September). *Cyber-Physical Systems Security: Limitations, Issues and Future Trends*. Microprocessors and Microsystems. https://www.ncbi.nlm.nih.gov/pmc/articles/PMC7340599/.

Zaid, T., Garai, S. (2024, April 28). *Emerging Trends in Cybersecurity: A Holistic View on Current Threats, Assessing Solutions, and Pioneering New Frontiers*. Blockchain in Healthcare Today. https://doi.org/10.30953/bhty.v7.302.

1

Why Medical Device Cybersecurity Cannot Be Ignored

More than 90% of hospitals worldwide rely on networked medical devices to deliver critical care, yet nearly half of these devices have known security vulnerabilities that remain unaddressed. The integration of advanced technology into healthcare systems has created vast attack surfaces, exposing patients and providers to risks beyond data breaches—risks that can directly threaten human life. This chapter examines why ignoring the cybersecurity of medical devices is no longer an option and underscores the urgent need for robust, transparent defenses in the evolving landscape of connected healthcare.

Lives on the Line: The Human Cost of Device Exploits

The stakes in medical device cybersecurity are as urgent and concrete as the risk to human life. Failures in securing healthcare technology can result in immediate harm, such as incorrect therapy administration or sudden loss of critical function. Attackers who exploit medical devices move beyond data theft and directly threaten patients' safety, sometimes with irreversible consequences. In one well-documented case, insulin pumps with unsecured wireless communication protocols proved vulnerable to remote exploitation. Hackers demonstrated that, using rudimentary radio equipment and software flaws in the device firmware, it was possible to transmit unauthorized commands and trigger the pump to deliver dangerously excessive or insufficient doses of insulin. These attacks bypassed authentication due to the absence of strong encryption and session management, exposing patients to life-threatening hypoglycemia or hyperglycemia. Had such methods been used maliciously in a real clinical environment, the result could have been coma, organ failure, or fatality, especially for patients relying on these pumps for daily glycemic control (Aldosari, 2025).

Pacemakers have also been targeted. In a high-profile security demonstration, researchers exploited outdated, unpatched firmware in a commonly deployed brand of implantable cardiac defibrillator. The researchers reverse-engineered communication protocols, exploiting the lack of cryptographic verification between the programming device and implant. By sending malicious radio frequency transmissions, they triggered the device to deliver unnecessary, potentially fatal electrical shocks or, conversely, disabled lifesaving therapy during cardiac events. Clinical compromise was possible because the wireless protocol enabled over-the-air updates without verifying origin or integrity, creating an open channel for attackers within range. While no deaths resulted from this demonstration, the technical route to harm was clear and has sparked widespread concern about similar vulnerabilities in thousands of deployed units worldwide (Elendu et al., 2024).

Large-scale hospital breaches illuminate how attacks on medical device infrastructure affect not just individuals but also entire critical care workflows. The WannaCry ransomware incident remains a prominent example. This attack infiltrated hospital networks using a widely known exploit targeting unpatched Windows systems, quickly spreading laterally to networked infusion pumps, diagnostic imaging devices, and life-support systems. Devices frozen by the ransomware could not be operated, monitored, or programed, resulting in the suspension of surgeries and the transfer of vulnerable patients to unaffected centers. Mechanisms involved included a combination of exploit chaining, lateral privilege escalation, and poor network segmentation, amplifying a local device compromise to affect hospital-wide operations and increasing the risk of death for critical inpatients (Aldosari, 2025; Elendu et al., 2024).

Vulnerabilities often begin with minor oversights that scale into catastrophic failures. A single device left with outdated firmware or a default password may function as an entry point. Attackers scan for such weaknesses, breach the device, and then use it as a foothold to probe the network for further vulnerabilities or launch denial-of-service attacks against interconnected systems. For example, a compromised infusion pump could be commanded to alter drug dosages, but it could also serve as a jump host, letting attackers map the internal hospital landscape and disable monitoring devices in intensive care units (ICUs). This chain reaction threatens not only single patients but entire wards, especially as device interconnectivity increases. The technical progression—from software bug to network compromise—means a minor misconfiguration can rapidly escalate to a critical care incident (Aldosari, 2025).

Legal consequences for neglecting medical device security are substantial. Healthcare organizations and device manufacturers found liable for preventable breaches face regulatory penalties from agencies such as the U.S. Food and Drug Administration (FDA) and the Department of Health and Human Services (HHS). Notable lawsuits have resulted in multi-million-dollar settlements, regulatory oversight, and mandatory product recalls. For instance, after vulnerabilities were discovered in a line of cardiac devices, the manufacturer was compelled to issue software updates and submit to ongoing FDA scrutiny. The obligations imposed in these settlements often require improved device monitoring, regular patching regimes, and more rigorous premarket security testing. Such legal precedents have influenced sector-wide policy, embedding cybersecurity by design as a regulatory requirement and establishing industry standards for device life cycle management (Elendu et al., 2024).

There is an equally pressing ethical dimension. Stakeholders in the development and deployment of medical devices hold a professional and moral obligation to ensure the safety and security of their products. Medical device manufacturers must integrate robust security measures from the earliest stages of product development, recognizing that any failing directly endangers patients. Technical professionals and clinicians share responsibility as advocates for implementing and updating secure technology in their institutions. Failure to address known vulnerabilities or allocate resources to security is not only a breach of patient trust but falls short of fundamental professional ethics, especially given the potentially fatal consequences of an exploited device (Aldosari, 2025).

Rapid technological evolution has transformed medical devices from standalone machines into components of complex, interconnected hospital networks. This shift has multiplied attack surfaces and amplified both the potential impact and obligation for strong device security. Cybersecurity must match this complexity, evolving as a core aspect of patient care and organizational accountability (Aldosari, 2025; Elendu et al., 2024).

The Evolution of Medical Devices: From Analog to Networked

Medical device security has rapidly evolved from an afterthought to a mission-critical concern due to the dramatic technological changes unfolding across the healthcare landscape. As frontline practitioners and technical stewards encounter persistent real-world cyber threats, it is vital to recognize that the challenges are not merely device-specific but rather a byproduct of a profound transformation in system design and connectivity, which has fundamentally altered risk profiles over time.

Initially, medical tools operated as standalone, electromechanical devices with minimal digital integration. These early machines—infusion pumps run by basic timers, manual defibrillators, or single-function monitors—provided essential care functions that were largely isolated from external influence. Cyber risk was almost nonexistent, as physical proximity and specialized technical access were necessary prerequisites for sabotage or malfunction.

The transformative turning point emerged with the embedding of microprocessors, memory, and software-based controls into medical hardware. Software-driven functionalities facilitated more advanced signal processing, improved diagnostics, and tailored therapy delivery. For instance, patient monitors evolved to provide real-time waveform analysis and customizable alarm parameters, while infusion pumps began supporting programmable dosing and interoperability with electronic medication orders. This new paradigm enabled clinical advancements but simultaneously introduced layers of hidden complexity and interdependence—each module, driver, and software library now represented a potential vector for error or exploitation.

These new capabilities set the stage for the next era: comprehensive system integration across local networks and enterprise IT environments. Medical devices, formerly independent, now share data with centralized workstations and electronic health record systems, allowing clinicians to monitor multiple patients in real time, share diagnostics instantaneously, and synchronize care interventions between modalities. The sophistication of such integrations is seen in ICUs, where multiparameter monitors, networked ventilators, and infusion stations deliver harmonized care routines. This interconnectedness advances safety and efficiency, allowing collaborative intervention and seamless escalation of care, but at the expense of new and consequential cybersecurity dependencies (Ahmed et al., 2023).

The expansion of wireless networking—using protocols like Wi-Fi and Bluetooth—ushered in continuous, location-independent data flow. Hospital settings now support mobile carts, wearable telemetry units, and remote-controlled imaging systems, while ambulatory and home-based applications rely on Internet of Medical Things (IoMT) architectures to relay vital data between patient, provider, and caregiver. For example, wireless pacemakers transmit rhythm logs to cloud analytics platforms, and network-enabled infusion pumps receive remote programming and alerts from care teams. This heterogeneity in device make and network protocols enables distributed, proactive care but at the same time complicates risk modeling and incident response. With such pervasive connectivity, patient safety has become increasingly contingent on secure wireless communication, meticulous endpoint management, and system-wide visibility into network activity (Ahmed et al., 2023).

Each incremental network connection inherently increases the attack surface. Where once a device needed to be physically accessed to be compromised, now adversaries can potentially exploit any exposed communication channel. The multitude of interfaces—ranging from proprietary device protocols, standard service discovery mechanisms, and remote firmware update services—multiplies the opportunities for exploitation. Multimodal monitors that combine electrocardiogram (ECG), pulse oximetry, and blood pressure sensors rely on continuous network presence, creating multiple vectors for adversaries to disrupt or manipulate clinical data. Wireless-enabled imaging systems may expose authentication weaknesses or contain errors in third-party communication stacks, as seen in several high-profile vulnerability disclosures in recent years. Each new connectivity layer augments clinical capacity but also requires an equivalent increase in security diligence (*A Review on the Security Vulnerabilities of the IoMT against Malware Attacks and DDoS*, 2019).

The migration toward firmware and software-centric design not only is about flexibility but also forms the operational core of every contemporary medical device. The move away from fixed-function, hardware-driven controls allows remote update capability, the addition of new features, and rapid deployment of clinical improvements. However, this dynamic environment transforms risk management into an ongoing obligation; a single software bug or misconfiguration can render devices nonfunctional or, worse, dangerous to patients. In practice, an exploitable flaw in networked pacemaker firmware may require urgent patching across thousands of deployed units—highlighting the operational, clinical, and regulatory complexities tied to secure update mechanisms (*A Review on the Security Vulnerabilities of the IoMT against Malware Attacks and DDoS*, 2019).

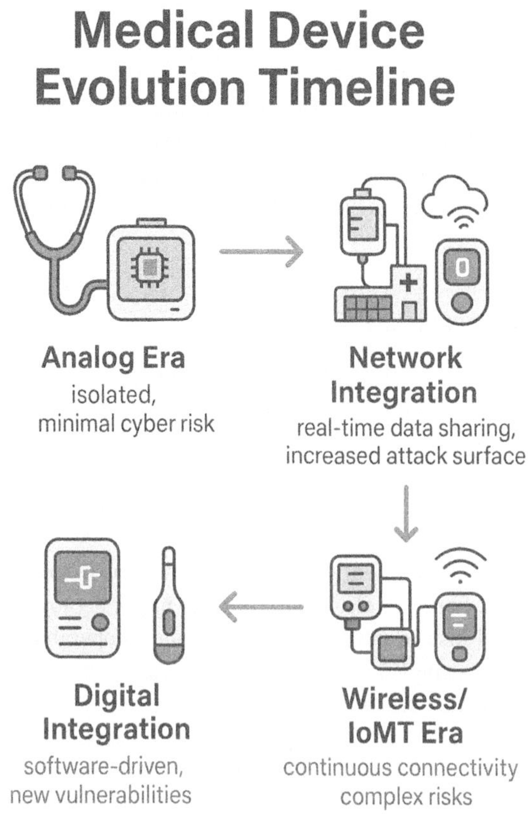

All these technological advances have contributed to the increasing intricacy of medical device ecosystems. The result is not just a growth in technical sophistication but a fundamental transformation in attacker strategy and defender responsibility. Cyber adversaries now target chains of interconnected endpoints, seeking vulnerabilities in software, configurations, or overlooked protocol layers. The attack surface has multiplied, bringing with it a diversity of threat actors—from opportunistic malware campaigns to targeted ransomware and sophisticated state-sponsored attacks.

The convergence of complexity, interconnectivity, and dynamic software controls has indelibly changed the risk environment for medical devices. This ever-widening attack surface has not only heightened the need for specialized incident response but also opened the door to a more diverse array of adversaries and motives, which will further inform the evolving approaches to healthcare cybersecurity.

Hidden Threats: Attackers in the Shadows

Advances in digital technology have blurred the boundaries between clinical care and computational systems. Medical devices once isolated from the outside world are now deeply woven into hospital networks, managing everything from vital monitoring to automated drug delivery. However, with increased network integration comes an enlarged attack surface. Adversaries with varied motivations can exploit device vulnerabilities, making it critical to recognize the distinct groups and ideological forces that drive such attacks—and the practical risks they pose to patient safety and healthcare operations.

Malicious Hackers and Cybercriminal Groups

Malicious hackers—ranging from independent actors to highly organized collectives—are often drawn by the lure of financial profit, widespread notoriety, or a simple urge to disrupt. Their attacks frequently begin with identifying weak points in device integration: default passwords on infusion pumps, unpatched vulnerabilities in MRI machines, or unsecured wireless telemetry. These avenues allow unauthorized access, which can be weaponized for data theft, operational disruption, or extortion.

Financially motivated attacks frequently focus on ransomware, leveraging the urgent, life-critical nature of hospital services. For example, cybercriminals might deploy specialized ransomware to target a fleet of networked insulin pumps, rendering them inoperable and demanding payment for restoration. Other attacks employ sophisticated phishing schemes to steal credentials granting access to imaging or diagnostic devices, ultimately allowing attackers to exfiltrate protected health information to sell on dark web marketplaces (He et al., 2021; Riggi, 2020). There are also disruptive campaigns designed to demonstrate the hacker's technical prowess: disabling ventilators during a public health crisis or manipulating device firmware to call attention to security gaps. Even when such actions are not politically motivated, they carry profound consequences for patient health and trust in medical technology (He et al., 2021).

State-Sponsored Threats and Advanced Persistent Adversaries

State-sponsored malicious actors—often referred to as advanced persistent threats (APTs)—operate under government directives, seeking rewards beyond mere profit. Their objectives might include sabotaging rival nations' healthcare infrastructure, collecting sensitive health data for intelligence purposes, or undermining public confidence in national health systems during pandemics. State actors can invest significant resources into uncovering zero-day vulnerabilities or crafting custom malware designed to evade detection.

The infamous WannaCry outbreak, attributed to North Korean sponsorship, stands as a striking example of this class of adversary: hospitals across Europe and Asia endured widespread device outages, diagnostic delays, and forced patient diversions when ransomware disabled everything from imaging devices to basic patient records systems (Riggi, 2020). Russian-backed adversaries have unleashed ransomware with destructive payloads, lacking even the possibility of decryption, while Iranian groups have targeted research facilities to pilfer intellectual property for government use (Riggi, 2020). In each case, the underlying motivations are geopolitical, whether to inflict harm, gain strategic advantage, or retaliate for sanctions. These actors often work methodically over long periods, embedding spyware or backdoors in devices, gathering intelligence over months or years with carefully staged operations.

The Insider Threat

Insider threats compound the risk landscape, as those with legitimate access—such as clinical staff, IT consultants, or contract technicians—can intentionally or inadvertently trigger catastrophic vulnerabilities. Insider actions may be rooted in grievances, a desire for financial gain, or simple carelessness. For example, an employee might download unauthorized software onto a hospital workstation, unintentionally introducing malware that propagates to dozens of connected medical devices. Inadequate security training and insufficient oversight allow such risks to persist (He et al., 2021).

At a more grave extreme, disgruntled insiders with elevated privileges can intentionally sabotage device functionality. By exploiting poorly managed access controls or overlooked audit logs, they might disable alarm features in life-sustaining ventilators or alter medication delivery settings to dangerous levels. Mistakes can also have devastating outcomes: a contractor who accidentally deletes a critical firmware update from a shared server could prevent essential devices from receiving patches, leaving known vulnerabilities open to exploitation. As the medical cyber-physical environment grows more intricate, human error and negligence become increasingly consequential (He et al., 2021).

Criminal Enterprises and Extortion Specialists

Organized crime syndicates have discovered that hospitals represent high-value, vulnerable targets. Unlike traditional enterprises, healthcare organizations cannot easily pause operations. Criminal enterprises exploit this operational pressure by locking down critical assets, including medical devices, through ransomware and threatening to release or destroy sensitive data unless large ransoms are paid. These attackers operate with the knowledge that disruptions can jeopardize patient lives, using this leverage to maximize financial returns (He et al., 2021; Riggi, 2020).

Criminal groups may operate on a "ransomware-as-a-service" basis, offering sophisticated malware toolkits to lesser-skilled actors who wish to extort healthcare institutions. In some cases, ransomware campaigns are so thoroughly orchestrated that criminals systematically disable redundant systems, backup devices, and network segments to ensure compliance with their demands. The result is a systemic risk not just to individual hospitals but to the continuity of entire health systems, as witnessed during coordinated attacks that paralyzed emergency care and canceled thousands of appointments worldwide (He et al., 2021; Riggi, 2020).

The prevalence of these adversaries is not simply the product of growing connectivity but of misconceptions that system architecture or proprietary technology will deter sophisticated attackers. In reality, such assumptions foster complacency, leaving critical vulnerabilities exposed to all categories of motivated adversaries, setting the stage for deeper exploration into why secrecy alone offers little true protection.

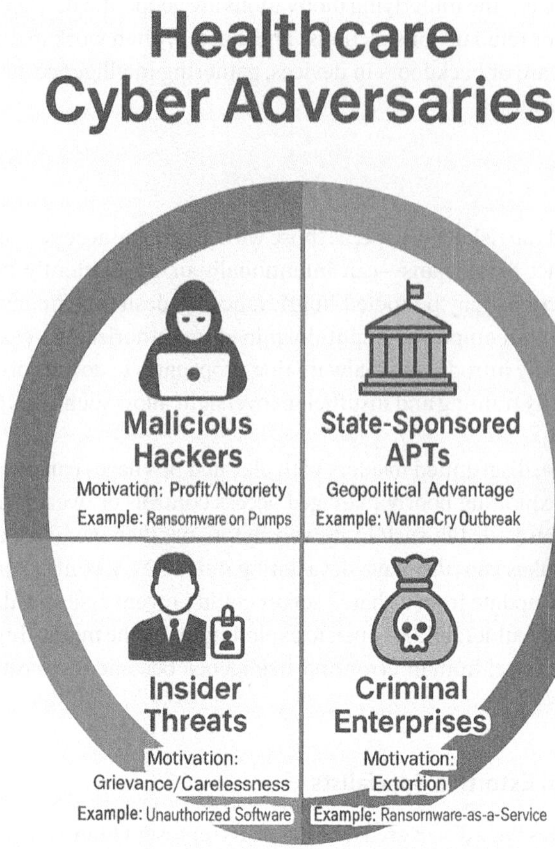

The Myth of Security by Obscurity in Medical Devices

For many years, manufacturers of medical devices relied on the belief that simply hiding the workings of their products or keeping software proprietary would provide sufficient security. This meant that as attacks grew more sophisticated, the industry continued to depend on secrecy and obscure code as its main line of defense, assuming that attackers would struggle to exploit what they could not see. However, this reliance on obscurity as a means to achieve security has created fertile ground for risky complacency. Advanced adversaries consistently exploit weaknesses that stem from overconfidence in proprietary systems rather than real, tested defense mechanisms.

The notion that security is guaranteed through an undisclosed or nonpublic design often leads engineers and developers to overlook rigorous testing or comprehensive protective strategies. Proprietary systems can mask unexamined vulnerabilities simply because they are hidden from public scrutiny. Companies sometimes view security more as a marketing claim than a design imperative, since the internal workings remain invisible to most users and regulators unless a serious incident occurs. This approach not only stifles meaningful evaluation and improvement but also enables dangerous gaps to persist. For example, device manuals and certification documents routinely divulge technical details about frequency use and transmission protocols, providing adversaries with enough data to mount informed attacks, even when the underlying source code remains closed (Williams & Woodward, 2015).

In practice, the myth of obscurity fails because attackers do not respect the artificial walls set up by proprietary designs. They deploy sophisticated techniques to reverse engineer devices, even without official access to design files. Extraction and analysis of firmware is one of the most common approaches. Attackers frequently use tools such as Joint Test Action Group (JTAG) debuggers and disassemblers to pull binary code directly from hardware or memory chips. Once extracted, this code can be decompiled and painstakingly teased apart to reveal how the device operates. Security vulnerabilities—such as hardcoded credentials, unencrypted data transfers, or weak authentication protocols—often become apparent after just a few hours or days of analysis by a skilled researcher or malicious hacker. Public reports of vulnerabilities routinely surface through independent discovery, not through voluntary disclosure by manufacturers. This makes clear that hiding technical details does little to delay or prevent determined adversaries from uncovering flaws.

Concrete examples of failed obscurity abound within the healthcare sector. Medical infusion pumps, which regulate lifesaving medications, have been found with embedded web services that completely lack authentication or encryption. Attackers have demonstrated how networked pumps can be manipulated remotely, even from outside hospital boundaries (Williams & Woodward, 2015). In another public case, researchers at the University of Michigan's Archimedes Center successfully compromised pacemakers and insulin pumps by exploiting radio interfaces intended for routine device management. In these scenarios, vulnerabilities such as default passwords, poorly protected wireless links, and undocumented maintenance functions were exposed through sheer determination and technical skill—none of which required original design documents.

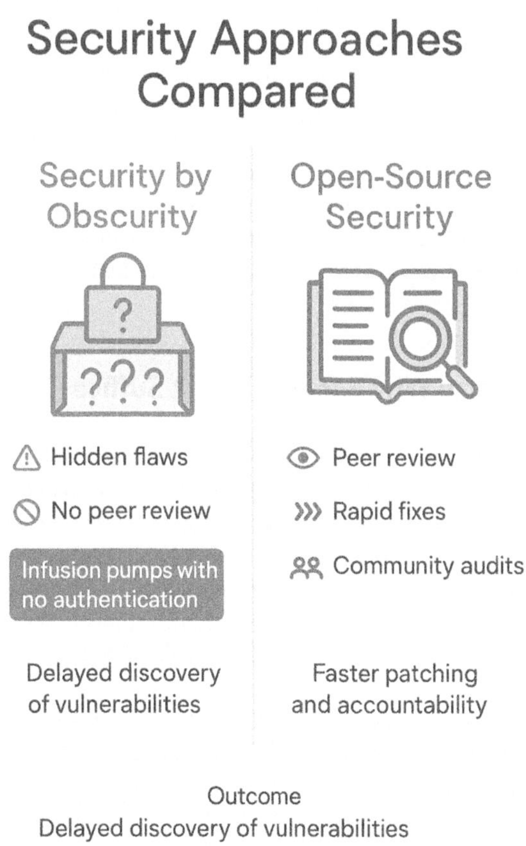

Historically, manufacturers have sometimes discovered faults in their proprietary devices but failed to share them with clinicians or regulators. The tragic case of a 21-year-old man who died when his implantable defibrillator short-circuited highlights this problem. The manufacturer, Guidant, first noticed the flaw several years earlier and made undisclosed corrective changes in production but neglected to notify the public or medical professionals. This incident triggered the recall of 70,000 devices—only after fatal consequences brought the problem to light (*Killed by Code: Software Transparency in Implantable Medical Devices—Software Freedom Law Center*, 2010). Such secrecy delays both warnings and solutions, exposing patients to unacceptable risks.

Open-source security strategies present a sharp contrast. When code is visible and available for independent analysis, researchers, developers, and even concerned end users can collectively examine and test devices for vulnerabilities. Peer review provokes rapid detection and correction of flaws that might otherwise remain dormant in closed systems. Open standards create uniformity and allow different products to work together, improving safety and simplifying compliance. Community-driven security audits and independent bug reports foster a culture of accountability beyond what closed, proprietary design can achieve. Industry regulators and advocacy organizations have increasingly pointed to transparent models as offering defense-in-depth and more timely patching of discovered vulnerabilities. Large open-source projects have established quality controls to ensure that only trusted code is adopted, stopping malicious or untested changes before they can endanger end users (*Killed by Code: Software Transparency in Implantable Medical Devices—Software Freedom Law Center*, 2010).

As reliance on embedded medical technology deepens, the limitations of obscurity in ensuring patient safety become more obvious and dangerous. Lessons from the past—including preventable fatal incidents—demonstrate that the ethical and professional requirements of device security cannot be met by secrecy alone. The stakes are high: unreliable devices put lives at risk, and when patient safety depends on technology, every aspect of protection must be open to verification, challenge, and improvement. Accountability begins not with the promise of hidden defenses but with clear, responsible, and community-supported security principles. The ongoing evolution of medical devices compels those who design, test, and deploy them to prioritize transparency—the only enduring foundation for trust and safety in modern healthcare (*Killed by Code: Software Transparency in Implantable Medical Devices—Software Freedom Law Center*, 2010; Williams & Woodward, 2015).

Unethical Complacency: Why Excuses Won't Suffice

Failures in the security of medical technology create a silent but dangerous gap between what designers intend and what patients and clinicians truly experience. When engineers and organizations tacitly believe that obscurity is enough to protect network-connected medical devices, they risk more than technical vulnerabilities. They invite moral and legal consequences that stretch far beyond the boundaries of code. Trust in "hidden" systems, instead of verified protections, can mask deep flaws in design, exposing patients to harm and organizations to liability. A misplaced confidence in obscurity reflects a deeper breach—a lapse in professional duty to the people whose lives depend on these technologies.

As medical devices link to hospital networks or cloud platforms, threats multiply and new vulnerabilities emerge. Device manufacturers who dismiss the urgency of patch management, encryption, or routine risk assessment place patients at risk of harm to both privacy and property. The harm goes beyond technical failure: compromised devices can interrupt life-sustaining therapy, expose sensitive health records, or offer an entryway for ransomware threats. Neglecting available security updates violates not only industry best practices but also legal mandates like the General Data Protection Regulation's Article 32, which requires immediate action to mitigate known vulnerabilities (Dhirani et al., 2023). Inaction is not a victimless oversight—it can lead to massive regulatory fines and, more critically, patient endangerment.

The professional responsibility to secure medical devices is not theoretical; it is rooted in the lived reality of modern medicine. Every software update, configuration decision, and incident response plan can mean the difference between continuity of care and catastrophic failure. Engineers and developers hold a duty of care, accountable not solely to their employer or the letter of the law but to those relying on every breath, heartbeat, or dose a device controls. This ethical obligation cannot be contracted away or diminished by ignorance (Augusta University, 2024). Medical device security is a direct extension of the Hippocratic principle: do no harm. Skirting these responsibilities is not merely a lapse in technical judgment but a breach of a professional standard that prioritizes human life.

Healthcare organizations and device manufacturers are required to embrace a proactive stance that treats security as fundamental to patient care. Such a stance involves regular vulnerability assessments, robust firewalls, and continuous monitoring for signs of intrusion (Augusta University, 2024). It also demands swift and transparent disclosure of discovered breaches. The Consolidated Appropriations Act of 2022, for example, instructs companies to report known cyber incidents within 72 hours of identification, reinforcing the significance of rapid and open communication for mitigating harm (Augusta University, 2024). Transparency and disclosure—whether about vulnerabilities or successful attacks—are acts of moral leadership. They empower clinicians, patients, and technical teams to take informed, risk-reducing actions.

The idea that lack of expertise excuses poor security simply does not hold up under scrutiny. The professional duty to learn is as essential as the obligation to care. Ongoing education and systematic professional development in cybersecurity are basic requirements, not aspirational goals, for responsible participation in developing or operating life-critical medical devices (Dhirani et al., 2023). Real-world cases have shown that organizations often face severe penalties or lawsuits when they fail to patch known flaws in devices, especially when those lapses result in patient harm or widespread data exposure. For instance, when a manufacturer decided to defer installing a critical patch in hospital software to avoid immediate inconvenience, they violated both the spirit and the letter of safety regulations—a choice explicitly condemned under General Data Protection Regulation (GDPR) statutes (Dhirani et al., 2023).

Concrete accountability mechanisms are necessary to demand and ensure safe practices in medical device security. Manufacturers and healthcare providers face product liability claims, where courts have held that reasonably foreseeable vulnerabilities must be addressed or else be grounds for legal action. Regulatory authorities impose fines and can even revoke approvals when organizations conceal incidents or neglect mandated security controls (Augusta University, 2024; Dhirani et al., 2023). Strong incident reporting systems and clear internal channels for raising security concerns allow organizations to learn from mistakes instead of repeating them. A culture encouraging transparency and learning—rather than blame and silence—becomes the bedrock of trust and continual safety improvement.

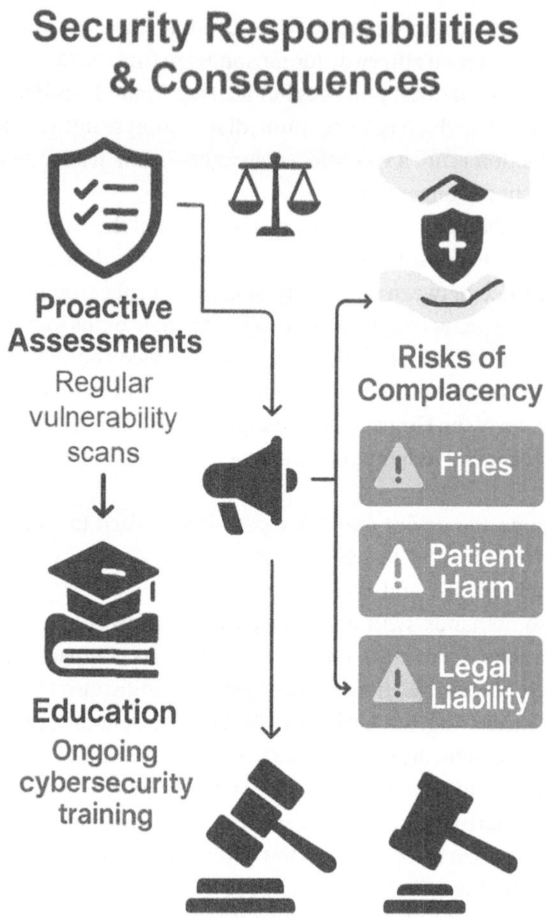

Patients trust that every step has been taken to secure the technologies their lives depend on. Upholding this trust means medical device professionals must reject complacency. They must act promptly to address known risks and commit to lifelong learning about emerging threats (Augusta University, 2024; Dhirani et al., 2023). Promoting ethical practices in cybersecurity is not merely about compliance, it is about protecting dignity, privacy, and life itself. Each decision—action or indecision—contributes to a shared culture that defines whether medicine will remain a source of safety and hope in an increasingly digital world.

Final Insights

Recognizing the profound human risks posed by medical device vulnerabilities and the complex, interconnected nature of modern healthcare technology compels us to adopt a comprehensive, transparent approach to cybersecurity. Now that we understand the limitations of security through obscurity and the diversity of adversaries targeting these critical systems, it is clear that safeguarding patient safety demands more than reactive fixes—it requires embedding robust security practices throughout device design, deployment, and ongoing management.

Healthcare professionals, engineers, regulators, and researchers must collaborate closely to develop resilient architectures, prioritize timely vulnerability disclosure and patching, and foster a culture of continuous learning and ethical responsibility. By doing so, we can build medical device ecosystems that not only withstand evolving cyber threats but also uphold the highest standards of accountability, ensuring trust and protection for patients in an increasingly digital healthcare environment.

References

A Review on the Security Vulnerabilities of the IoMT Against Malware Attacks and DDoS. (2019). Arxiv.org. https://arxiv.org/html/2501.07703v1.

Ahmed, S. F., Alam, M. S. B., Afrin, S., Rafa, S. J., Rafa, N., Gandomi, A. H. (2023, September 29). *Insights into Internet of Medical Things (IoMT): Data Fusion, Security Issues and Potential Solutions*. Information Fusion. https://doi.org/10.1016/j.inffus.2023.102060.

Aldosari, B. (2025, May 6). *Cybersecurity in Healthcare: New Threat to Patient Safety*. Media LLC; Cureus; Springer Science and Business. https://doi.org/10.7759/cureus.83614.

Augusta University (2024). *Cybersecurity Ethics: What Cyber Professionals Need to Know*. Cybersecurity Ethics: What Cyber Professionals Need to Know. https://www.augusta.edu/online/blog/cybersecurity‐ethics.

Dhirani, L. L., Mukhtiar, N., Chowdhry, B. S., Newe, T. (2023, January 19). Ethical Dilemmas and Privacy Issues in Emerging Technologies: A Review. Sensors. https://doi.org/10.3390/s23031151.

Elendu, C., Omeludike, E. K., Oloyede, P. O., Obidigbo, B. T., Omeludike, J. C. (2024, September 27). *Legal Implications for Clinicians in Cybersecurity Incidents: A Review*. Medicine; Wolters Kluwer. https://doi.org/10.1097/md.0000000000039887.

He, Y., Aliyu, A., Evans, M., Luo, C. (2021). Healthcare Cyber Security Challenges and Solutions Under the Climate of COVID19: A Scoping Review. Journal of Medical Internet Research; ncbi. https://doi.org/10.2196/21747.

Killed by Code: Software Transparency in Implantable Medical Devices—Software Freedom Law Center. (2010, July 22). Softwarefreedom.org. https://softwarefreedom.org/resources/2010/transparent-medical-devices.html.

Riggi, J. (2020). *Ransomware Attacks on Hospitals Have Changed | Cybersecurity | Center | AHA*. Www.aha.org. https://www.aha.org/center/cybersecurity-and-risk-advisory-services/ransomware-attacks-hospitals-have-changed.

Williams, P., Woodward, A. (2015, July). *Cybersecurity Vulnerabilities in Medical Devices: A Complex Environment and Multifaceted Problem*. Medical Devices: Evidence and Research. https://doi.org/10.2147/mder.s50048.

2

Understanding the Medical Device Cyber Landscape

"Did you verify the device's firmware before deployment?" The question lingered in the room, a simple query loaded with implications. In healthcare settings around the world, such questions are not just routine—they are vital checkpoints in an intricate network of technology, safety, and compliance. Every connected medical device, from the simplest wearable to the most complex implantable system, operates within a vast and dynamic ecosystem where security vulnerabilities can ripple through patient care environments with far-reaching consequences.

Understanding this interconnected landscape requires more than surface knowledge; it demands a detailed grasp of how medical devices communicate, the roles of diverse stakeholders, and the evolving tactics of those who seek to exploit weaknesses. It also involves navigating a complex global regulatory framework designed to ensure safety without hindering innovation. This environment poses unique challenges, especially when legacy equipment must coexist with cutting-edge technologies, blurring traditional boundaries between privacy and security.

The dialogue about medical device cybersecurity is ongoing and multifaceted, reflecting the critical balance between technological advancement, patient safety, and regulatory oversight. Professionals engaged in securing these systems must continuously adapt to emerging threats, shifting regulatory expectations, and operational realities. Their work is essential to maintaining the integrity and trustworthiness of healthcare delivery in an increasingly connected world.

Defining the Medical Device Ecosystem

Medical devices in healthcare span a wide spectrum, from basic fitness trackers that monitor heart rate and movement to high-risk, surgically implanted equipment like pacemakers or insulin pumps. Each category serves distinct clinical and operational purposes, and each brings its own unique set of connectivity needs and security requirements. Wearables, such as smartwatches and activity bands, often use Bluetooth or Wi-Fi for regular data uploads to mobile devices or cloud platforms. Their primary risks involve unauthorized access to personal health data during wireless transmissions and while stored in mobile applications (*The Ultimate Guide to Seamless Medical Device Integration*, 2024). In comparison, advanced devices—such as pacemakers or infusion pumps—are critical to life support and frequently incorporate long-range connectivity options for remote diagnostics and firmware updates, thus making them susceptible to manipulation or denial-of-service attacks if security measures are inadequate (FDA, 2020).

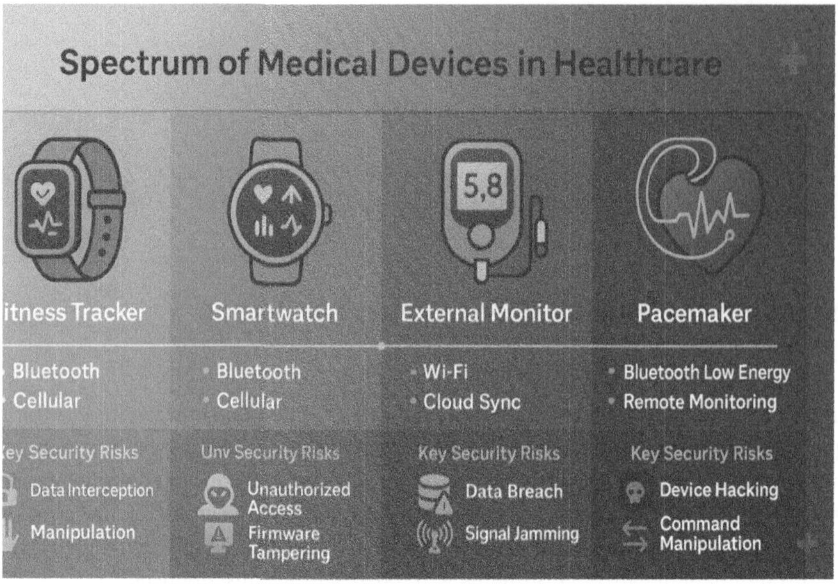

Context of use greatly influences vulnerabilities. In home environments, a networked blood pressure monitor might transmit readings to a caregiver or clinic through unsecured Wi-Fi, making it open to interception or malicious data injection. In contrast, a hospital-based infusion pump integrated into the institution's network faces complex risks: if device authentication is weak or segmentation is poor, an intruder might gain access, potentially altering dosages with direct consequences for patient safety (FDA, 2020). The operational context demands tailored protective measures: for consumer wearables, threats focus on data confidentiality and robust app ecosystem hygiene, while for implantables and hospital equipment, the priorities are real-time integrity protection, strong access controls, and comprehensive audit logging.

Stakeholder Roles and Influence on Security

Stakeholders in this ecosystem form a critical web for device security. Manufacturers design and build devices, embedding software and hardware controls that define the baseline exposure to risk. Their choices around secure boot, encryption, remote telemetry, and software updatability determine the device's lifelong resilience to threats (FDA, 2020). Healthcare delivery organizations, particularly hospitals, bear responsibility for secure integration. Their duties include inventory management, rigorous network segmentation, and regular vulnerability assessments as they operate mixed fleets of devices with varying protection levels. For hospitals, maintaining compliance with interoperability standards—such as Health Level 7 (HL7) or Integrating the Healthcare Enterprise (IHE)—complicates the security equation, increasing the danger of weakly integrated endpoints (*The Ultimate Guide to Seamless Medical Device Integration*, 2024).

Clinicians serve as both users and custodians of devices. Their ability to recognize anomalies, handle device interfaces securely, and report irregularities impacts operational safety. Patients, meanwhile, are stakeholders whose use and management of connected devices, especially in outpatient settings, affect the potential attack surface and likelihood of unauthorized access (*The Ultimate Guide to Seamless Medical Device Integration*, 2024). Regulators, like the FDA, define safety and cybersecurity standards, enforcing premarket and postmarket oversight. They require thorough documentation, rigorous testing, and continuous risk mitigation, all of which shape deployment timelines and operational procedures (FDA, 2020).

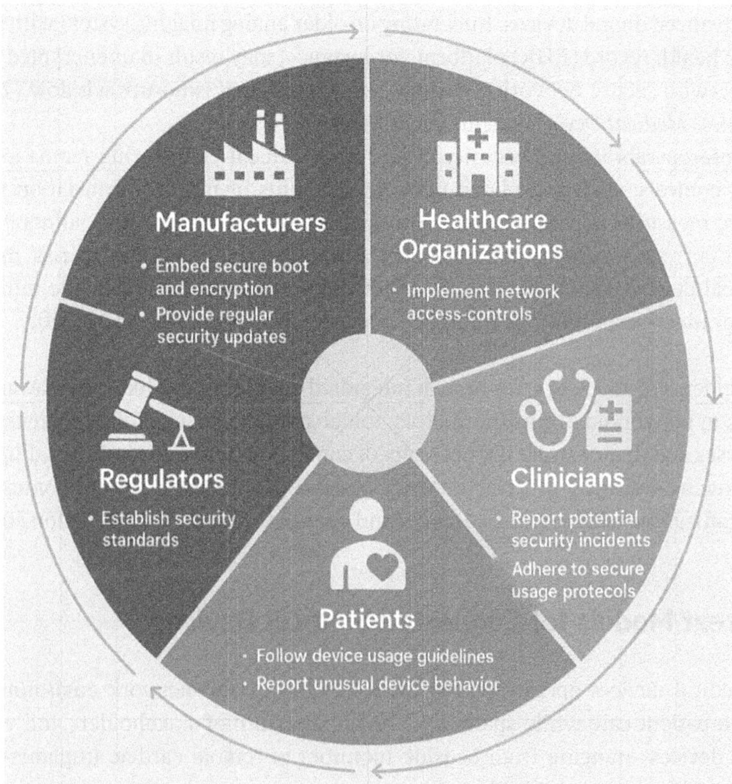

Data and Control Flow Pathways

Data and control signals travel complex paths in contemporary medical device networks. A wearable glucose monitor, for example, collects patient data and relays it via Bluetooth to a smartphone, which may then transmit the data over the internet to a cloud server for processing and aggregation. Results, analytics, or alerts might then flow to a clinician's secure dashboard, enabling timely medical intervention (FDA, 2020). Each phase—device transmission, mobile app handoff, cloud processing, and clinician interface—represents a potential point of compromise if authentication and encryption are insufficient.

For hospital-based networked devices, such as ventilators or infusion pumps, bidirectional flows often extend between on-premises gateways and cloud systems for predictive analytics and software updates. When remote updates are delivered, they must be securely validated and applied, ensuring that a compromised update service does not serve as an entry point for attackers. Examples of compromised data integrity in these flows—such as altered drug dosage commands—demonstrate high-risk scenarios that require stringent access controls and real-time monitoring (FDA, 2020).

Legacy Integration and Security Complexities

The prevailing reality in healthcare settings is the coexistence of legacy medical devices alongside modern, internet-enabled equipment. Legacy systems, often lacking inherent cryptographic protection, secure network protocols, or updatable firmware, frequently operate within the same hospital networks as advanced digital devices. Integrating an older analog imaging system with a contemporary electronic health record (EHR) platform, for instance, may result in unencrypted patient data traversing otherwise secure networks, introducing a significant exposure window (*The Ultimate Guide to Seamless Medical Device Integration*, 2024).

Balancing interoperability and security often forces clinical engineering teams to implement compensating controls, such as isolated network segments or protocol translation middleware. However, these measures may themselves introduce points of failure or new vulnerabilities if not maintained to current security standards. This creates continual tension in risk management: ensuring critical devices can communicate effectively for patient benefit, while minimizing the risk that integration—especially of outdated systems—jeopardizes confidentiality, integrity, or availability.

Concrete incidents of attack or error in such integrated environments, like ransomware exploiting vulnerabilities in networked diagnostic devices, underscore the far-reaching repercussions of network weaknesses and demonstrate the necessity of continuous, cross-disciplinary diligence (FDA, 2020). The growing interplay of diverse device generations and operational contexts demands persistent, organization-wide threat awareness and adaptive, robust risk mitigation strategies.

Typical Threat Models for Connected Medical Devices

Connected medical devices operate within intricate clinical and network environments, filling critical roles in patient care while spanning a broad spectrum of stakeholders and vendors. The integration of devices—ranging from bedside monitors to remote cardiac implants—underpins healthcare's technological evolution but also generates opportunities for exploitation. With each connection and dependency, attackers gain a new avenue, motivated by a mix of financial, ideological, and strategic objectives.

Attacker Motivations and Typologies

Financially motivated actors are the most prevalent and persistent threat. Healthcare environments are lucrative for extortion-focused attackers because operational disruption leads to immediate, high-stakes consequences. Ransomware campaigns, for example, have targeted imaging devices and infusion pumps, encrypting control software and forcing hospitals to pay to restore life-sustaining equipment. Attackers frequently exploit this urgency to demand substantial ransoms, knowing that downtime can paralyze surgical suites or intensive care operations. Devices involved in complicated clinical workflows, if rendered inaccessible, have a cascading impact, affecting everything from diagnostics to electronic medical record (EMR) systems.

Ideologically driven adversaries seek to advance social, political, or hacktivist causes. These actors choose targets to maximize visibility during critical events or to further agendas by disrupting public health infrastructure. Hospitals become focal points during times of civil unrest or contentious political climates, with operations deliberately hampered to attract public attention or erode trust in institutional stability. The deliberate targeting of networked ventilators or critical monitoring devices during periods of heightened patient need exemplifies tactics rooted in ideological disruption rather than personal gain.

Strategic attackers, often associated with nation-states or sophisticated organized groups, operate with broader objectives. Healthcare systems, viewed as part of a nation's critical infrastructure, are valuable components in campaigns designed to degrade public morale, destabilize essential services, or extract sensitive medical research data. For instance, attempts to interfere with device communication protocols have been observed during escalating geopolitical tensions, with adversaries seeking long-term, systemic disruption rather than immediate, visible chaos.

Common Entry Points and Vulnerabilities

The complex web of connected medical devices is only as secure as its weakest element. Inadequate device authentication remains a prevalent vulnerability, with many devices still reliant on default, hard-coded, or weak credentials. This deficiency allows unauthorized users to gain privileged access to diagnostic tools or medication dispensers. Unsecured endpoints further expose devices to exploitation, particularly when boundary controls separating clinical and administrative networks are lax or inconsistently enforced.

Legacy communication protocols such as HL7, Digital Imaging and Communications in Medicine (DICOM), or proprietary vendor technologies often lack robust encryption or support for modern authentication frameworks. These weaknesses permit attackers to intercept, manipulate, or redirect device data streams and commands. External networks represent frequent gateways for intrusion. Patient monitoring portals, telehealth platforms, and remote management interfaces extend trust boundaries outside the physical perimeter of hospitals and clinics, increasing the exposure of critical devices to internet-based threats.

Insider threats present significant challenges, as employees with elevated privileges may misuse access, either intentionally or accidentally. A disgruntled technician can reprogram or disable safeguards on life-support systems, causing harm or operational delays. Third-party vendors, responsible for routine maintenance or software updates, add another attack surface; compromised laptops, remote-access tools, or stolen credentials can serve as entry points for wider infiltration campaigns.

Plausible Threat Scenarios

In real-world incidents, attackers have remotely altered dosing regimens on network-connected analgesia pumps, directly endangering patient safety by administering excessive medication. Disabling alarm functionality on central monitoring systems introduces dangerous response delays, as staff may remain unaware of deteriorating patient conditions. Other attacks leverage access to device-generated data. In a frequently cited breach, attackers extracted high volumes of diagnostic images and patient identifiers, reusing this information for identity theft, blackmail, or to manipulate clinical outcomes.

The operational impact extends beyond data breach. Ransomware targeting ventilators or surgical robots creates widespread chaos, causing mass cancellation of elective procedures, diverting emergency admissions, and potentially resulting in patient morbidity. The interconnectedness of devices means that a single compromised segment can threaten broader clinical workflows, compounding risk across entire healthcare networks.

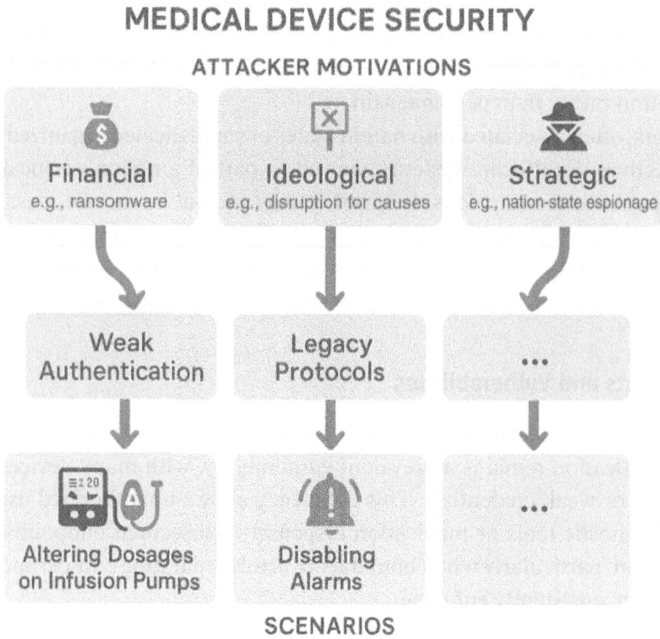

Evolving Attacker Tactics

Adversaries adapt as defenders strengthen technical and procedural safeguards. Advanced phishing schemes target clinical staff with messages crafted to mimic authentic system alerts or procedural updates. Social engineering bypasses technical controls altogether by tricking personnel into divulging credentials or approving malicious software installation. Attackers also search hospital networks for zero-day vulnerabilities—previously unknown software flaws lacking vendor-supplied patches. Tools leveraging AI and process automation scan hospital IP spaces, rapidly detecting devices running outdated firmware or exposed management interfaces.

Sophisticated malware now targets the unique features of medical device operating systems and communication protocols. Recent trends include the development of self-propagating threats engineered to exploit DICOM vulnerabilities or to install persistent backdoors in device firmware. Malware authors employ automated reconnaissance bots to map hospital device topologies and identify clusters of similar vulnerabilities, enabling coordinated and large-scale exploitation campaigns.

Continuous advancement of attacker methodologies, resourcefulness in discovering unprotected networks, and the availability of specialized, purpose-built malware accentuate the urgent need for industry-wide mechanisms that prioritize, standardize, and enforce countermeasures against both established and emerging threats within the expanding digital healthcare landscape.

Regulation Roundup: Navigating the Frameworks

The sophistication of modern cyber threats against medical devices creates a landscape where precise and harmonized regulatory controls are essential. Attackers now target not only individual devices but also the wider clinical environments in which they operate, exploiting vulnerabilities that can compromise patient safety, damage institutional reputations, and lead to significant financial loss (Ostermann et al., 2025). This hostile climate underscores the nonnegotiable need for rigorous and well-enforced security frameworks. Effective regulation provides a crucial baseline that supports the technical and ethical obligations of manufacturers and healthcare organizations, setting clear expectations to address emerging risks with consistency and technical depth (*Cybersecurity Requirements for Medical Devices*, 2025).

Regulatory oversight represents the backbone of medical device risk management. National and international authorities have established mandatory requirements encompassing the entire product life cycle, from conception through sustained use in clinical settings. Each jurisdiction espouses a framework of interlocking obligations, mandating that manufacturers proactively design, validate, and continually monitor the security of their devices. These controls do more than meet a legal minimum—they safeguard the well-being of patients and institutional systems.

FDA Guidance (United States)

In the United States, the Food and Drug Administration (FDA) sets specific expectations for cybersecurity both before and after a device reaches the market. Manufacturers must incorporate a comprehensive, risk-based approach to security in every phase of development. Premarket submissions require explicit documentation of threat modeling, identifying potential avenues of attack and corresponding mitigation strategies. Submissions must also include a software bill of materials, which allows evaluators to understand the full scope of integrated software components and their individual vulnerabilities (Ostermann et al., 2025). These measures enable regulators to assess whether manufacturers have anticipated and managed foreseeable threats.

After a device has entered the market, postmarket guidance compels manufacturers to establish ongoing mechanisms for vulnerability monitoring, patch development, and incident response. If a vulnerability is discovered, manufacturers must notify the FDA and affected users, develop remediation plans, and document all incident management steps. For example, a cardiac device company could demonstrate compliance by regularly monitoring third-party software components for newly disclosed weaknesses, issuing timely software patches, and maintaining thorough records of incident communications and risk reassessments (*Cybersecurity Requirements for Medical Devices*, 2025).

EU MDR/IVDR (European Union)

Europe's regulatory framework centers around the Medical Device Regulation (MDR) and the In Vitro Diagnostic Regulation (IVDR). These regulations view cybersecurity as integral to the overall safety and performance of a device. Manufacturers must conduct exhaustive risk assessments that cover technical and organizational threats throughout the device life cycle. Documentation must clearly outline identified risks, the likelihood and severity of harm, and the specific mitigation controls employed. Regulatory authorities expect ongoing, periodic cybersecurity reassessments to ensure newly discovered vulnerabilities or attack methods are addressed without delay (Ostermann et al., 2025).

A biomedical startup releasing a new diagnostic device in the European Union would need to maintain a cybersecurity file. This documentation details hazard identification, regular risk scoring, and implemented countermeasures such as encrypted communications or user authentication mechanisms. These records are subject to review during device conformity assessments and factory audits.

ISO/IEC Standards

Internationally harmonized standards play a foundational role in unifying practices across borders. ISO 14971 sets out the principles of risk management for medical devices. It requires structured risk analysis, evaluation, and control processes, which help manufacturers methodically identify hazards and develop appropriate safeguards. IEC 62304 addresses the software development life cycle, mandating clear processes for requirements definition, design, hazard analysis, verification, and maintenance. Adhering to these standards supports compliance across regulatory regions, enabling organizations to demonstrate the thoroughness of their design and monitoring activities (*Cybersecurity Requirements for Medical Devices*, 2025).

For example, incorporating a routine software hazard analysis into each development sprint promotes early identification of vulnerabilities, helping teams avoid the propagation of weak controls into finished products. This harmonized quality assures regulators and end users of consistent, risk-aware engineering practices regardless of product origin.

Consequences of Noncompliance

The operational and legal ramifications of failing to meet these frameworks can be severe. Regulatory authorities may impose sanctions, suspend or withdraw device approvals, and levy significant monetary penalties. Health organizations relying on noncompliant technology may face costly recalls or be barred from certain markets, while manufacturers can face increased liability if compromised devices result in patient injury or death (Ostermann et al., 2025). For instance, a company failing to patch a known vulnerability in an infusion pump may be subject to a global market recall and millions in fines following a cyber incident linked to delayed patient care and data breach.

Noncompliance erodes trust, damages reputations, and can force organizational restructuring or exit from the medical technology market. Industry professionals must therefore view regulatory vigilance not as a periodic task but as a core aspect of product stewardship and clinical partnership.

As connected technologies advance and medical devices age, the challenge of maintaining compliance intensifies. New technology introduces fresh attack surfaces, while legacy devices may lack features needed for current controls. Sustained risk mitigation remains critical, requiring ongoing adaptation and vigilance from all stakeholders.

The Legacy Device Dilemma

Achieving compliance with regulatory and standards-based cybersecurity requirements is fundamental for healthcare organizations, but this proves challenging when a significant share of the device inventory consists of legacy systems. Regulatory frameworks, such as those enforced by the FDA, demand manufacturers and users to demonstrate that devices are safe, effective, and capable of meeting evolving security threats throughout their operational life. Cyclical audits, ongoing risk assessments, and traceable records of patching or remediation activities are mandatory. However, these best practices and requirements frequently clash with reality: many medical devices remain in active duty for well beyond their original support windows, leaving organizations to navigate enforcement risks when patches or technical controls are missing. The persistent presence of outdated equipment often results not from neglect but from complex operational dependencies, large capital outlays for device procurement, and the mission-critical roles these devices play in patient care (Charter Communications, 2024; FDA, 2020).

Legacy devices, such as early-generation infusion pumps and diagnostic imaging equipment, lack the necessary support from manufacturers for software updates and security patches. When a vendor ceases maintenance, vulnerabilities that emerge remain unaddressed, creating ongoing attack surfaces within clinical networks. Attackers may exploit unpatched firmware flaws or exploit weak authentication schemes that do not adhere to current industry standards. Unsupported devices frequently run obsolete operating systems or embedded software incapable of sustaining modern encryption or access control protocols. The continued use of these technologies allows lateral movement by threat actors, often making the weakest link a strategic point of compromise in a hospital's security posture (Charter Communications, 2024).

The technical constraints on patching legacy devices are acute. In many cases, these devices require physical access for firmware updates, including use of proprietary diagnostic cables, legacy media, or on-site technical support. This disrupts clinical workflows and can temporarily take critical devices out of service. For example, updating a critical care cardiac monitor may require hours of downtime, making it unavailable for emergent patient care. Hospital IT teams face additional challenges: some devices must undergo extensive postupdate validation to comply with regulatory recertification, while others risk voiding their remaining warranties if unauthorized patches are applied (FDA, 2020). In practice, the cost and resource demands frequently exceed IT staff bandwidth, especially in organizations with sprawling device inventories.

Even if patching were logistically feasible, delayed or absent updates can have grave consequences. Attackers may obtain knowledge of known vulnerabilities from manufacturer advisories or public exploit databases and weaponize them far faster than healthcare organizations can implement mitigations. This risk is not hypothetical—multiple advisories focus on vulnerabilities in insulin pumps, intracardiac defibrillators, or mobile cardiac telemetry units. Unpatched devices may expose protected health information through unauthorized access, triggering regulatory action, or, in extreme scenarios, put patient safety at risk due to manipulated settings or unauthorized commands (Charter Communications, 2024).

Interoperability introduces further layers of difficulty. Many legacy devices communicate using outdated protocols lacking essential security features, such as device and user authentication, encryption, or proper segmentation. Integrating older diagnostic tools to an EHR platform may require custom middleware, network segmentation, or use of security bridges—each introduces new configuration complexities and lurking risks. For example, network translation gateways that make legacy infusion pumps visible to modern systems may themselves have vulnerabilities or lack timely updates, forming new attack vectors and points of failure. Efforts to overlay modern security solutions, such as network-level firewalls or microsegmentation, help but rarely fully compensate for the absence of device-level controls (FDA, 2020).

Balancing the replacement or retrofitting of legacy equipment complicates decision-making. Full device replacement provides the most definitive resolution but comes at high financial cost and may disrupt care delivery. Procurement cycles can take years, and existing capital budget constraints frequently slow adoption of replacement technologies. On the other hand, retrofitting through network segmentation, authentication overlays, or secure gateways may offer incremental improvements, but these measures seldom bring devices up to parity with modern security standards and may have hidden compatibility hurdles. Organizations must conduct rigorous risk assessments to weigh the likelihood and impact of threats, then decide if retrofitting yields sufficient risk mitigation or if the business case for replacement is justified. These evaluations further depend on factors like vendor roadmaps, reimbursement models, and overall impact on clinical operations (Charter Communications, 2024; FDA, 2020).

Throughout these technical, operational, and financial dilemmas, patient data protection remains at the heart of every risk calculation. Legacy device vulnerabilities and operational challenges contribute directly to exposure of sensitive patient information and disruption of health services. As healthcare organizations parse the risks and benefits of maintaining legacy assets, they confront complex trade-offs that touch upon not only cybersecurity best practices but also the broader mandates for privacy and uninterrupted patient care. The pressing need to secure aging medical devices therefore becomes inseparable from the ongoing responsibility to safeguard the privacy and security of patient data across the healthcare continuum (Charter Communications, 2024; FDA, 2020).

Data Privacy vs. Data Security in Healthcare

Legacy medical devices have often made healthcare networks more vulnerable because their aging hardware and software lack modern protections, and many cannot be upgraded to meet current standards. Hospitals and clinics must deal with exposure to attacks that can exploit unpatched code or weak authentication routines. Even as these devices continue to play vital roles in treatment and diagnostics, their integration with newer systems extends the attack surface and increases risk. For example, when an older infusion pump connects with EHRs, it becomes a potential entry point for attackers seeking sensitive patient data or system control.

Privacy and security, while intimately linked, require careful distinction in this context. Privacy focuses on who is allowed to access or share patient health information. It reflects ethical obligations to protect individuals' rights, addresses patient trust, and follows legal principles about consent, access, and confidentiality. Regulatory frameworks, like the Health Insurance Portability and Accountability Act (HIPAA) and General Data Protection Regulation (GDPR), underscore that keeping patient data private is not just a technical matter but an ethical duty for clinicians, administrators, and technology suppliers (Conduah et al., 2025; McGraw & Mandl, 2021). Security provides the technological basis for enforcing privacy rules. It covers all the safeguards—physical, procedural, and technical—that prevent unauthorized access, ensure data cannot be manipulated, and shield systems from interruption or destruction.

The overlap comes into focus through global regulations that blend privacy and security mandates. HIPAA in the United States requires organizations to limit access to protected health information, give patients control over who sees their data, use encryption, and notify authorities about breaches. The GDPR in Europe sets even stricter rules around data minimization (collect only what is needed), purpose specification, and the right to erasure, alongside demanding robust technical controls such as encryption and pseudonymization (Conduah et al., 2025). The California Consumer Privacy Act (CCPA) and South Africa's POPIA add to this global web, requiring organizations to report breaches and safeguard data across jurisdictions. Hospitals adopting new medical technologies must verify that device log data, for example, cannot be easily linked to a patient, and that any patient identifiers stored locally or in the cloud meet regional consent and encryption standards.

Tension often arises between the aims of privacy and the technical needs of security. Rigorous authentication or access controls, implemented to prevent data loss or system tampering, can slow down clinicians who need swift access to critical patient records at the bedside. On the other hand, privacy rules may require organizations to anonymize or limit access to logs for oversight, reducing the ability of security teams to spot insider threats or respond to incidents in real time (McGraw & Mandl, 2021). An example can be drawn from a hospital's network team establishing surveillance tools to spot malware outbreaks. If these tools capture too much detail, they can risk violating privacy requirements. Conversely, if restricted too tightly, the system may miss vital signs of compromise.

Striking this balance requires layered solutions, such as role-based access controls. A hospital administrative assistant might have clearance to see scheduling data but not clinical notes, while an oncologist can view lab results relevant to their patients but not unrelated individuals in the system. Regular audit logging lets the institution monitor for inappropriate access, while privacy-aware policies ensure logs mask or tokenize patient identifiers. Manufacturers updating device firmware for security must test workflows so updates don't inadvertently increase the risk of unauthorized data disclosure.

Consider health data flow within typical organizations. Patient vitals and medication orders travel from bedside devices, across wireless hospital networks, to local servers or the cloud. Backups and analytics may transmit data elsewhere, and external researchers or specialists could request access for care coordination or studies. Each handoff or connection point introduces potential vulnerabilities, such as exposure to intercepts, unauthorized access, or accidental disclosure. Practical defenses must address each link: encrypting data in transit and at rest, applying multifactor authentication for remote device management, and ensuring information sent outside the organization is de-identified or minimized as required (Conduah et al., 2025; McGraw & Mandl, 2021).

A compliance officer overseeing a multiregion deployment must reconcile overlapping but not identical requirements. For example, she might have to respond to a data breach differently in Europe versus the United States, as each jurisdiction imposes unique timelines and reporting obligations. A failure to clarify whether a security safeguard also upholds privacy obligations, or vice versa, can lead to fines, damaged reputation, or compromised patient trust.

Recognizing and properly managing the distinct but interdependent nature of privacy and security is essential for protecting individuals and institutions in today's connected healthcare ecosystem. Clear definitions and practices make it possible to deploy and use innovative devices while meeting regulatory demands and maintaining public confidence. In rapidly changing environments, maintaining this clarity is not optional but foundational for ethical, safe, and effective healthcare.

Bringing It All Together

Having established a comprehensive understanding of the medical device ecosystem, associated threat models, regulatory frameworks, legacy device challenges, and the nuanced relationship between data privacy and security, we are now equipped to address the multifaceted risks inherent in connected healthcare technologies. This foundational knowledge enables cybersecurity professionals, engineers, and compliance officers to anticipate vulnerabilities, design effective safeguards, and implement policies that balance safety, functionality, and regulatory compliance. As medical devices continue to evolve and integrate into complex clinical environments, ongoing collaboration, adaptive risk management, and adherence to rigorous standards will be essential to protecting patient well-being and maintaining trust in healthcare systems worldwide.

References

Charter Communications. (2024). *Reducing cybersecurity risk from legacy medical devices | Spectrum Business*. Spectrum.com. https://enterprise.spectrum.com/insights/blog/legacy-medical-devices-cybersecurity-risk.html.

Conduah, A. K., Ofoe, S., Siaw-Marfo, D. (2025, May). *Data Privacy in Healthcare: Global Challenges and Solutions*. DIGITAL HEALTH. https://doi.org/10.1177/20552076251343959.

Cybersecurity Requirements for Medical Devices. (2025). TÜV SÜD. https://www.tuvsud.com/en-us/industries/healthcare-and-medical-devices/medical-devices-and-ivd/medical-device-testing/medical-device-cybersecurity.

FDA (2020, September 22). *Cybersecurity*. FDA. https://www.fda.gov/medical-devices/digital-health-center-excellence/cybersecurity.

McGraw, D., Mandl, K. D. (2021, January 4). *Privacy Protections to Encourage Use of Health-relevant Digital Data in a Learning Health System*. NPJ Digital Medicine.

Ostermann, M., Mathias, R., Jahed, F., Parker, M. B., Hudson, F. D., Harding, W. C., Gilbert, S., Freyer, O. (2025, July 15). Cybersecurity Requirements for Medical Devices in the EU and US—A Comparison and Gap Analysis of the MDCG 2019-16 and FDA Premarket Cybersecurity Guidance. Computational and Structural Biotechnology Journal; Elsevier. https://doi.org/10.1016/j.csbj.2025.07.024.

The Ultimate Guide to Seamless Medical Device Integration. (2024, August 30). Virtelligence. https://virtelligence.com/blog/ultimate-guide-seamless-medical-device-integration/.

3

Architectures and Technologies: The Building Blocks of Secure Devices

How can medical devices maintain trust when every component—from the moment they power on to the data they send across networks—faces potential threats? What architectural choices and technologies ensure that life-critical systems remain safe, reliable, and resistant to tampering or unauthorized access? How do embedded software and hardware work together to defend against sophisticated attacks while meeting stringent regulatory requirements? As healthcare increasingly relies on connected devices, these questions become vital not only for engineers designing the technology but also for cybersecurity professionals, regulators, and clinical staff responsible for protecting patients. Exploring the underlying structures and mechanisms that form the backbone of secure medical devices sheds light on how these challenges are addressed in practice.

Hardware Security: The Root of Trust

Hardware-based trust mechanisms are crucial for medical devices due to their role in safeguarding patient health, meeting regulatory obligations, and maintaining system reliability. Within healthcare, the integrity of a device directly affects patient outcomes, driving the need for an immutable foundation that can attest to its trustworthiness. Hardware-level security features mitigate risks by establishing a clear, auditable baseline from the moment a device powers on. These features form the basis for subsequent software protections, meeting stringent safety and compliance standards.

Secure Boot

Secure boot serves as the initial barrier against unauthorized firmware by ensuring only signed and verified code executes during device startup. At power-on, a securely embedded root-of-trust module checks the integrity and authenticity of the firmware against cryptographic signatures stored in tamper-resistant hardware. The process starts when the device loads an immutable bootloader from its hardware root of trust. This bootloader carries out cryptographic verification on the firmware image, validating it using a digital signature created by the device manufacturer. If this check fails, the boot process halts or enters a recovery state to prevent any further execution. Each subsequent component, such as the operating system, is chained to the previous verification outcome, maintaining a continuous and auditable trust chain. This architectural choice prevents the installation of malware or persistent rootkits that could otherwise compromise clinical workflows or allow attackers to tamper with device behavior undetected (*Hardware Based Security for Government—Purism*, 2024). Medical devices employing secure boot protocols thus address concerns identified in regulatory frameworks, such as FDA requirements for device integrity and safety.

Trusted Platform Modules (TPMs) and Secure Elements

TPMs and secure elements are dedicated microcontrollers embedded within medical devices to provide hardware isolation for critical security tasks. TPMs protect sensitive cryptographic keys, passwords, and digital certificates within a physically shielded area, offering resistance to both software-based attacks and physical extraction attempts (*Trusted Platform Module (TPM) in Embedded System Security*, 2025). These keys are used for device authentication, secure boot processes, and storing attestations of device state. For instance, when a medical device connects to hospital networks or cloud-based platforms, its TPM proves its identity by signing cryptographic challenges with a private key held within the chip. This prevents attacks that leverage cloned or counterfeit devices to gain unauthorized access.

Secure elements go further by offloading cryptographic computations from the main processor, minimizing the device's attack surface. They safeguard transactions such as secure wireless updates, payment operations, or patient data encryption, crucial for mobile medical devices that require wireless authentication or transmit electronic protected health information (ePHI). Their secure memory ensures that, even in the event of a breach in the operating system or application layer, attackers cannot retrieve authentication material or secret keys, maintaining compliance with privacy-focused regulations like the Health Insurance Portability and Accountability Act (HIPAA) (*Hardware Based Security for Government—Purism*, 2024).

Physical Tamper Resistance

Robust physical security features are vital to prevent direct hardware attacks, such as attempts to extract cryptographic keys or manipulate firmware. Medical devices integrate tamper-evident packaging and custom enclosures made from materials engineered to foil drilling, chemical exposure, or microprobing. On-board sensors can detect physical intrusions—such as excessive temperature, voltage, or attempts to open the casing—and will trigger protective responses. For example, the system may erase sensitive cryptographic information or lock down critical functionality, rendering the device inert if tampering is detected. These actions inhibit attackers seeking to reverse engineer firmware or access restricted data. Design strategies include circuit mesh barriers, epoxy potting on critical modules, and encrypted secure storage that self-destructs key material in response to sensor events. This level of tamper response fulfills requirements highlighted in various regulatory guidelines and reassures clinical users regarding device trustworthiness.

Hardware Random Number Generators

Strong cryptographic systems depend on true random numbers for secure key generation, authentication, and encrypted communications. Hardware random number generators rely on unpredictable physical processes, such as electronic noise, oscillator jitter, or radioactive decay, to produce entropy that is virtually impossible to reproduce or guess. In contrast, software-based pseudorandom generators often use deterministic algorithms that can be predicted if their internal state or seed is exposed. Attackers have exploited weak random number generator (RNG) implementations to reconstruct session keys or credentials, enabling unauthorized access in security-sensitive systems.

In medical environments, robust hardware RNGs ensure that keys and authentication tokens used to secure patient data or device communications are not susceptible to replay or brute-force attacks. By drawing on entropy from nondeterministic processes embedded directly in secure hardware modules, medical devices can provide provably unpredictable key material, satisfying security audits and cryptographic certification standards (*Hardware Based Security for Government—Purism*, 2024).

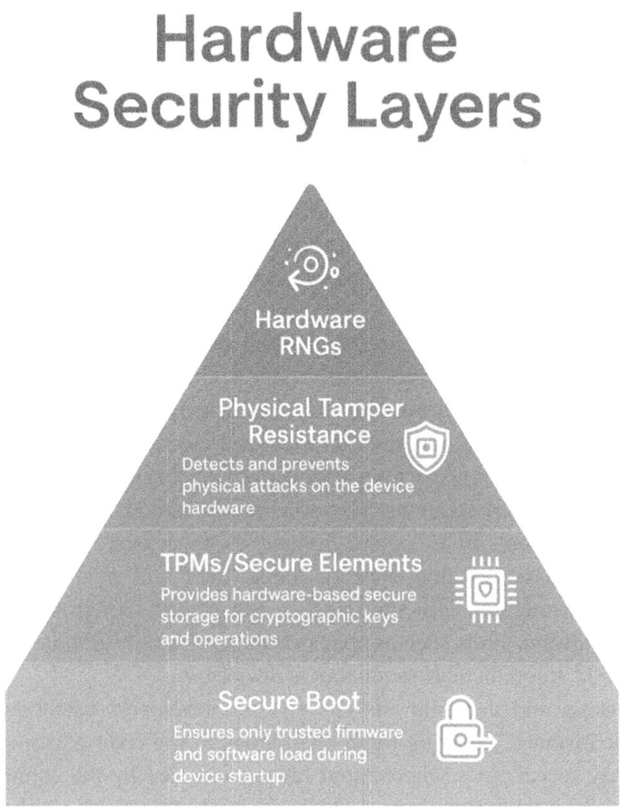

A strong hardware security foundation ensures the trustworthiness of medical devices from initial power-on through to secure network integration and encrypted patient data handling. By anchoring software security controls in transparent, provable hardware mechanisms, device makers provide high-assurance protections that regulatory bodies and clinical stakeholders demand, supporting the safe and reliable delivery of patient care.

Software Security Fundamentals for Medical Devices

Establishing trustworthy embedded medical device systems necessitates a steadfast focus on secure coding techniques and robust software architectural principles. High code quality lies at the foundation of resilient medical software, directly reducing the prevalence of latent vulnerabilities while also improving maintainability across a device's life cycle (Espinosa, 2024; Gupta, 2025). The adoption of memory-safe programming languages, such as Rust or modern iterations of Java, exemplifies this approach by significantly lowering the propensity for buffer overflow vulnerabilities and memory corruption. In contrast, unsafe or legacy C code has repeatedly served as a vector for severe exploitation, especially in resource-constrained embedded environments.

For instance, a single unchecked buffer copy operation could result in a remote code execution flaw that might allow unauthorized modification of infusion pump dosages, placing patient safety at risk. Selecting languages and frameworks with inherent memory management safeguards, and consistently enforcing their use across all subsystems, helps ensure these hazardous classes of errors are systematically excluded from codebases.

Maintaining a consistent, modern code style across the development team further mitigates security-critical errors. Leveraging language features such as bounds-checked data structures, strong type definitions, and exception handling reduces ambiguity and eliminates many traps that can result in subtle vulnerabilities. For example, explicit variable initialization policies and the avoidance of implicit conversions decrease the risk of unintentional logic errors, such as those undermining access control or input processing routines. Standardizing on advanced static analysis tools within continuous integration pipelines supports adherence to these software quality guidelines, enabling developers to detect nonconformant code early and with minimal additional effort (Gupta, 2025).

Static code analysis and dynamic runtime verification serve as mutually reinforcing safeguards against the introduction of vulnerabilities. Static analysis tools, such as Coverity and CodeSonar, scrutinize source code before execution and excel at identifying patterns associated with null pointer references, misuse of cryptographic libraries, and race conditions. Static analysis proves especially effective for preemptively catching unsafe input parsing in routines that establish device settings or actuate hardware components. However, static methods alone cannot always expose runtime behaviors dependent on execution context or asynchronous hardware interactions. Dynamic analysis methods, including fuzzing frameworks and memory sanitizers, evaluate compiled software as it interacts with simulated or live external environments. This approach excels at uncovering runtime faults such as heap overflows or input validation gaps that may only manifest under specific, unpredictable sequences of device use.

Deploying both static and dynamic analysis ensures developers catch both structural and emergent flaws. For instance, applying these techniques to the code governing infusion pump rate adjustments may reveal cases where input ranges accepted by the command parser differ between code paths, potentially exposing unsafe device states that evade single-method detection (Espinosa, 2024). Comprehensive adoption of both forms of analysis across the entire software life cycle creates a feedback loop that not only addresses immediate security defects but also strengthens institutional knowledge of common failure patterns.

Sound memory management remains a cornerstone for embedded medical safety. Devices are vulnerable to exploitation via out-of-bounds writes or heap overflows, and the consequences in a clinical setting are profound. A single malformed network packet or corrupt file system operation can cause unpredictable device crashes during patient monitoring or diagnostic imaging, compromising data integrity or availability. Furthermore, poorly managed memory may lead to the unintentional retention of sensitive patient information beyond its intended scope, enabling data leakage through subsequent memory reuse or system diagnostics dumps. Developers should apply boundary checking at every layer of data handling, ensuring that all input and internal buffers are sized and validated to prevent overruns. Utilizing safe libraries, like those designed specifically for embedded contexts and subject to rigorous open-source review, limits exposure to well-known flaws. All use of deprecated or explicitly unsafe routines—such as "strcpy" in C—must be prohibited within code standards and enforced via automated review tools, since these functions are responsible for numerous catastrophic device failures.

The software supply chain introduces an external risk dimension that cannot be overlooked. Third-party libraries and commercial off-the-shelf (COTS) software constitute a large portion of embedded stacks, often providing vital cryptographic or communications functionality. These dependencies must be rigorously vetted before integration by scanning all new and updated packages for known vulnerabilities with dedicated software composition analysis tools, such as Black Duck or Snyk, and maintaining strict version controls to ensure timely patching. Organizations should establish clear criteria for trusted suppliers, requiring detailed provenance and change history documentation for every library or binary artifact imported into production builds. Workflow mandates—such as only allowing code with signed attestations of origin and preapproved security reviews—further reduce the risk that a compromise enters via external channels (Gupta, 2025).

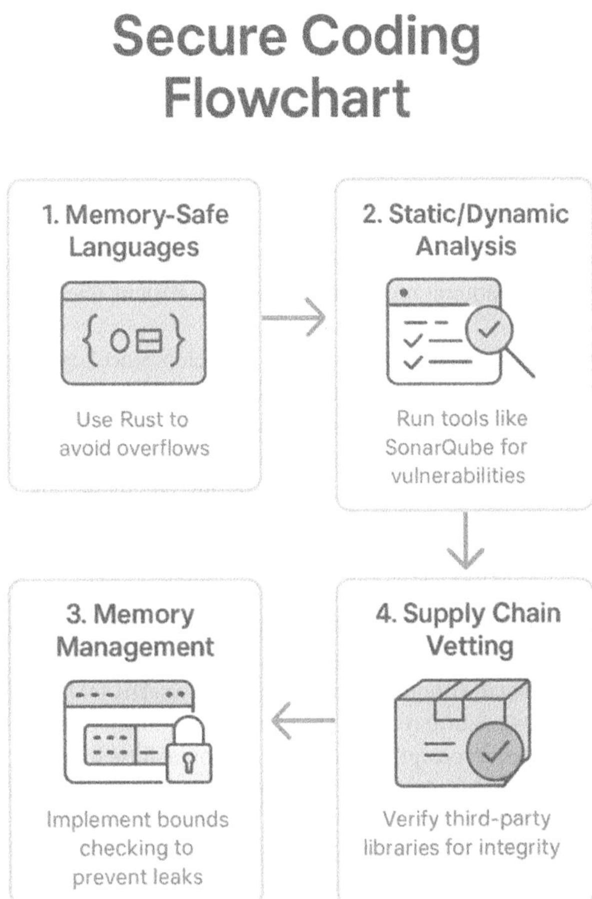

As device interconnectivity increases across clinical settings, attention must turn to the secure movement of data between heterogeneous components within these networks. Ensuring that software interfaces and communication endpoints are developed on foundations grounded in secure coding principles lays the groundwork for robust, regulated exchanges of clinical and operational information, facilitating the transition to more specific analysis of secure network protocol options and their role in safeguarding healthcare communications (Espinosa, 2024; Gupta, 2025).

Networking Protocols in Healthcare

Standardized communication protocols establish the foundation for interoperability across diverse medical devices and information systems in healthcare. Protocols such as Health Level 7 (HL7) and Digital Imaging and Communications in Medicine (DICOM) serve as cornerstones, enabling consistent formatting, integration, and reliable transmission of clinical and imaging data between laboratories, radiology units, and electronic health records. HL7 structures the grammar and syntax for exchanging lab results, admission records, and medical orders. DICOM encapsulates imaging metadata, enforcing rules for transferring MRI, CT, and ultrasound images within and between hospital networks (*20 Medical Device Protocols: Descriptions, Purposes, Security Flaws, FDA Guidance, and Examples—Blue Goat Cyber*, 2024). These protocols maintain data fidelity and clinical accuracy, yet their efficacy depends on rigorous attention to security throughout their life cycle.

Wireless protocols have become essential in connecting a new generation of medical devices that focus on portability, low power consumption, and patient comfort. Bluetooth Low Energy (BLE) powers communication across wearable monitors, continuous glucose meters, and smart inhalers, allowing periodic health data uploads to central aggregators or clinician dashboards. Zigbee offers a mesh networking architecture, linking implantable devices, bedside telemetry, and smart infusion pumps with redundant nodes to ensure resilience against local interruptions. Hospitals increasingly employ Wi-Fi as the backbone network, supporting vendor-agnostic communication and high-throughput integrations among mobile imaging carts, patient trackers, and medical workstations (Koutras et al., 2020).

Context-driven security approaches are mandatory, as each protocol operates amid distinct risk landscapes. BLE's short-range design does not obviate the need for encryption or secure pairing, given the proximity of attackers in shared hospital spaces. Zigbee's mesh model increases attack surface due to greater numbers of connecting devices, necessitating robust node authentication and periodic key refresh. Wi-Fi bridges clinical and administrative domains, often carrying both protected health information and routine network traffic. Security policies segment devices, but vulnerability in one zone can cascade across the system without careful boundary enforcement.

Many communication protocols adopted for medical environments stem from standards conceived before the contemporary focus on cyberthreats, leading to inherent gaps. HL7, for instance, often lacks provisions for native payload encryption or endpoint authentication, allowing for eavesdropping or spoofing if deployed without compensating controls. DICOM traditionally provides no mandated encryption of image data at rest or in motion; plain transmission over hospital networks permits not only unauthorized interception but also injection of false imaging, potentially altering diagnoses (*20 Medical Device Protocols: Descriptions, Purposes, Security Flaws, FDA Guidance, and Examples—Blue Goat Cyber*, 2024).

Wireless protocol vulnerabilities gained notoriety after proof-of-concept attacks on unencrypted BLE transmissions, where adversaries captured and replayed blood glucose readings, generating false alarm events. Hospitals reported malware intrusions via inadequately secured Wi-Fi-connected devices, sometimes resulting in ransomware attacks that disrupted patient monitoring and forced diversion of acute care (Koutras et al., 2020). These failures occur when protocols transmit data in the clear or implement weak, static credentials for authentication. Insufficient cryptographic protections can result in the disclosure of patient identities or, more critically, manipulation of pacemaker or infusion pump settings.

End-to-end encryption forms the backbone of defending data in motion within medical device networks. Protocols like Transport Layer Security (TLS) implement handshake mechanisms, negotiating symmetric session keys using public-key cryptography and ensuring confidentiality and message integrity across hospital LANs and wireless segments. Embedded systems often apply TLS profiles tailored for constrained resources, such as TLS_ECDHE_ECDSA with lightweight curves and hardware-accelerated cryptographic modules, to maintain performance within wearable devices (Koutras et al., 2020).

Strong key management sustains secure encryption. Manufacturers provision devices with unique, factory-generated keys stored in tamper-resistant modules. Key rotation schedules—using ephemeral keys issued with each authentication session—reduce exposure if attackers gain transient access. Cryptographic key exchanges leverage Elliptic Curve Diffie-Hellman (ECDH) to securely derive session keys between diagnostic workstations and imaging servers.

Modern authentication surpasses static passwords and introduces mutual verification at the protocol layer. Mutual TLS (mTLS) uses X.509 certificates installed in device secure elements or hardware keystores. Certificate pinning locks a device to accept only a specific server certificate, preventing man-in-the-middle attackers from presenting counterfeit credentials. For implantable devices and critical monitoring solutions, hardware tokens based on TPMs provide cryptographically signed nonces—unique, device-bound identity assertions validated by backend servers. This ensures the sender is authentic and prevents replay or impersonation attacks, even if the network is compromised.

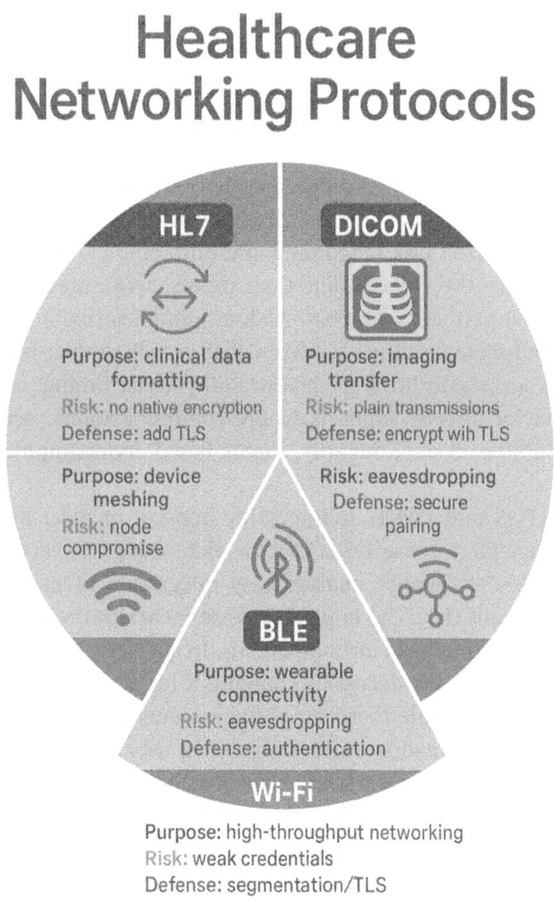

Cryptographic credentials circulate within a layered trust architecture: devices prove their identity to gateway servers, which provision session keys and broker connections to hospital data stores. Central policy engines manage certificate revocation, audit credential usage, and trigger isolation in the event of compromise. Routine updates and vulnerability assessments help sustain defensive posture amid evolving threats.

Secure implementation of communication protocols does not exist in isolation. Embedded operating systems enforce process isolation for protocol handlers, restrict device file access, and apply real-time policy checks on all communications. Access controls ensure only authorized processes initiate protocol stacks, and mandatory integrity checks monitor for tampering or unsafe stack modifications. These operational controls form the link from cryptographic defense at the protocol level to systemic enforcement at the operating system layer (Koutras et al., 2020; *20 Medical Device Protocols: Descriptions, Purposes, Security Flaws, FDA Guidance, and Examples—Blue Goat Cyber*, 2024).

Operating Systems and Real-Time Constraints

Modern medical devices increasingly depend on secure, real-time communication protocols to connect with clinicians, hospital networks, and remote monitoring platforms. The security and assurance of these protocols, especially in life-critical scenarios, are fundamentally tied to the capabilities and architecture of the device's underlying operating system. Embedded systems in medical technology must support both communication security—like robust encryption and authentication—and the uncompromising reliability expected in environments where patient safety is on the line.

Real-Time Operating Systems (RTOSs) differ sharply from general-purpose operating systems (GPOS) such as Windows or Linux in their architectural priorities. An RTOS is purpose-built to provide deterministic, prioritized task scheduling and ultra-predictable response to time-critical stimuli (*RTOS: A Comprehensive Guide—Blue Goat Cyber*, 2024). Medical scenarios frequently depend on guarantees that particular actions—such as administering a precise drug dose—will occur within microsecond-scale deadlines. Unlike GPOS, which emphasize broad versatility and feature breadth, RTOSs are streamlined for predictability, strict timing, and resource isolation. This deterministic behavior directly supports secure communications, where delays or missed deadlines can lead to failures in data integrity, timing-based cryptographic protocols, or urgent alarm notifications.

Choosing between RTOS and GPOS architectures impacts medical device security in several important ways. General-purpose OSes include rich sets of drivers, network stacks, and management utilities. This abundance enables easy integration of new connectivity features but also introduces significant risk. The larger codebase increases the attack surface, making it difficult to verify the trustworthiness of all components. In addition, nondeterministic scheduling in GPOS can lead to unpredictable delays or contention for resources, which is unacceptable in critical real-time scenarios. While these operating systems tend to offer more sophisticated patch management, update automations, and fine-grained user controls, their complexity often outweighs these advantages in life-critical medical environments (*RTOS vs Bare-Metal*, 2023).

RTOSs, in contrast, execute a tightly controlled scheduler that only allows high-priority medical processes to preempt less critical tasks. By avoiding unnecessary background tasks and minimizing the code footprint, RTOSs reduce not only performance-related risk but also opportunities for attackers to exploit vulnerabilities in less-used software modules. A small attack surface and predictable task execution enable security controls to be more effectively audited and verified. Although this comes at the cost of general flexibility, it is an essential foundation for secure medical workflows—especially where any unpredictability could have direct safety consequences (*RTOS: A Comprehensive Guide—Blue Goat Cyber*, 2024).

Security patch management presents a crucial balancing act for medical device manufacturers. Vulnerability remediation must occur promptly to protect against threats, but maintenance windows are often limited, and device downtime can have grave implications for patients. Effective update mechanisms address this with several technical strategies. First, all updates must be cryptographically signed and verified at deployment, ensuring only authentic, trusted firmware or software is installed. Second, staged update processes, where updates are downloaded and verified in advance but only applied during safely scheduled periods, can prevent disruption. Dual-partition storage designs offer further assurance: the active device firmware is isolated from the area storing the new update, and the device can revert to a known-good previous version if the update fails or interrupts critical operations. These mechanisms together help patch vulnerabilities while upholding ongoing device integrity and real-time performance—the heart of safe medical device operations (*RTOS: A Comprehensive Guide—Blue Goat Cyber*, 2024).

Process scheduling and robust task isolation are key architectural features for defending against system compromise. Real-time operating systems are often configured so that diagnostic modules, data logging tasks, and communication processes are strictly separated from core therapy delivery components. In an infusion pump, for example, isolating the therapy module ensures that even if a noncritical diagnostic process were compromised, malicious commands cannot affect the device's drug delivery functions. Enforcing least-privilege principles—that is, giving each process only the access rights it strictly needs—and using mandatory access controls over inter-process messages further limit the lateral movement of threats inside the device. Minimal, clearly defined inter-process communications not only streamline security oversight but also contain potential breaches to noncritical parts of the system (*RTOS vs Bare-Metal*, 2023).

Trusted Execution Environments (TEEs) take this separation even further by carving out protected processor regions for ultra-sensitive operations. TEEs securely manage cryptographic keys, patient credentials, and secure boot processes, shielding them both from the rest of the operating system and from any dubious third-party code. This is especially critical when transmitting confidential patient data wirelessly or verifying the authenticity of device software at startup. While the integration of TEEs brings new challenges—like stricter development requirements and potential platform compatibility issues—their benefit is substantial in safeguarding high-value data streams and establishing trust in device operations. For example, a TEE might securely sign patient records as they are transferred to a clinician's tablet, protecting against both interception and manipulation (*RTOS: A Comprehensive Guide—Blue Goat Cyber*, 2024).

As medical devices move toward remote management, over-the-air updating, and cloud-integrated decision support, robust foundation in specialized operating systems and real-time architectures becomes even more central. The ability to reliably and securely support advanced connectivity, while meeting the unyielding demands of patient safety, requires careful choice and technical execution at the OS level—setting the stage for addressing new challenges and opportunities in the evolving landscape of connected health technology.

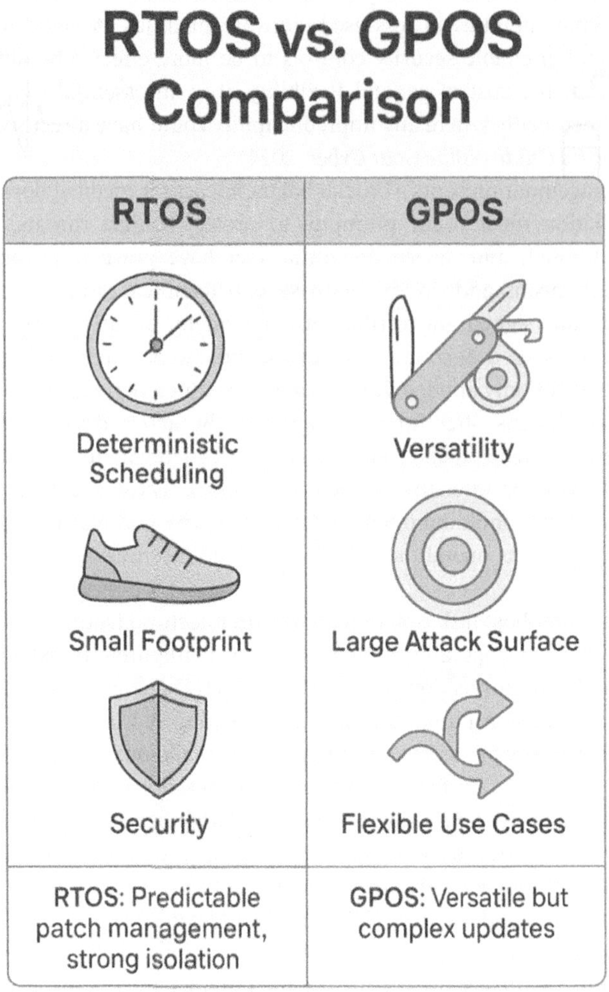

Cloud and Remote Access Mechanisms

Medical device security extends from the integrity of real-time operating systems to the broader challenge of cloud integration and remote management. Once devices achieve reliable local execution and data handling, the architecture must adapt to systematic connectivity with cloud services. This transition enables scalable data aggregation, external analytics, and continuous support but simultaneously increases the attack surface and regulatory burden.

Cloud integration models for medical devices fall into two categories: direct device-to-cloud and gateway-mediated communication. Direct integration allows devices like infusion pumps to transmit telemetry or software logs straight to a cloud platform using encrypted network channels. This architecture supports efficient scaling of device fleets and minimizes latency for time-sensitive analytics. Its major risk is exposing endpoint devices directly to the open internet or wide-area networks. A breach of device credentials or misconfiguration can place sensitive data or even device controls at risk.

Gateway-mediated models interpose a local aggregation point—either onsite or within a protected network. Implantable devices, for example, may relay periodic state data to a hospital-managed gateway, which then enforces additional security and policy controls before forwarding to the cloud. This reduces risk by containing initial exposure in a more tightly controlled domain. However, gateways also become critical trust anchors whose compromise can affect all connected assets. In both patterns, designers must ensure mutual authentication and transport encryption protocols—such as TLS 1.2 or above—enforce confidentiality and integrity for medical data traversing external networks (Mehrtak et al., 2021; Pradeep Kumar et al., 2023). Regulatory frameworks like HIPAA or General Data Protection Regulation (GDPR) require that protected health information is not only encrypted in transit but also subject to access auditing and breach notification controls.

Remote monitoring enables real-time observation and troubleshooting of complex medical devices. For an imaging scanner or cardiac implant, remote telemetry empowers specialists to detect anomalies from anywhere. This accessibility, if not precisely managed, introduces new vectors for unauthorized control, data theft, or sabotage. Threat actors have used compromised remote access systems to manipulate device settings or stage ransomware attacks on healthcare infrastructure. To counteract this, effective systems implement strong transport encryption, commonly using mutual TLS. All authentication attempts should require multiple factors, such as possession-based digital certificates plus time-limited codes. Every control session and data access must be logged with tamper-evident records to satisfy forensic and regulatory requirements (Pradeep Kumar et al., 2023).

Concrete scenarios demonstrate this necessity: a biomedical field engineer remotely diagnosing a CT scanner uses a vendor portal that enforces device certificate checks and records all configuration changes. Similarly, a hospital's cardiac telemetry aggregation feeds cloud dashboards that are access-controlled using rotating credentials and monitored for anomalous access patterns. Secure APIs play a central role, requiring rigorous definition and ongoing verification (Mehrtak et al., 2021).

Keeping edge software current requires a robust update infrastructure. Devices must validate update payloads through cryptographic signatures, ensuring that only authentic firmware from trusted sources is installed. Updates travel over prescreened channels like mutually authenticated HTTPS or VPNs. Any update process must verify both sender and recipient cryptographically; mutual authentication prevents attackers from introducing malicious code or exploiting replay attacks (Pradeep Kumar et al., 2023). In practice, examples from infusion pump fleets or surgical robots show that cryptographic code signing and staged(rollout) procedures have averted large-scale compromise attempts. When errors occur—such as faulty updates—the device must be capable of rolling back to the last known good software state, thus limiting operational downtime or clinical risk.

As devices gain richer connectivity, their value can be enhanced by integrations with third-party analytics providers, research databases, or remote service partners. Each external integration, though, introduces potential for new vulnerabilities. Analytics engines could siphon protected patient data; insufficiently hardened partner platforms may become lateral entry points for broader network compromise. To mitigate these threats, organizations need strict partner vetting, requiring external platforms to comply with established security certifications and frequent audits (Mehrtak et al., 2021). Data exchanges must operate with the principle of least privilege, granting minimum access necessary for each integration. Continuous security monitoring is mandatory, with automated alerts for anomalous data flows or privilege escalations.

Examples illustrate these risks. Unvetted cloud analytics tools have, in some cases, introduced pathways for large-scale data exfiltration or platform compromise. On the control side, least-privilege design prevents analytics providers from reconfiguring device firmware or accessing raw sensor feeds not strictly needed for their purpose. Integrated monitoring systems draw on both network telemetry and device logs to trigger incident responses swiftly.

By intertwining secure local design, encrypted communications, credentialed remote access, and meticulous update ecosystems, medical device systems support clinical safety and operational resilience, even as their networks and use cases expand across cloud platforms and remote domains (Mehrtak et al., 2021; Pradeep Kumar et al., 2023).

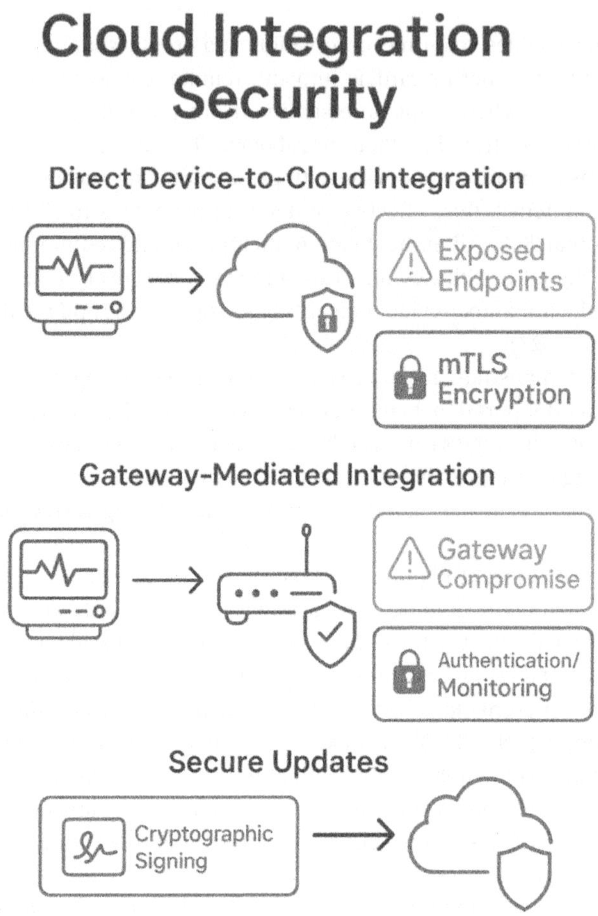

Bringing It All Together

Understanding the core architectures and technologies that underpin secure medical devices is essential for advancing patient safety and regulatory compliance in an era of increasing connectivity. By grounding device security in hardware roots of trust, rigorous software development practices, robust networking protocols, and specialized real-time operating systems, stakeholders can build resilient systems capable of defending against evolving cyber threats. As cloud integration and remote access become integral to healthcare delivery, implementing strong encryption, multifactor authentication, and vigilant update mechanisms will be critical to maintaining trust and operational integrity. Armed with this foundational knowledge, healthcare professionals, engineers, and regulators are better equipped to design, manage, and oversee medical device ecosystems that not only meet today's security demands but also adapt proactively to future technological and threat landscapes.

References

20 Medical Device Protocols: Descriptions, Purposes, Security Flaws, FDA Guidance, and Examples—Blue Goat Cyber. (2024, January 15). Blue Goat Cyber. https://bluegoatcyber.com/blog/20-medical-device-protocols-descriptions-purposes-security-flaws-fda-guidance-and-examples/.

Espinosa, C. (2024, February 9). *Secure Coding for Medical Devices—Blue Goat Cyber*. Blue Goat Cyber. https://bluegoatcyber.com/blog/secure-coding-for-medical-devices-ensuring-safety-and-protection-in-healthcare/.

Gupta, D. (2025, April 25). *Secure Coding Practices: Essential Principles to Prevent Vulnerabilities in 2025*. Deepak Gupta | AI Cybersecurity Innovation Leader | Founder's Journey from Code to Scale; Deepak Gupta | AI Cybersecurity Innovation Leader | Founder's Journey from Code to Scale. https://guptadeepak.com/secure-coding-practices-guide-principles-vulnerabilities-and-verification/.

Hardware Based Security For Government—Purism. (2024, August). Purism. https://puri.sm/posts/hardware-based-security-for-government/.

Koutras, D., Stergiopoulos, G., Dasaklis, T., Kotzanikolaou, P., Glynos, D., Douligeris, C. (2020, January 1). *Security in IoMT Communications: A Survey*. Sensors. https://doi.org/10.3390/s20174828.

Mehrtak, M., SeyedAlinaghi, S., MohsseniPour, M., Noori, T., Karimi, A., Shamsabadi, A., Heydari, M., Barzegary, A., Mirzapour, P., Soleymanzadeh, M., Vahedi, F., Mehraeen, E., Dadras, O. (2021, August). Security Challenges and Solutions Using Healthcare Cloud Computing. Journal of Medicine and Life; NCBI. https://doi.org/10.25122/jml-2021-0100.

Pradeep Kumar, K., Prathap, B. R., Thiruthuvanathan, M. M., Murthy, H., Jha Pillai, V. (2023, December 14). *Secure Approach to Sharing Digitized Medical Data in a Cloud Environment*. Data Science and Management. https://doi.org/10.1016/j.dsm.2023.12.001.

RTOS: A Comprehensive Guide—Blue Goat Cyber. (2024, February 27). Bluegoatcyber.com. https://bluegoatcyber.com/blog/rtos-a-comprehensive-guide/.

RTOS vs Bare-Metal. (2023). Embeddedrelated.com. https://www.embeddedrelated.com/thread/5762/rtos-vs-bare-metal.

Trusted Platform Module (TPM) in Embedded System Security. (2025, July 29). Sysgo.com. https://www.sysgo.com/blog/article/trusted-platform-module-tpm-in-embedded-system-security.

build resilient systems capable of defending against ever-evolving threats. As cloud, IoT, and 5G networks [illegible] layered security implementing [illegible] multifactor authentication, and vigilant update mechanisms will be critical to maintaining trust and operational integrity. Armed with this foundational knowledge, healthcare professionals, engineers, and regulators are better equipped to design, manage, and oversee medical device ecosystems that not only meet today's security demands but also adapt proactively to future technological and threat landscapes.

References

[illegible]

4

Cyber Threats Targeting Medical Devices

"The ventilator stopped responding just as the patient's oxygen levels began to drop." The voice over the emergency pager was tense, urgent. In a busy hospital unit, every second counts, yet technology meant to save lives had suddenly become an obstacle. Behind the scenes, unseen and silent, malicious software had infiltrated the device's core systems, locking clinicians out of critical controls. This was not a failure of medicine or human error but a breach of digital trust.

Medical devices have evolved from standalone tools into interconnected components of vast healthcare networks. While this connectivity offers enhanced monitoring and treatment capabilities, it also exposes these devices to complex cyber threats. From hidden malware quietly altering therapy settings to unauthorized intrusions exploiting weak security measures, medical technologies face persistent risks that can compromise their operation and, ultimately, patient safety.

Understanding the nature of these threats is essential for those who design, manage, and protect medical systems. It requires insight into how malicious actors exploit vulnerabilities at multiple layers—from software infections and supply-chain manipulations to wireless communication weaknesses. Addressing these challenges demands a comprehensive approach combining technical defenses with informed clinical practices.

The following discussion delves into the various cyber risks affecting medical devices today. It highlights the mechanisms of attack and the implications for healthcare environments, emphasizing the need for proactive strategies to safeguard these vital technologies against evolving digital dangers.

Malware Infections: A Growing Plague

Medical devices embedded in connected healthcare environments are prime targets for increasingly sophisticated malware variants. Ransomware, Trojans, rootkits, and worms each employ unique operational tactics to compromise device integrity, reliability, and clinical safety in ways that threaten patient outcomes and hospital operational continuity. Understanding the technical behaviors and risks associated with each threat is critical as medical device fleets grow in both scope and connectivity.

Ransomware Disruption of Medical Devices

Ransomware poses an acute threat to medical technologies by encrypting essential configuration files, application data, and, in some cases, entire device operating systems. Once encrypted, devices such as ventilators, infusion pumps, or MRI scanners become inaccessible to clinical staff. In the healthcare context, ransomware campaigns have specifically targeted Windows-based embedded systems and workstations integral to patient monitoring, radiology, and medication delivery platforms (Riggs et al., 2023). The immediate consequence is the denial of access to life-sustaining interventions. For example, a ventilator locked by ransomware prevents technicians from adjusting oxygen delivery or monitoring a patient in respiratory distress.

Restoring device functionality under these conditions often requires isolating infected hardware, purging malicious files, restoring from backups, and rigorously validating the integrity of system software before reintroducing equipment to clinical workflow. Even with rapid incident response, patient care delays become unavoidable, as contingencies like manual medication administration or temporary device replacement are rarely as safe or efficient as intended workflows. The period of operational unavailability dramatically elevates patient risk during emergencies, with delayed responses posing life-threatening consequences for critically ill individuals (Riggs et al., 2023).

Trojan Infections and Device Hijacking

Trojans represent another highly effective class of malware, exploiting trust in legitimate updates or device applications to stealthily gain unauthorized access. In the medical domain, scenarios have emerged where compromised software updates, sometimes installed directly from a vendor's cloud service or USB media, introduce Trojan executables masquerading as device-management utilities. Once inside, these Trojans can harvest administrative credentials, exfiltrate sensitive patient data, or alter embedded logic controlling therapy parameters.

A subtler but profound risk arises when a Trojan subverts device authentication or authorization mechanisms at the application layer. For example, a compromised drug infusion system might permit silent reprogramming, modifying dose schedules or drug libraries without alerting clinicians. Surveillance Trojans can capture keystrokes or intercept serial communications between devices, ferrying data out over persistent covert channels. The consequences extend beyond immediate device manipulation, threatening patient privacy and undermining clinical trust in technological infrastructure (Riggs et al., 2023; Smmarwar et al., 2024).

Rootkits and Stealth Manipulation

Rootkits present a deep and persistent menace by operating at the most fundamental layers of medical devices—either at the operating-system level or the firmware level. Sophisticated rootkits exploit vulnerabilities in legacy hospital operating systems or undersecured device firmware. Once deployed, these threats hide their existence from standard antivirus software by intercepting calls to system utilities and masking altered code. Embedded monitoring assets, designed to detect tampering or performance anomalies, can be blinded by rootkit activity.

A clinical scenario illustrating rootkit impact involves long-term manipulation of ventilator performance metrics or sensor calibration without visible alerts to operators. The rootkit may persist through device reboots, collecting data on therapy patterns or sabotaging specific patient treatments by delivering incorrect dosages over extended periods. By evading even thorough manual inspections, rootkit-infected devices compromise care quality in ways that clinicians may never directly detect. Long dwell times allow attackers to accumulate operational intelligence, posing ongoing risks to both patients and clinical workflows (Smmarwar et al., 2024).

Worm Propagation in Networked Healthcare Infrastructures

Worms exploit the densely connected ecosystems of contemporary clinical environments. Their ability to independently scan network segments, exploit commonly shared service vulnerabilities, and autonomously propagate between devices renders them especially disruptive. An initial infection within a single infusion pump's communication module, for example, can rapidly traverse unsegmented hospital networks, leapfrogging to imaging equipment, telemetry monitors, and laboratory analyzers.

Worms exploit weak network segmentation and outdated systems lacking current patches, rapidly overwhelming clinical workflows. A notable case involved a ransomware-carrying worm disabling not only medical equipment but also hospital communications, pharmacy dispensing, and administrative systems, causing cascading failures in coordinated care delivery (Riggs et al., 2023). Legacy hardware and inconsistent network-access controls provide persistent infection corridors, magnifying the challenge of containing widespread operational disruptions.

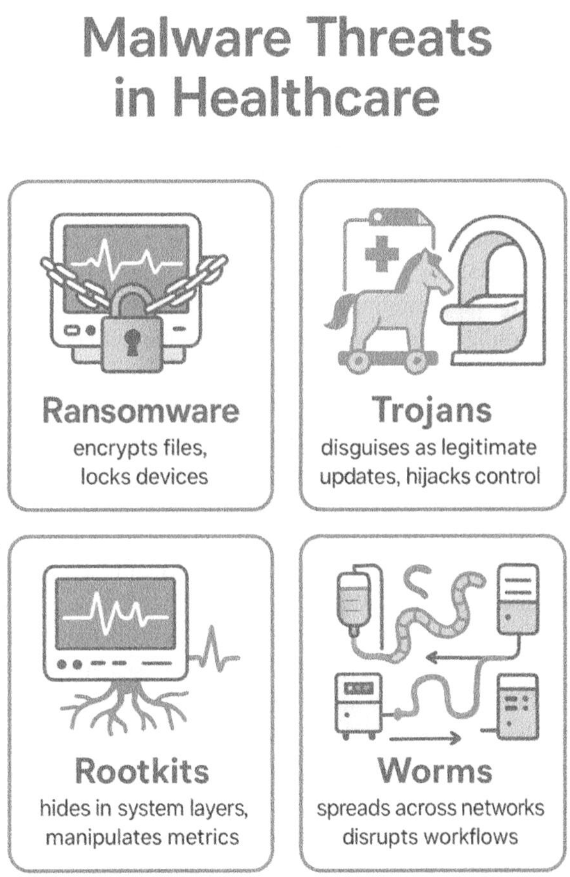

Tailored Countermeasures for Medical Malware

Protecting medical devices against these threats requires endpoint protection tailored for clinical-hardware environments rather than generic IT endpoints. Robust whitelisting denies execution of unapproved binaries or scripts. Continuous security scanning, appropriate for real-time and embedded systems, detects suspicious files or configuration changes. Rapid anomaly detection systems identify deviations from expected device activity, flagging early-stage compromises before patient care can be affected.

Incident response plans must include network-level isolation procedures and validated system-restore images to quickly replace compromised firmware or software. Regular cybersecurity training for clinical and technical staff heightens awareness of phishing vectors, social-engineering attempts, and improper device usage. Stringent patch management is vital, especially for devices running aging or custom operating systems. These countermeasures work together to reduce attack surfaces and boost resilience, but adversaries continually seek to bypass these layers, targeting weaknesses in credential storage, authentication practices, and remote-access policies. Their ongoing pursuit of privilege escalation and unauthorized access shapes the next frontier of medical-device cybersecurity.

Unauthorized Access: Breaking the Digital Locks

Attackers seeking to compromise medical devices often focus their efforts on authentication and authorization mechanisms, exploiting technical oversights and operational weaknesses to defeat digital safeguards. Weak passwords remain one of the most persistent vulnerabilities in clinical technology environments. Medical devices, both legacy and newly deployed, frequently retain manufacturer-default credentials or employ simple, easily guessed passwords for the sake of setup convenience and operational continuity. This issue often arises when clinical staff or device installers prioritize quick configuration over the implementation of complex credential schemes. Additionally, password-management policies in many healthcare facilities fail to prevent the use of shared accounts or enforce unique, strong passwords for each user. Instances where the same login is shared among nurses, technicians, and even external support personnel can be observed. Such credentials are rarely updated or rotated—a result of workflow pressures and concerns about disrupting patient care delivery. The absence of stringent password complexity standards further compounds the problem; attackers can leverage automated tools that deploy brute-force or dictionary attacks, rapidly cycling through common phrases and manufacturer-default combinations until device access is obtained (Grimes & Wirth, 2021).

Once access is gained, attackers often escalate privileges to defeat limited-user protections and unlock deeper layers of device functionality. Privilege escalation attacks target flaws in device software, such as buffer overflows, unpatched firmware vulnerabilities, or misconfigured access-control rules. Medical-device operating systems, particularly those based on legacy or proprietary architectures, may lack routine security updates and robust boundary checks. For example, a vulnerable infusion pump or patient monitoring device might grant a successful attacker system-level privileges following the exploitation of a firmware bug. With elevated rights, malicious actors can modify device configurations, disable security services such as encrypted communications, or install persistent backdoor programs. These changes can go undetected by users focused on immediate patient care tasks, allowing the adversary to maintain long-term control or pivot further into hospital networks. Another fervent risk emerges if attackers disable alarms or modify therapy settings, potentially endangering patient safety or clinical outcomes (Suleski et al., 2023).

Authentication protocols, particularly older or poorly implemented ones, introduce their own set of weaknesses that attackers readily exploit. Certain networked medical devices continue to rely on outdated authentication mechanisms lacking modern safeguards. For example, the absence of cryptographic protections may render session tokens or passwords vulnerable to interception or replay by adversaries monitoring network traffic. Credential replay attacks—where intercepted authentication information is reused by an attacker—are particularly dangerous when communication lacks session binding or freshness guarantees. Attackers can also exploit session-hijacking vulnerabilities, injecting malicious packets or spoofing authentication messages to masquerade as authorized users. Medical-device communication standards, such as Digital Imaging and Communications in Medicine (DICOM) for imaging devices or Health Level 7 (HL7) for patient records, have been targets for such protocol attacks, especially if the integration with clinical IT infrastructure is managed without comprehensive security assessments. The result is a breach of device integrity—attackers not only access but may alter or corrupt vital medical data, threatening the reliability of connected clinical systems (Grimes & Wirth, 2021; Suleski et al., 2023).

The healthcare sector has begun integrating multifactor authentication (MFA) to strengthen device-access controls, yet significant challenges persist in practice. While the adoption of MFA—combining factors such as passwords (what you know), smartcards or tokens (what you have), and biometrics (what you are)—offers theoretically robust protection, real-world deployment is often hindered by legacy-device limitations and clinical-workflow constraints. Many older devices simply do not support modern MFA mechanisms without costly hardware upgrades or disruptive changes. Even where MFA is technically feasible, integration may burden clinical staff with additional steps, potentially slowing down critical patient interventions and causing resistance to adoption. For instance, requiring a fingerprint scan and a one-time passcode before reprogramming an emergency defibrillator could delay urgent care. Usability considerations are paramount in healthcare, where security mechanisms must not impede lifesaving treatments. The balance between operational security and seamless care delivery defines the ongoing struggle in MFA implementation. Technical barriers—such as resource-constrained device platforms or inconsistent interoperability between security systems—further complicate large-scale MFA rollouts (Grimes & Wirth, 2021; Suleski et al., 2023).

Attackers rarely limit themselves to exploiting a single weakness. Initial access gained through weak passwords may be leveraged for privilege escalation, while flaws in authentication protocols expose avenues for impersonation and lateral movement across interconnected clinical systems. MFA, when improperly deployed or omitted due to practical constraints, can leave supposedly protected devices vulnerable to layered, multistep attacks. The cumulative effect of these vulnerabilities widens the attack surface, making it possible for malicious actors to subvert device integrity, disrupt critical healthcare services, and threaten patient safety on a systemic scale (Grimes & Wirth, 2021; Suleski et al., 2023).

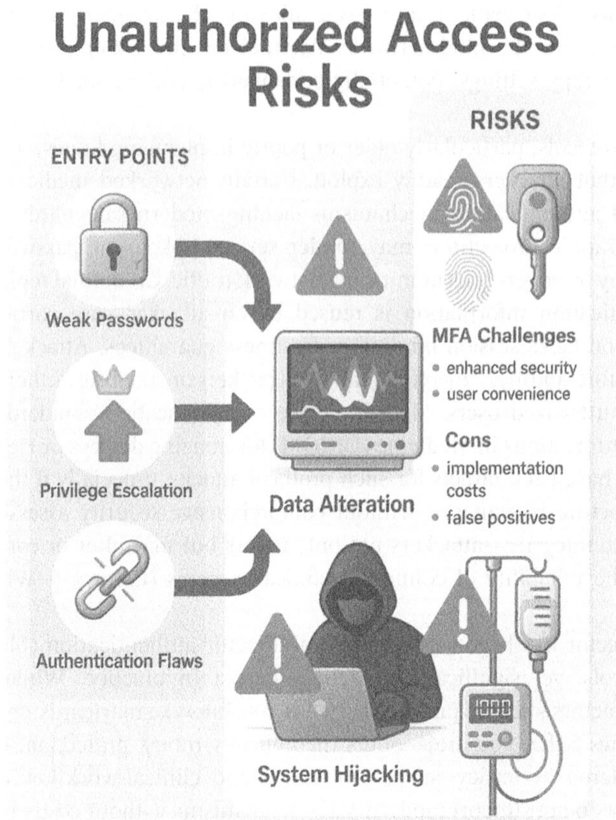

Denial of Service and Its Consequences

A pattern of unauthorized access in healthcare settings reveals deeper systemic flaws. Compromised credentials and weak segmentation offer threat actors persistent footholds, exposing entire device fleets and hospital networks to attacks that extend beyond data theft. One of the most dangerous consequences of these vulnerabilities is the loss of device and network availability in clinical environments, where continuous operation is critical for patient survival and effective care delivery (Koutras et al., 2020). Denial of service (DoS) and distributed denial of service (DDoS) attacks weaponize these weaknesses, impeding access to life-sustaining systems. The rupture of service, even for seconds, can result in acute clinical risk and cascading operational crises.

Risks from DoS Attacks on Individual Medical Devices

A direct DoS attack focuses on disrupting the function of a critical device—such as an infusion pump, ventilator, or cardiac monitor—by overwhelming it with traffic or forcing it into repeated reboots. In these scenarios, a single vector of attack can halt monitoring or delivery of therapy. If a networked infusion pump is forced offline through a DoS event, medication delivery to a patient is interrupted. Restoration often requires manual intervention: staff must locate the affected device, execute a shutdown or forced reboot, and sometimes revert to manual dosage calculation. This process introduces delays with tangible consequences. A patient on a ventilator whose device is incapacitated due to a DoS attack might face interrupted oxygen delivery. Staff must act quickly using manual resuscitation equipment, but even brief downtime raises the risk of hypoxia or fatal outcomes. Each step—reboot, replacement, or manual override—adds time and operational complexity, heightening staff stress and diverting scarce attention from other patients (Koutras et al., 2020).

These attacks do not only threaten direct patient outcomes. They also erode workflow integrity and damage clinical confidence in technology. For example, in scenarios where telemetry data cannot reach central monitoring stations due to a DoS event, nurses and physicians lose visibility into patient status for the duration of the outage. Even after restoration, the risk of data desynchronization, time-stamp misalignment, and loss of event logs complicates post-incident care review.

Escalation to Hospital Network Disruption

Widespread network attacks amplify disruption far beyond a single device. Attackers flood hospital communication infrastructure with illegitimate traffic, saturating shared network segments that host medical, administrative, and building-management systems (Koutras et al., 2020). This can be achieved through packet flooding or malformed packet injection targeting common communication protocols used in medical environments, such as HL7, DICOM, or message queuing telemetry transport (MQTT) (Hernandez-Jaimes et al., 2023). Systems designed to handle finite data rates buckle under the load, causing delays or total loss of communication between devices and control hubs.

A paralyzed hospital network stops vital telemetry feeds from reaching nurses' stations. Central monitoring dashboards may lose real-time information from dozens of wireless infusion pumps and telemetry units. In high-acuity settings such as intensive care or emergency departments, these lapses eliminate remote oversight and force workflows back onto manual checks. Staff scramble to patrol more rooms and verify device status in person, slowing critical response. This surge in demands on human resources increases the probability of missed events, medication errors, or overlooked deterioration.

The interdependence of systems means that the impact of a saturated network radiates outward. Radiology, pharmacy, clinical decision support, and laboratory information updates may all stall. Even after attacks subside, the process of safe network restoration—inspection for device corruption, sequence checks, and reauthentication—complicates and delays the return to normal operations.

Botnets and the Scale of DDoS

The most devastating DDoS events draw on global networks of compromised devices, many of which are IoT-based and include vulnerable medical and nonmedical endpoints (Hernandez-Jaimes et al., 2023). Attackers discover unpatched devices by scanning network ranges for known vulnerabilities—such as outdated firmware or weak default credentials—and then place malware agents. These infected devices receive attack instructions from remote command servers. When activated, the botnet generates massive, coordinated surges of malicious traffic pointed at the target institution. The volume easily surpasses the handling capacity of standard network infrastructure, causing hospital-wide paralysis.

Detecting and responding to such events presents practical challenges. Attack traffic may blend with legitimate communications, making it hard to distinguish friend from foe. Hospital IT teams must rapidly analyze traffic flows, identify attack sources, and segment afflicted portions of the network. Delays can prolong outages, risking ongoing patient care.

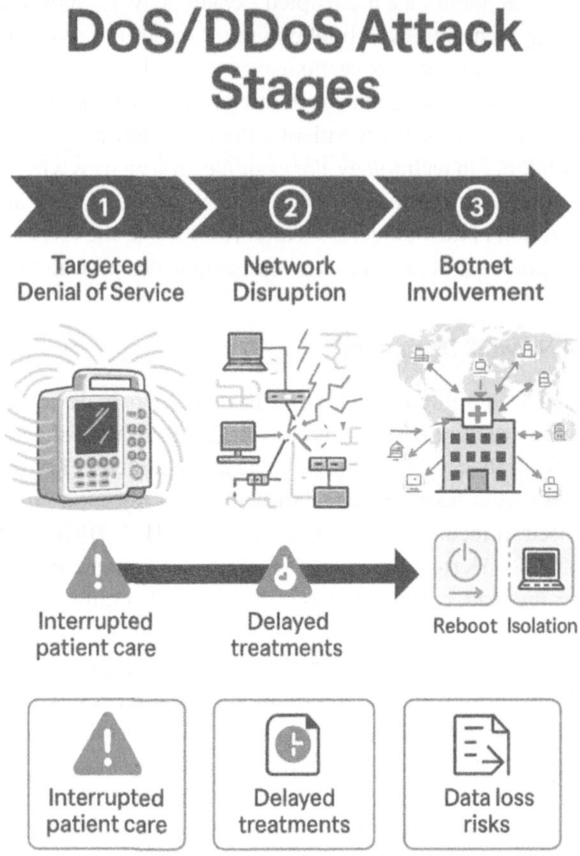

Engineering Resilience and Defense

Technical defenses prioritize maintaining operational continuity. Hospitals implement redundant network links, failover routes, and segmentation barriers so local outages do not cascade. Rate limiting and deep packet inspection block or reduce malicious bursts. Automated filtering based on source addresses, protocol types, and traffic behaviors diminishes attack impact before it overwhelms devices (Koutras et al., 2020). Incident response frameworks involve ongoing staff training, table-top exercises, and live drills. Rapid isolation protocols and tested routines for restoring device operation form the backbone of effective preparedness.

The distinct characteristics of DoS, DDoS, and network saturation attacks demand layered protection and ongoing vigilance. Yet even with robust countermeasures, vulnerabilities in the supply chains of devices and software present additional footholds for attackers—threats that can infiltrate well before devices reach clinical use. The next challenge lies in understanding how component and update risks underpin broader systemic exposure.

Supply-Chain Attacks: Poisoning at the Source

Modern clinical environments depend on highly available and reliable medical devices, but this assurance increasingly hinges on the integrity of complex, globally-sourced supply chains. As adversaries become more sophisticated, they exploit trust relationships, dependencies, and weak points embedded throughout the production, delivery, and maintenance processes of medical technology. Efforts to secure system availability and service continuity require recognizing that device compromise can begin long before deployment. Trust in vendors, updates, and manufacturing is essential but often misplaced, exposing clinical networks to disruptive attacks.

Medical devices are rarely built from scratch by a single entity. Instead, they integrate microprocessors, drivers, firmware, and application code produced by a variety of third parties. This interconnectedness brings the threat of compromised components into sharp relief. A malicious chipset, for example, may be manufactured to include hidden logic that remains dormant during standard screening, triggering only in response to specific signals or operational patterns. Similarly, a software library imported at build-time may carry an intentionally crafted vulnerability, introduced either through insufficient oversight or as a deliberate act by a supply-chain adversary. The case of the Avaya 9600 desk phone vulnerability, attributed to outdated open-source firmware contained within a proprietary product, illustrates how insecure or unmaintained components can expose millions of devices to root-level compromise and eavesdropping within highly sensitive networks (Breaking Trust: Shades of Crisis across an Insecure Software Supply Chain, 2020). Dormant threats like these can quietly evade traditional endpoint security measures, remaining invisible until actors activate them, resulting in device malfunction, unauthorized data access, or patient safety incidents.

Supply-chain interdiction represents another persistent threat. Attackers may intercept shipments of critical hardware, replacement parts, or installation media. During this breach of the logistics chain, devices can be physically modified to include hidden backdoors, covert chips, or preinstalled malware—alterations that standard post-delivery inspections are often unable to detect. An illustrative scenario involves malicious actors rerouting medical-device shipments, briefly accessing the packaging, and embedding hardware implants that later transmit confidential patient data when connected to clinical networks. Even advanced technical assessments can fail to reveal such modifications, since tampering may occur at the circuitry or firmware level without leaving visible traces. Industry reports have documented instances of "shipment interdiction" targeting network appliances, where infected firmware or hardware was only discovered after extensive post-compromise forensics (Yahya, 2024). These events highlight the limitations of visual inspection and checksum validation when facing advanced adversaries working at the supply-chain's physical layers.

Malicious updates constitute a third, frequently exploited vector in medical device compromise. Regular software and firmware updates introduce new code, making them attractive opportunities for adversaries to inject malicious payloads, provided update mechanisms lack robust cryptographic verification and end-to-end integrity checks. When code-signing processes are undermined, tampered updates can appear authentic to systems and administrators, facilitating silent infiltration (Breaking Trust: Shades of Crisis across an Insecure Software Supply Chain, 2020). Instances such as trojanized applications distributed via app stores or unauthorized patches deployed to unsegmented networks demonstrate how attackers leverage weak signature verification or outdated delivery channels to subvert device controls. Time pressure to resolve vulnerabilities further elevates risk if organizations prioritize speed over thorough validation. Firmware updates signed with obsolete algorithms or delivered without attestation increase the likelihood that devices will accept unauthorized code, potentially resulting in DoS, false readings, or remote access for threat actors.

Compromises at the source resonate throughout the lifecycle of medical devices. Robust vetting of vendors and ongoing risk assessments are foundational defenses. Systematic evaluation of supplier cybersecurity practices, adherence to contractual obligations, and active due diligence beyond initial procurement are cardinal elements of device security. Frameworks such as NIST SP 800-161 offer criteria for assessing supplier risk, addressing requirements for vulnerability notification, secure-development life cycles, and continuous monitoring. The U.S. Food and Drug Administration (FDA) provides further guidance for device manufacturers, emphasizing the importance of ensuring supply-chain transparency, update security, and ongoing vulnerability management. Establishing contractual obligations requiring software bills of materials (SBOMs), regular independent audits, and incident-notification protocols directly enhances device security posture (Breaking Trust: Shades of Crisis across an Insecure Software Supply Chain, 2020; Yahya, 2024).

Technical assurance and organizational diligence must integrate seamlessly into broader network and interface defenses. Compromised components, firmware, or updates often serve as launching pads, providing attackers with footholds to pivot into wider digital exploitation. Wireless interfaces and network-connected diagnostic modules become channels for lateral movement or data exfiltration. As a result, organizations must view supply-chain risk as an amplifier of broader cybersecurity concerns, connecting origins of compromise with downstream vulnerabilities that threaten every level of device operation and patient-care integrity.

Targeting Wireless Interfaces: Physical Meets Digital

Medical devices in clinical environments rely on a broad array of wireless and physical interfaces to deliver patient care, but each interface introduces opportunities for adversaries to compromise cybersecurity and patient safety. These vulnerabilities persist even as healthcare institutions implement controls within the supply chain, emphasizing the need to understand and defend every access point across the operational landscape.

Bluetooth and BLE Interfaces

Bluetooth and Bluetooth Low Energy (BLE) remain some of the most prevalent wireless protocols for medical devices requiring short-range communication. Many hospital assets, such as infusion pumps, patient monitors, and wearable sensors, use BLE for simplified pairing and effortless data exchange. However, weaknesses arise due to insecure pairing procedures, poor cryptographic implementations, or improper authentication configurations. Attackers can intercept or manipulate traffic by exploiting these weaknesses, leading to unauthorized data access and device control. Eavesdropping becomes possible when attackers capture pairing exchanges or use packet-sniffing tools to monitor unencrypted data transmitted between devices and central controllers, such as nurse workstations or smartphones (Barua et al., 2022).

Inadequate pairing mechanisms, such as those relying on static PINs or poorly implemented cryptographic routines, can expose sensitive data or allow malicious actors to perform Man-In-The-Middle (MITM) attacks. Once communication is hijacked, attackers may inject rogue commands to manipulate therapy settings or inhibit alarm functions. BLE's limited transmission range provides no guarantee of security in crowded clinical spaces where adversaries can gain physical proximity. Real-world incidents have demonstrated that attackers, even while situated within the same building, can initiate battery-draining denial-of-service attacks or silently capture patient data flows without direct contact with the medical device or its user (Barua et al., 2022).

Wi-Fi and Proprietary Wireless Protocols

Wi-Fi has become a mainstay for connecting critical medical systems to broader hospital networks and electronic health records. Poorly secured devices—those lacking updated firmware, strong authentication, or adequate segmentation—are highly susceptible to remote exploitation attempts. Threat actors can leverage weak Wi-Fi credentials, lack of encryption, or outdated protocols to gain a foothold within the clinical environment. This could result in unauthorized monitoring, data breaches, or manipulation of device operation, with significant repercussions for both patient safety and data privacy (Yaacoub et al., 2020).

Numerous medical device manufacturers employ proprietary wireless protocols designed for specific use cases, such as telemetry or real-time monitoring. These protocols sometimes forego standardized security measures in favor of performance or backward compatibility. A proprietary protocol lacking robust mutual authentication or encryption might allow attackers equipped with inexpensive radio hardware to reverse engineer communications, inject malicious signals, or impersonate authorized controllers within the hospital setting. The result is elevated exposure to data interception, unauthorized command injection, or covert data leaks, especially when hospital staff are unaware of the vulnerabilities inherent in these "closed" protocols (Yaacoub et al., 2020). Regular evaluation and strict access controls on wireless network infrastructure are required to monitor and limit traffic, reducing the likelihood of lateral movement or exfiltration attacks.

Physical Ports and On-Site Intrusion

Physical interfaces such as USB or SD card slots open an entirely different vector for attack, sometimes bypassing network-level defenses altogether. Malicious insiders—such as disgruntled employees or contract personnel—may utilize these accessible ports to introduce malware-laden storage media, compromising device integrity. In one scenario, a technician performing routine maintenance unknowingly connects a compromised diagnostic tool, resulting in malware propagation to connected systems.

To defend against such exposure, device-lockdown policies that disable unused physical ports, mandate strict access logging, and implement read-only controls for permitted interfaces are essential. Consideration of workflow realities is crucial; maintenance activities must be secured by scanning all external media for malware and ensuring that port usage is carefully authorized and monitored. Automated alerts for unauthorized port usage, layered with MFA for administrative access, provide further security without compromising operational needs (Yaacoub et al., 2020).

Advanced Attack Techniques: Jamming and Signal Injection

Attackers may pursue more disruptive strategies by targeting the wireless communication channels themselves. Signal jamming represents a significant danger, particularly in mission-critical

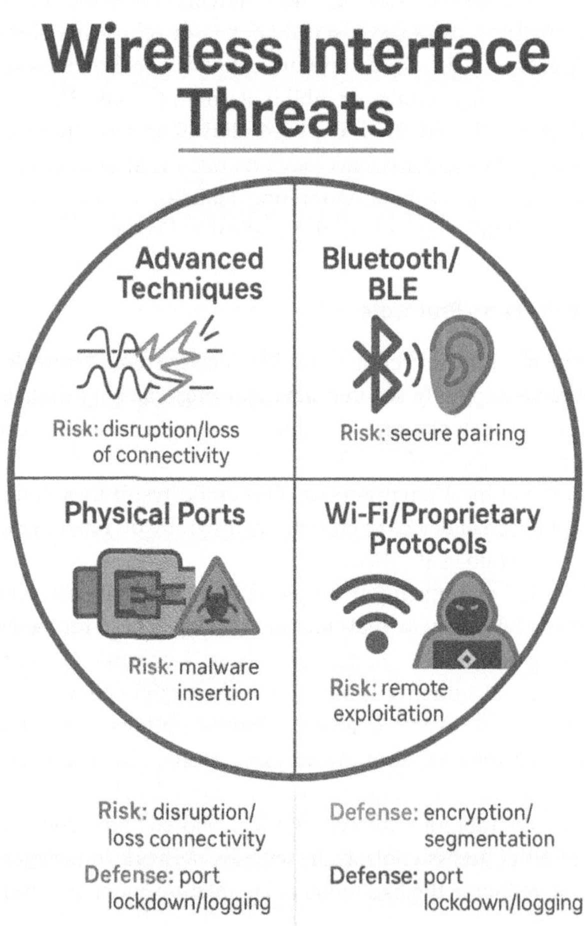

device clusters reliant on wireless coordination. By transmitting high-powered noise over frequencies allocated to medical-device communication, adversaries can cause targeted devices to lose connectivity, enter fault states, or default to less secure backup modes.

Signal injection presents another form of attack, with adversaries generating crafted signals or commands that deliberately alter device behavior. A practical illustration involves injecting false telemetry into a vital-signs monitor, resulting in erroneous clinical decisions or unnecessary interventions. Detection of these attacks requires continuous monitoring of radio-frequency activity, anomaly detection algorithms for unusual communication patterns, and robust fallback procedures to ensure clinical operations remain unaffected. Implementation of frequency-hopping techniques, authenticated command structures, and layered wireless encryption enhances resilience, ensuring that medical devices can withstand sophisticated jamming or injection attempts (Barua et al., 2022; Yaacoub et al., 2020).

By rigorously securing each interface—wireless or physical—healthcare organizations safeguard critical medical devices against the evolving array of cyber threats.

Bringing It All Together

Now that we have detailed the multifaceted cyber threats targeting medical devices—from malware and unauthorized access to DoS, supply-chain risks, and wireless vulnerabilities—it is clear that securing these critical systems demands a comprehensive and proactive approach. Healthcare organizations, device manufacturers, and regulators must collaborate to implement robust protections that address each attack vector while balancing clinical usability and safety. Emphasizing continuous risk assessment, integrating advanced security controls tailored to medical environments, and fostering ongoing education for all stakeholders will strengthen device resilience and protect patient outcomes. As technology evolves and connectivity expands, vigilance and innovation in cybersecurity practices are essential to anticipate emerging threats and safeguard the integrity, availability, and confidentiality of life-sustaining medical technologies.

References

Barua, A., Al Alamin, M. A., Hossain, M. S., Hossain, E. (2022). Security and Privacy Threats for Bluetooth Low Energy in IoT and Wearable Devices: A Comprehensive Survey. IEEE Open Journal of the Communications Society. https://doi.org/10.1109/ojcoms.2022.3149732.

Breaking Trust: Shades of Crisis Across an Insecure Software Supply Chain. (2020, July 27). Atlantic Council. https://www.atlanticcouncil.org/in-depth-research-reports/report/breaking-trust-shades-of-crisis-across-an-insecure-software-supply-chain/.

Grimes, S. L., Wirth, A. (2021, July 1). *The Case for Medical Device Cybersecurity Hygiene Practices for Frontline Personnel*. Biomedical Instrumentation Technology. https://doi.org/10.2345/0890-8205-55.3.96.

Hernandez-Jaimes, M. L., Martinez-Cruz, A., Ramírez-Gutiérrez, K. A., Feregrino-Uribe, C. (2023, October 1). Artificial Intelligence for IoMT Security: A Review of Intrusion Detection Systems, Attacks, Datasets and Cloud–Fog–Edge Architectures. Internet of Things. https://doi.org/10.1016/j.iot.2023.100887.

Koutras, D., Stergiopoulos, G., Dasaklis, T., Kotzanikolaou, P., Glynos, D., Douligeris, C. (2020, January 1). *Security in IoMT Communications: A Survey*. Sensors. https://doi.org/10.3390/s20174828.

Riggs, H., Tufail, S., Parvez, I., Tariq, M., Khan, M. A., Amir, A., Vuda, K. V., Sarwat, A. I. (2023, January 1). Impact, Vulnerabilities, and Mitigation Strategies for Cyber-secure Critical Infrastructure. Sensors. https://doi.org/10.3390/s23084060.

Smmarwar, S. K., Gupta, G. P., Kumar, S. (2024, March 1). Android Malware Detection and Identification Frameworks by Leveraging the Machine and Deep Learning Techniques: A Comprehensive Review. Telematics and Informatics Reports; Elsevier BV. https://doi.org/10.1016/j.teler.2024.100130.

Suleski, T., Ahmed, M., Yang, W., Wang, E. (2023, May 22). *A Review of Multi-factor Authentication in the Internet of Healthcare Things*. Digital Health. https://doi.org/10.1177/20552076231177144.

Yahya, F. (2024, May 17). *Vendor Risk Management: Securing Your Supply Chain*. Faisal Yahya. https://faisalyahya.com/cybersecurity-leadership/vendor-risk-management-securing-your-supply-chain/.

Yaacoub, J.-P. A., Salman, O., Noura, H. N., Kaaniche, N., Chehab, A., Malli, M. (2020, September). *Cyber-physical Systems Security: Limitations, Issues and Future Trends*. Microprocessors and Microsystems. https://www.ncbi.nlm.nih.gov/pmc/articles/PMC7340599/.

5

Regulatory and Legal Landscape: Raising the Bar

Medical device cybersecurity is no longer a peripheral concern; regulatory and legal frameworks governing this space have grown considerably complex and demanding. Compliance now extends beyond traditional safety and efficacy to encompass rigorous security requirements throughout the entire device life cycle. The stakes are high, as failure to meet evolving standards can result in significant legal liabilities, financial penalties, and interruptions in market access. Understanding how these regulations have developed, the key compliance expectations across multiple jurisdictions, and the interplay with privacy laws is essential for professionals involved in healthcare security, device design, and regulatory oversight. This chapter examines the shifting landscape that shapes current obligations, the challenges posed by global regulatory divergence, and the legal consequences of inadequate cybersecurity measures in medical devices.

The Evolution of Regulatory Demands

Early medical devices, designed before the widespread acknowledgment of cyber risk, operated with minimal regulatory oversight regarding cybersecurity. For much of the twentieth century, regulatory frameworks were focused on functional safety, efficacy, and basic patient protection rather than addressing the new frontier of data flows and digital interfaces. The introduction of software-driven medical devices and, later, network connectivity, fundamentally changed the threat landscape. Legacy devices were frequently shipped without authentication requirements, encrypted data transfer capabilities, or secure update protocols, leaving basic system configurations exposed to both intentional and accidental compromise.

Concrete examples demonstrate the consequences of these oversight gaps. In the early 2000s, researchers demonstrated that pacemakers and insulin pumps could be wirelessly accessed and manipulated due to weak or nonexistent security. In 2011, Jay Radcliffe, a diabetes researcher, highlighted the ability to remotely access his own insulin pump, raising awareness around the potential for hackers to deliver fatal doses. Researchers also exposed vulnerabilities in internet-connected CT scanners and MRI machines, showing these devices lacked even rudimentary protections. These incidents underscored a reality where insufficient regulation did not just create data privacy risks but introduced acute dangers to patient health.

Legislators and regulators began responding to these risks as both the volume and interconnectedness of devices expanded. The United States enacted the Health Insurance Portability and Accountability Act (HIPAA) in 1996, initially orienting itself toward protecting health data privacy and security. While HIPAA's Security Rule represented an early attempt to address electronic protected health information (ePHI), its focus was squarely on data management rather than device-specific risks. This meant that many networked devices, acting as both data collection points and clinical instruments, fell through the cracks of regulatory protection (*Advancing Medical Device Cybersecurity beyond Compliance: Managing Risk with Governance*, 2023; Elendu et al., 2024).

A decisive turning point came during the 2000s and early 2010s, reflecting the recognition that mere data confidentiality was insufficient and a comprehensive approach to device security was necessary. The Health Information Technology for Economic and Clinical Health (HITECH) Act of 2009 drove the adoption of electronic health records (EHRs) while mandating breach notification requirements and expanding penalties for noncompliance (Elendu et al., 2024). Around this period, international bodies such as the International Organization for Standardization introduced standards like ISO/IEC 27001, which created formal frameworks for managing information security risks and encouraged device manufacturers and healthcare organizations to shift toward structured, repeatable cybersecurity management programs.

From the mid-2010s onward, expectations for medical device manufacturers grew sharply. Regulators and industry groups called for the integration of cybersecurity throughout the device life cycle, from design and development through postmarket surveillance. The U.S. Food and Drug Administration (FDA) began issuing specific guidance, such as the 2014 and 2016 premarket and postmarket cybersecurity guidance documents, making it clear that security was as fundamental as functionality and safety. Manufacturers now bear explicit responsibility to identify risks, document threat modeling efforts, and provide evidence of secure design in their submissions. Ongoing obligations include timely software patching, vulnerability management, coordinated disclosure with cybersecurity researchers, and real-time threat monitoring for devices already deployed in the field (*Advancing Medical Device Cybersecurity beyond Compliance: Managing Risk with Governance*, 2023).

In the European Union, the General Data Protection Regulation (GDPR) set stringent data protection standards in 2018, requiring privacy by design and rigorous data handling protocols for devices that process health data. The Medical Device Regulation (MDR) reinforced these expectations, imposing cybersecurity requirements as a component of safety and performance. Enforcement was no longer an abstract concept. National agencies, such as the FDA in the United States, the European Medicines Agency, and various data protection authorities across Europe and Asia, began auditing compliance, issuing fines, and even suspending device approvals in cases of egregious nonconformance (Elendu et al., 2024).

Countries adopted distinct regulatory models prioritizing either proactive or reactive strategies. The United States and much of Europe increasingly favor a proactive model, requiring manufacturers to build secure-by-design principles into devices and implement regular risk assessments. For example, Japan's Act on the Protection of Personal Information (APPI) and Australia's Privacy Act mandate the implementation of data security measures prior to incidents, reflecting a preventive orientation. In contrast, regulatory regimes in other jurisdictions, such as China's Cybersecurity Law, emphasize penalizing noncompliance and mandating breach response plans, favoring incident-driven improvements and corrective actions (Elendu et al., 2024).

These regulatory differences flow from national priorities and legal traditions. Highly developed markets with robust regulatory institutions and histories of major data breaches often choose anticipatory frameworks, while markets with less regulatory capacity or rapid technological adoption may emphasize enforcement and remediation. Ultimately, the evolution of regulatory demands shows a clear progression from minimal oversight and repeated vulnerability exposure to a multifaceted expectation that encompasses device architecture, organizational governance, and continuous life cycle accountability. This trajectory leads directly into questions about the specifics of compliance guidance and enforcement, shaping present-day obligations for all stakeholders.

FDA Cybersecurity Guidance and Expectations

Achieving FDA clearance for networked medical devices today requires far more than demonstration of physical safety or effectiveness. The agency maintains rigorous expectations for cybersecurity, directly integrated into both premarket and postmarket regulatory structures. First, device manufacturers seeking premarket approval must provide comprehensive documentation addressing the cybersecurity posture of their products. This core expectation includes a cybersecurity risk assessment that systematically identifies reasonably foreseeable threats and vulnerabilities, maps their potential impacts, and evaluates mitigations. The FDA expects the submission to detail threat modeling processes, highlighting how developers anticipate, categorize, and address security risks from the earliest design phases (Health, 2022).

In parallel, manufacturers are required to establish security-focused design controls as part of their overall Quality System. This means incorporating specific security requirements within device architecture, such as authentication mechanisms, secure update capabilities, access controls, data encryption, and mechanisms to ensure data integrity. FDA reviewers commonly expect diagrams or detailed descriptions of the device's cybersecurity architecture, listing trusted computing bases, isolation boundaries, and secure boot logic. The documentation must provide evidence that the product can detect, resist, and recover from common attack scenarios relevant to its connectivity context (Health, 2022). For instance, a wireless infusion pump's submission might include detailed penetration test results, explain how firmware integrity is checked on startup, and provide technical descriptions mapping each system component's exposure to networked threats.

FDA guidance also mandates that manufacturers provide a Software Bill of Materials (SBOM) and that all off-the-shelf software components be inventoried, tracked, and monitored. Documentation should define processes for vulnerability management across both proprietary and third-party software, underscoring the need for prompt responsiveness as new vulnerabilities emerge in the field (FDA, 2020). This holistic approach reflects the agency's understanding that security assurance is continuous and dynamic, not static.

After market entry, manufacturers face robust ongoing obligations designed to maintain device security across its operational life. FDA postmarket guidance makes clear that security responsibilities do not end with launch. Manufacturers must actively monitor for emerging threats, maintain open channels for vulnerability disclosure, and rapidly produce software or firmware patches when new risks are identified (FDA, 2020). A comprehensive postmarket surveillance plan should detail how the manufacturer collects cybersecurity intelligence, evaluates incoming reports, and prioritizes response actions.

Timely software updates are expected—guidance stresses that delay in patch distribution, once a viable fix is available, may expose patients to unnecessary risk. FDA encourages the integration of autonomous or semiautomated update mechanisms to reduce dependence on manual intervention. Manufacturers should document patching protocols, from development and testing through deployment, and provide clear communication channels for affected healthcare facilities to receive and apply updates efficiently (Health, 2022). Monitoring practices must also include deployment of runtime security monitors and robust incident detection systems for high-risk devices. Long-term life cycle management requires formalized incident response procedures, including playbooks for triage, containment, eradication, and recovery. These protocols are supported by coordinated communication strategies with healthcare providers and the public.

Prompt and transparent reporting is critical to FDA's incident response paradigm. Manufacturers are legally required to report certain cybersecurity-related adverse events—those that could lead to patient harm—on an expedited timeline. Documentation of each incident must include affected device models, nature and origin of the vulnerability, incident timeline, mitigation status, and communication efforts with end users. In some cases, where an identified vulnerability cannot be adequately addressed through patching or where patient safety is at risk, manufacturers may be compelled to initiate recalls or corrective actions under FDA supervision (FDA, 2020). For instance, a manufacturer might contact the FDA, healthcare organizations, and affected users within days of discovering a critical credential management flaw, simultaneously distributing a mitigation script and removing impacted products from the market. These actions, guided by established regulatory protocols, enhance both patient safety and public trust while limiting liability.

Case history continues to shape the practical boundaries of compliance. A recurring theme in FDA enforcement is the insufficiency of generalized security claims unsupported by concrete evidence. Enforcement actions commonly follow from incomplete documentation of threat modeling, lack of responsive patching, or late reporting of significant events. In one noted case, a manufacturer failed to update default login credentials embedded within widely deployed clinical devices, exposing hospital networks to exploitation until regulatory intervention forced swift remediation. Such cases clarify expectations and promote a culture of continuous improvement, providing instructive reference points for firms seeking to align their own practices with regulatory demands. Recognizing and analyzing patterns in enforcement helps organizations understand the implementation of FDA guidance as a living set of standards, constantly calibrated against new risks and lessons learned (FDA, 2020; Health, 2022).

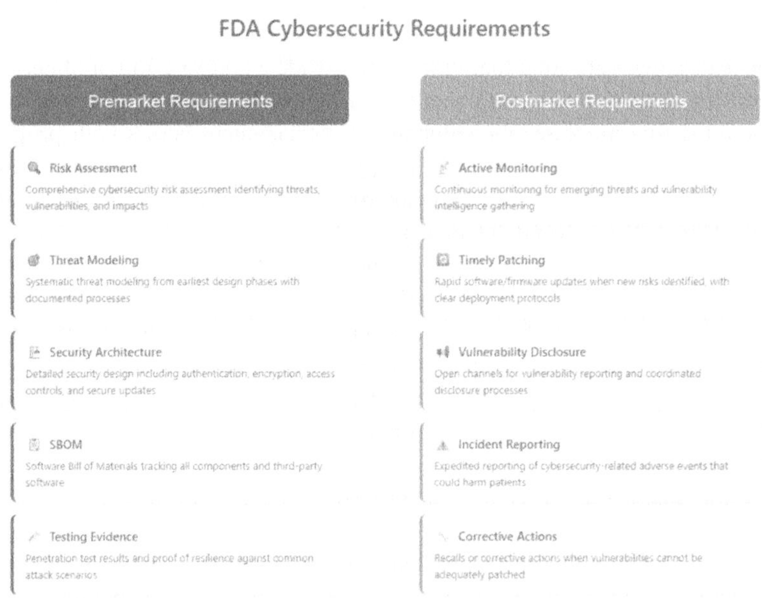

As the landscape of device connectivity expands and multinational deployment becomes standard practice, regulatory demands increasingly reflect the interplay of regional, national, and international expectations for cybersecurity. The convergence of standards and compliance approaches hints at new layers of complexity and opportunity for coordinated global risk reduction, broadening the horizon beyond what any individual regulator can achieve alone.

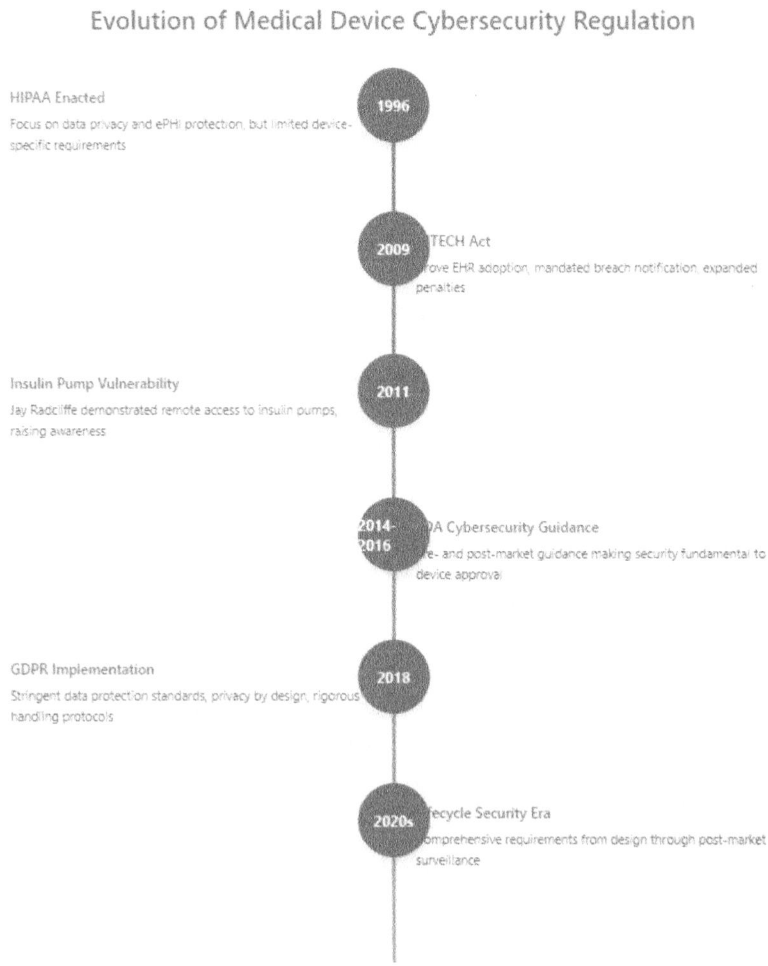

International Compliance Challenges for Medical Device Cybersecurity

Medical device cybersecurity demands expertise that spans regulatory landscapes in multiple regions. European controls, formalized through the MDR and in vitro Diagnostic Regulation (IVDR), set strict security standards for medical technologies. These regulations require that manufacturers actively address cybersecurity from product conception through ongoing life cycle management, integrating risk management and protection against both anticipated and emergent threats. Manufacturers must maintain detailed technical documentation, which includes comprehensive risk analyses, detailed assessment of threats and vulnerabilities, verification of mitigation strategies, and explicit evidence of secure product design. Regulators demand that this evidence supports claims of device safety and effectiveness, emphasizing periodic reviews and necessary updates when new risks surface.

During audits, European notified bodies may scrutinize documents and test evidence demonstrating that a device's operating system or embedded software resists cyber threats. A manufacturer may be required to show log files, threat models, and incident response plans tailored to preserve patient data and critical system functionality. Several compliance failures have resulted in market withdrawals or mandatory recalls. Public cases, such as regulatory warnings over inadequate software update mechanisms or cryptographic flaws in networked diagnostic tools, show that lack of sustained vigilance leads to regulatory intervention. Under MDR and IVDR, failure to maintain compliance exposes companies to suspension of CE marking, effectively barring products from EU markets.

The situation changes in other regions. Japan, for example, integrates cybersecurity within its Pharmaceuticals and Medical Devices Agency (PMDA) review. Japanese regulators focus on whether device manufacturers incorporate security risk management alongside safety and performance. However, enforcement varies; the PMDA may require localized testing and evidence that devices conform to Japan-specific cryptographic and privacy standards, even when devices meet international certifications. Variations in audit rigor and local language requirements for technical documentation can cause repeated delays or reapproval cycles.

In China, regulations evolve rapidly, impacting both imported and domestic products. The National Medical Products Administration (NMPA) enforces new standards intended to align with broader national cybersecurity laws. These laws mandate data localization and heightened controls on information crossing borders. A device that transmits clinical data from China to global servers may require infrastructure changes and explicit government review. Recent amendments compelled several global manufacturers to overhaul remote update functions and guarantee local storage of clinical event logs. Policy shifts often accompany these legal changes. For instance, following a security breach traceable to imported laboratory analyzers, the NMPA escalated device audits, suspending import licenses until revised architectures addressed identified issues (*Regulatory Updates*, n.d.).

Australia takes a risk-based approach led by the Therapeutic Goods Administration (TGA), which classifies devices according to their seriousness and cyberattack exposure. The TGA publishes guidelines, periodically amending them in reaction to emerging vulnerabilities. After high-profile ransomware attacks against radiology devices, the TGA ramped up demand for threat modeling, robust authentication, and postmarket surveillance. Unlike Europe, the TGA grants conditional approvals with the expectation of rapid field updates for new threat vectors. This approach means manufacturers face more frequent compliance reviews but gain flexibility to react to fast-moving threats (*Regulatory Updates*, n.d.).

A major challenge for multinational device manufacturers lies in the lack of harmonization between standards. The EU, Japan, China, and Australia each imposes its own requirements for technical documentation, postmarket vigilance, and reporting timelines. For example, Europe requires immediate notification of certain cyber incidents, while China's process is less transparent and subject to government review before disclosure. Differences in definitions of what constitutes a reportable cyber incident lead to confusion and compliance risk. A global manufacturer may create multiple incident response and documentation workflows to satisfy each jurisdiction, increasing operational burden and engineering complexity.

Regulatory disparities also affect device approval timelines. A low-severity software update rolled out in the EU might require full reassessment and local certification in Japan or China, resulting in product unavailability for months in critical markets. In the event of a widespread cyber incident, overlapping regulatory authority generates confusion. Following an attack on an internet-connected infusion pump with sales across Europe, China, and Australia, the investigation brought simultaneous requests from European Data Protection Authorities, NMPA officials, and the Australian TGA for incident logs, forensic images, and patient notification records. Responding to asynchronous, sometimes contradictory requests, alternate timelines for incident disclosure, and incompatible technical standards, the manufacturer risked partial compliance and further enforcement action in several countries.

Despite some progress in inter-governmental cooperation, complete harmonization remains elusive. Technical standards—such as secure encryption, event logging, or backup requirements—continue to diverge. The exposure to uneven regulatory scrutiny is growing as postmarket surveillance intensifies and as governments demand more transparency and resilience from manufacturers. The global compliance environment for medical device cybersecurity tests not only technical expertise but also cross-border agility in regulatory affairs and risk management, signaling a growing need for robust data privacy practices and comprehensive security governance to avoid legal liability and protect patient safety (*Regulatory Updates*, n.d.).

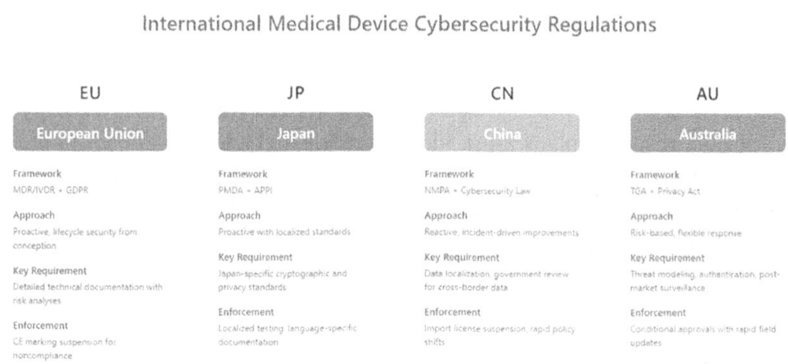

Privacy Laws and Their Impact on Device Security

Privacy regulations, particularly HIPAA and GDPR, embody some of the most pronounced differences and operational hurdles for healthcare stakeholders managing medical device ecosystems. These frameworks frame privacy as a legal and technical necessity, enforcing nonnegotiable standards for the management of ePHI and personally identifiable information (PII). For medical device manufacturers, compliance officers, and IT professionals, the detailed demands of these laws drive how device design, deployment, and support processes unfold across multiple jurisdictions.

HIPAA sets exacting access and authorization controls for U.S. entities handling ePHI. Device developers and health systems must implement strict policies to ensure that only appropriately authorized staff can access ePHI, based on job need and documented supervision. The law enforces "minimum necessary" use, prompting technical implementations such as granular role-based access controls and routine access reviews to prevent overexposure of sensitive data. Security policies must outline who may view or modify patient information, and under what circumstances those abilities are conferred and monitored, with mandatory security awareness training for all workforce members involved in device management or support (*Summary of the HIPAA Security Rule*, 2024).

The GDPR creates different expectations for entities encountering the data of European residents. Consent is central: device manufacturers must secure explicit, intelligible consent before collecting or processing personal data, including biometric measurements or device-generated clinical signals. Withdrawal of consent must be as straightforward as granting it, meaning device firmware, mobile apps, and related interfaces must support prompt, transparent opt-out mechanisms (Robb Taylor-Hiscock, 2022). Beyond consent, organizations must allow patients direct access to their health data, prompt deletion upon request unless legal justification exists, and account for every form of data usage—mandating real-time audit logs and meticulously tailored privacy configurations.

Both laws treat data security as a multifaceted technical discipline. Devices that capture, transmit, or store health data must implement encryption for data at rest and in transit, unique user authentication, and tamper-evident logging to deter and detect unauthorized activity. For example, an implantable cardiac monitor designed for smartphone synchronization must encrypt telemetry from device to app and restrict data uploads to authorized cloud endpoints. IT departments require audit mechanisms that trace every access attempt and categorize activities for routine forensic review, a process essential for HIPAA's and GDPR's security standards compliance. Incident response plans must address anticipated and unplanned events, including natural disasters or targeted cyberattacks, and set clear escalation protocols for both internal and external communications (*Summary of the HIPAA Security Rule*, 2024).

Patient rights remain at the core of global privacy mandates. Under GDPR, individuals can demand access to, correction of, and deletion of their health records. When a patient using a home dialysis device submits a request to see all retained metrics, device vendors and hospital administrators must rapidly verify identity, generate a complete data export, and deliver information securely while logging every step for compliance review (Robb Taylor-Hiscock, 2022). These processes fuel operational challenges—engineers must balance robust authentication with usability, facilitate exports that comply with privacy-by-design, and prevent unauthorized disclosure during transfer or storage.

Reporting obligations in the event of a data breach offer a lens into the rigor of enforcement. GDPR imposes a hard 72-hour deadline for notifying supervisory authorities and data subjects regardless of breach scope, while HIPAA sets a 60-day cap for notifiable incidents involving 500 or more people. If a hospital discovers a flaw in infusion pump firmware that exposes patient logs to remote compromise, security teams must immediately secure the devices, contain the incident, document the extent of exposure, and draft notices for affected patients and, where applicable, regulators. In practice, this requires IT operations to maintain detailed incident-response runbooks, centralized breach documentation systems, and well-rehearsed communications workflows (Robb Taylor-Hiscock, 2022; *Summary of the HIPAA Security Rule*, 2024).

Data minimization is both a regulatory imperative and a risk mitigation principle. Device creators must justify the data elements collected and retained, ensuring the minimum required for device efficacy. For instance, a wearable glucose monitor may limit stored data to anonymized trend readings rather than detailed patient demographics, thus reducing attack surface and compliance obligations. This approach curbs exposure if compromise occurs and simplifies cross-border data transfers and regulatory reporting, especially where localization mandates and deletion rights apply.

HIPAA vs. GDPR: Key Privacy Requirements

Category	HIPAA (United States)	GDPR (European Union)
Scope	Electronic Protected Health Information (ePHI) for covered entities	All personal data of EU residents, including health data
Consent	Implied consent for treatment, payment, and operations	**Explicit consent** required before data collection/processing
Access Controls	Role-based access, "minimum necessary" standard	Privacy by design, granular access controls with audit logs
Patient Rights	Right to access and request amendments to records	Right to access, correction, deletion, and data portability
Breach Notification	**60 days** for incidents affecting 500+ individuals	**72 hours** to notify authorities regardless of breach size
Data Encryption	Addressable requirement (recommended but not mandatory)	Required for data at rest and in transit
Data Minimization	Minimum necessary use and disclosure	Strict data minimization—only collect what's essential
Penalties	Up to $1.5M per violation category annually	Up to €20M or 4% of global revenue, whichever is higher

Medical device environments amplify these operational complexities. Regulators insist on sector-specific features, such as advanced device logging, region-locked data storage, and routine audit submission for high-risk platforms. Consider a global imaging device manufacturer: to market in the EU, it must maintain patient scans in local data centers, implement special audit protocols keyed to GDPR's requirements, and prove compliance with both host country and international standards during audits.

The layered demands of privacy regulations expose organizations to legal and organizational risk if device security lapses occur. Failures in compliance not only threaten patient trust but also trigger investigations, fines, and reputational harm—issues that demand further exploration of liability and financial exposure.

Legal Risks and Consequences of Negligence in Cybersecurity

Cybersecurity lapses in medical devices expose both practitioners and organizations to direct, significant legal liabilities. The legal framework has evolved amidst high-profile incidents, prompting courts and regulatory agencies to increase scrutiny over how healthcare entities protect connected technologies. When patient harm or data compromise occurs due to inadequate security, affected organizations and professionals may face lawsuits, regulatory sanctions, and even criminal investigations. The expectation is not only that reasonable safeguards are in place, but also that device makers and healthcare providers adapt rapidly to emerging threats, documenting compliance with up-to-date best practices at every stage.

Professional Liability and Negligence

Healthcare providers face growing risk of malpractice claims if patients experience injury or deteriorating health because of compromised devices. Courts and regulatory bodies have started to regard failure to secure devices as a form of negligent care, particularly when well-established security frameworks or manufacturer guidance is overlooked. Notably, the 2017 incident involving an unpatched insulin infusion pump that was compromised remotely resulted in the provider being sued for failing to implement recommended security controls. The court concluded that awareness of potential vulnerabilities, coupled with inaction, constituted a breach of the duty of care.

Malpractice claims increasingly reference the duty to maintain up-to-date knowledge of cyber risks and to respond appropriately to security advisories. If hospitals or clinics cannot demonstrate that physicians or IT staff received appropriate training and implemented basic safeguards, liability increases significantly. Moreover, the burden of proof often shifts to the defendant once it is shown that vulnerabilities were known but unaddressed, raising settlement and insurance payout costs.

Regulatory Precedent and Expanded Accountability

Legal trends reinforce that regulatory compliance alone is not always sufficient as a defense. Agencies such as the U.S. Food and Drug Administration (FDA) and the Office for Civil Rights (OCR) have imposed financial penalties and mandated corrective action plans in cases where organizations were found to have ignored disclosed vulnerabilities or failed to respond to security incidents in a timely manner. The OCR's $3 million penalty against a major hospital system in 2021, for failing to encrypt devices that ultimately contributed to a data leak, set a notable precedent for enforcement grounded in cybersecurity malpractice.

Manufacturers' Exposure to Litigation

Device manufacturers sit at the center of mounting legal and financial exposure linked to lax security postures. Lawsuits may originate from patients, hospital partners, or state attorneys general and increasingly take the form of class action lawsuits if multiple parties are affected. In a widely reported 2018 case, a manufacturer of cardiac defibrillators faced both federal regulatory fines and a class-action lawsuit when vulnerabilities were exploited, exposing patient data and risking device failure. Settlements included both financial penalties and requirements to pursue aggressive overhauls of security programs.

Manufacturers also face risks of export bans or product recalls, which can exceed direct legal costs. The FDA's authority to halt device distribution or issue mandatory field corrections amplifies both the financial and reputational stakes. Data from industry analysts demonstrates that remediation and crisis communications spending typically exceeds projected costs by 30–40% after major incidents, underscoring the need for proactive security investment and continuous compliance monitoring.

Whistleblower Protections and Organizational Transparency

Legal frameworks such as the U.S. False Claims Act and Department of Labor regulations directly protect employees who alert leadership or regulators to cybersecurity risks in medical devices. These protections extend to both internal and external reporting and are designed to reduce the risk of retaliation for exposing security shortcomings. Whistleblowers have historically played pivotal roles in uncovering concealed vulnerabilities, accelerating regulatory intervention, and preventing more severe patient or data harm.

Organizations with robust whistleblower protections are more likely to surface risks before they escalate. For example, a Midwestern hospital group avoided costly litigation when technicians reported a firmware flaw through a well-protected internal channel, enabling immediate remediation and prompt regulatory notification. Conversely, poor whistleblower response at a major diagnostics company led to suppressed reports, which ultimately resulted in public disclosure of a widespread vulnerability, triggering federal investigations and multimillion-dollar settlements.

Broader Financial and Operational Consequences

The financial fallout from cybersecurity incidents frequently extends beyond direct legal costs. Healthcare entities may file claims under cyber liability insurance, but a consistent trend of post-incident premium increases has emerged, affecting overall operating costs. Insurers increasingly analyze historical incident data before underwriting, and multiple claims within a short period commonly result in policy restrictions or significantly higher deductibles.

Indirect costs, including the loss of contracts, long-term reputational damage, and reduced patient or customer trust, have also multiplied. The annual Ponemon Institute cost report for healthcare consistently flags reputational impact and business loss as accounting for more than half of the total cost of a serious security event. Following a ransomware attack on a large integrated delivery network, several institutional payers suspended new agreements, leading to an estimated $20 million reduction in projected revenue for the fiscal year.

Mitigating these risks hinges on sustained investment in workforce training, ongoing vulnerability monitoring, and transparent disclosure practices. Failing to anticipate or respond effectively to emerging security risks no longer exposes organizations to reputational harm alone; it now brings significant legal, financial, and regulatory consequences, shaping the evolving landscape of healthcare cybersecurity.

Final Thoughts

As regulatory and legal requirements for medical device cybersecurity continue to evolve, it is clear that manufacturers, healthcare providers, and compliance professionals must adopt a proactive, life cycle-wide approach to security that integrates rigorous risk management, continuous monitoring, and transparent reporting. Understanding the complex interplay of international standards, privacy laws, and enforcement actions enables stakeholders to anticipate emerging challenges and align their practices with both current expectations and future developments. By embracing secure-by-design principles, fostering cross-sector collaboration, and investing in ongoing education and governance, organizations can not only reduce legal and financial liabilities but also enhance patient safety and trust in an increasingly connected healthcare environment. Moving forward, this comprehensive perspective will be essential for navigating the expanding regulatory landscape and advancing resilient, innovative medical technologies.

References

Advancing Medical Device Cybersecurity Beyond Compliance: Managing Risk with Governance. (2023, December 19). HIMSS. https://legacy.himss.org/resources/advancing-medical-device-cybersecurity-beyond-compliance-managing-risk-governance.

Elendu, C., Omeludike, E. K., Oloyede, P. O., Obidigbo, B. T., Omeludike, J. C. (2024, September 27). *Legal Implications for Clinicians in Cybersecurity Incidents: A Review*. Medicine; Wolters Kluwer. https://doi.org/10.1097/md.0000000000039887.

FDA (2020, September 22). *Cybersecurity*. FDA. https://www.fda.gov/medical-devices/digital-health-center-excellence/cybersecurity.

Health, C. for D. and R (2022, April 6). *Cybersecurity in Medical Devices: Quality System Considerations and Content of Premarket Submissions*. U.S. Food and Drug Administration. https://www.fda.gov/regulatory-information/search-fda-guidance-documents/cybersecurity-medical-devices-quality-system-considerations-and-content-premarket-submissions.

Regulatory Updates. (n.d.). ISPE I International Society for Pharmaceutical Engineering. https://ispe.org/initiatives/regulatory/updates.

Robb Taylor-Hiscock (2022, September 21). *HIPAA vs. GDPR Compliance: What's the Difference?* OneTrust. https://www.onetrust.com/blog/hipaa-vs-gdpr-compliance/.

Summary of the HIPAA Security Rule. (2024, December 30). U.S. Department of Health and Human Services. https://www.hhs.gov/hipaa/for-professionals/security/laws-regulations/index.html.

6

Risk Assessment: The Critical Starting Point

How can healthcare organizations systematically identify and manage the myriad cybersecurity risks embedded within medical devices? What strategies enable professionals to not only detect vulnerabilities but also translate that knowledge into effective protection tailored to clinical environments? How do risk assessment practices align with evolving regulatory requirements while keeping patient safety at the forefront?

This chapter addresses these questions by exploring the essential processes involved in risk assessment for medical device cybersecurity. It lays out methods for uncovering device weaknesses through both manual expertise and automated tools, highlights the application of specialized threat modeling frameworks, and explains how to measure and prioritize cyber risks in a way that reflects their clinical impact. Additionally, it emphasizes the importance of mapping cyber risks to real-world healthcare workflows and maintaining continuous monitoring to adapt defenses over time. Together, these coordinated efforts form the basis for developing safer medical technologies and more resilient healthcare systems.

Identifying and Disclosing Device Vulnerabilities

In medical device cybersecurity, the process of vulnerability identification starts with two complementary strategies: manual review and automated tools. Manual review depends on skilled professionals who combine deep technical knowledge with practical assessment of device use cases. Security experts analyze source code, inspect network traffic, and investigate device logic to uncover vulnerabilities that go beyond surface-level flaws. For instance, a manual assessor might detect a complex privilege escalation if an application erroneously grants administrative access when certain sequences of user actions occur, a scenario often missed by automation because it requires insight into business and clinical logic (Company, 2025). This method exposes hidden weaknesses, such as misuse of authorization checks or subtle logic bugs, but it is time-consuming, may suffer from inconsistencies in the expertise or focus of reviewers, and does not scale easily in environments with frequent software updates.

Automated tools like static application security testing (SAST), dynamic application security testing (DAST), and vulnerability scanners bring efficiency and consistency to vulnerability identification. Tools such as Burp Suite and ZAP systematically probe devices and their connected systems, searching for signs of common vulnerabilities like outdated libraries, open ports, and misconfigured HTTP headers (*Mastering Web Vulnerability Discovery: Manual Techniques and Powerful Tools Explained—ExamCollection*, 2025). These solutions excel at rapidly covering broad areas, ensuring that obvious flaws—such as missing patches—are flagged routinely with every release. However, scanners are limited by their predefined signatures; they cannot interpret the intent of bespoke workflow designs, and they often miss emerging threats or sophisticated attacks dependent on business logic, such as manipulations of healthcare transaction processing (Company, 2025). As a result, false positives are frequent, requiring manual validation to rule out nonissues and focus remediation efforts.

Healthcare organizations achieve the best results by integrating both approaches. Automation provides a baseline defense, identifying known problems and reducing manual labor, while manual review brings contextual understanding, creative thinking, and adaptability to novel or nuanced device architectures. For example, an automated scan might identify unencrypted data transmission on a wireless infusion pump, while a manual tester could exploit a gap in role-based access control to alter medication dosages, demonstrating a critical issue that rule-based scans would overlook. Collaborative workflows enable teams to review, validate, and address findings in a manner suited to the device's operational context, patient safety imperatives, and regulatory demands.

Certain vulnerabilities dominate device security audits. Outdated software and firmware remain one of the greatest risks, as attackers commonly exploit known flaws to gain unauthorized access or install malicious code. A widely publicized case involved hospital imaging devices running unsupported operating systems, which allowed ransomware to take control of patient data storage and critical imaging functions. Weak authentication practices, such as default or hardcoded credentials, are another persistent hazard. One case demonstrated unauthorized parties gaining access to connected cardiac monitors and altering alarm thresholds due to never-changed factory-set passwords. Risks extend to failed updates, as devices with insecure update mechanisms—such as lacking digital signature verification—invite attackers to inject malicious firmware under the guise of a routine patch (*Mastering Web Vulnerability Discovery: Manual Techniques and Powerful Tools Explained—ExamCollection*, 2025).

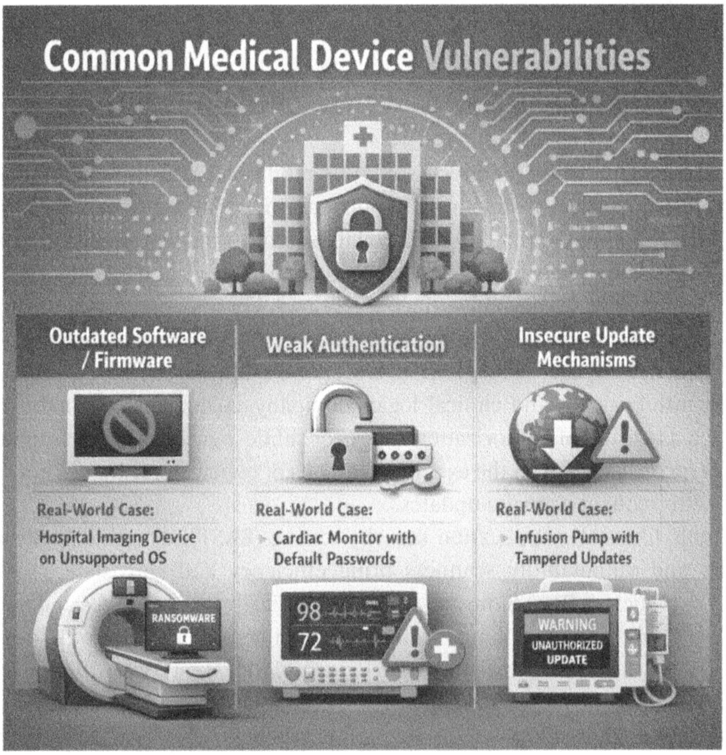

Unsecured communications present severe ramifications for patient privacy and safety. Devices that communicate over local hospital networks without encrypted protocols, for example, remain vulnerable to eavesdropping or injection attacks. Some networked infusion pumps have transmitted critical patient and device data in the clear, allowing adversaries to manipulate commands or access sensitive health information. Inadequate input validation also exposes devices to buffer overflows or injection attacks, which can lead to compromise of core device functions (Company, 2025).

Effective and ethical vulnerability disclosure practices form a cornerstone of medical device security. Security researchers and internal teams should follow established guidelines, ensuring vulnerabilities are reported privately to device manufacturers or responsible parties before making any information public. This minimizes the window of exposure for at-risk patients. Engaging directly with security response teams, providing clear technical documentation and proposed remediation steps, accelerates fixed deployments. Maintaining auditable records—including the discovery date, communications, remediation timeline, and final resolution—supports quality assurance and regulatory review. Additionally, ongoing communication with all stakeholders—ranging from device makers to healthcare providers—reinforces a collective responsibility to safeguard patient health. Responsible disclosure fosters trust, reduces the likelihood of panic or misinformation, and helps coordinate thorough, timely response actions (*Mastering Web Vulnerability Discovery: Manual Techniques and Powerful Tools Explained—ExamCollection*, 2025).

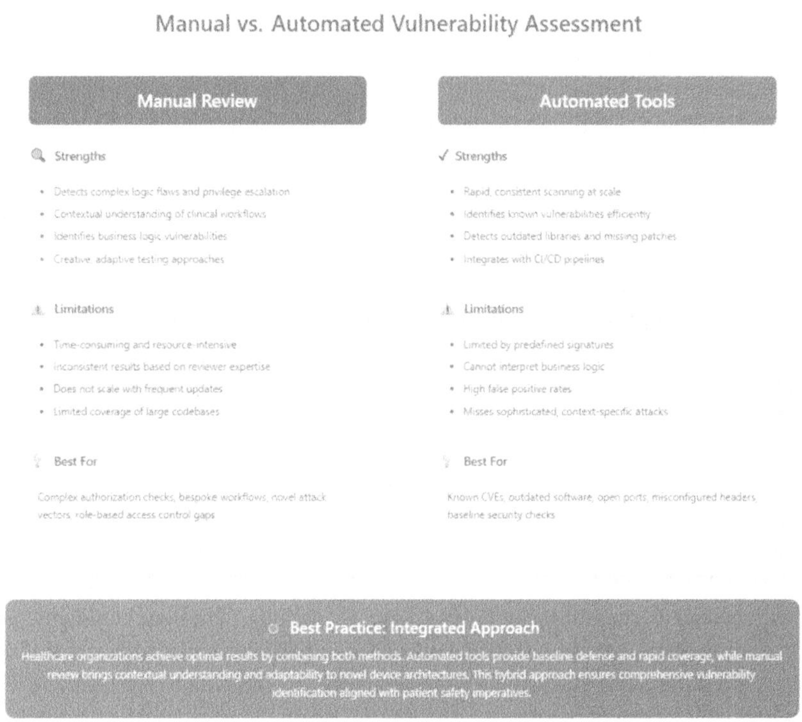

Independent, third-party security assessments add further resilience to device risk management efforts. External experts introduce specialized skills, unbiased scrutiny, and often leverage advanced testing techniques—including exploitation chaining or customized payloads—that internal teams may overlook under resource constraints (Company, 2025). Security firms and ethical hackers have revealed critical software or network-layer vulnerabilities previously undetected in internal reviews, directly contributing to stronger device designs and regulatory compliance. The validation of fixes through third-party retesting reassures manufacturers, healthcare organizations, and oversight bodies that mitigations have been effective, ultimately reinforcing system trustworthiness and market reputation.

From the process of vulnerability identification to the protocols of ethical disclosure and third-party evaluation, these practices collectively form the foundation for robust medical device security strategies. This groundwork supports a deeper understanding of how risk modeling and proactive defense measures can further refine the safeguarding of healthcare technologies.

Applying Threat Modeling to Healthcare Technology

Within healthcare, the act of identifying and disclosing device vulnerabilities is only the beginning of a truly effective cybersecurity strategy. The next critical step is threat modeling, a systematic process that transforms vulnerability knowledge into comprehensive risk awareness and actionable defense strategies. Threat modeling is not a generic cybersecurity exercise; it is deeply embedded in the context of patient safety and the realities of clinical care. As healthcare organizations continue to network life-critical devices and unlock new efficiencies through interoperability, threat modeling becomes the linchpin that connects technical insight to operational and regulatory mandates (Czekster et al., 2024; James, 2025).

Unlike broader cybersecurity assessments, medical device threat modeling considers a layered set of concerns. It incorporates not only cyber intrusions but also the unique patient impacts, such as treatment disruption or compromised data integrity, which can result from successful attacks. Regulatory frameworks like the FDA's requirements for premarket submissions and international standards for medical devices further shape threat modeling practices, demanding specificity, documentation, and validation at each step. Technical processes must align with these demands, beginning with the choice of methodology.

STRIDE and PASTA are two well-established methodologies that assist multifaceted healthcare teams in the rigorous identification and classification of threats. STRIDE provides a memorable, actionable framework organized around six categories: Spoofing, Tampering, Repudiation, Information Disclosure, Denial of Service, and Elevation of Privilege. For a hospital setting, Spoofing might include an attacker impersonating a clinician to remotely access patient ventilator settings, while Tampering could involve changes to infusion pump configurations after bypassing access controls. Repudiation concerns can arise when electronic health record (EHR) changes are not properly logged, allowing an actor to deny harmful alterations. Information Disclosure most often appears in the unauthorized access to sensitive patient data by exploiting wireless telemetry. Denial of Service is critical in hospital environments, as attackers can target wireless cardiac monitoring infrastructure, causing alarm failures or delayed care. Elevation of Privilege could emerge when a technician's account, intended for device diagnostics only, gains administrative capabilities across the network. Addressing each category in a structured way clarifies which risks demand immediate remediation and which present secondary concerns, enabling more systematic prioritization.

PASTA, a risk-oriented framework, guides healthcare cybersecurity teams through a layered, process-driven model. Starting from business objectives—such as continuous infusion monitoring, immediate clinical response, or compliance with the Health Insurance Portability and Accountability Act (HIPAA)—PASTA facilitates a mapping between technical threats and operational goals. This framework compels analysts to connect threat scenarios to the specific functions of medical devices and associated outcomes. The clarity and depth provided by PASTA can, for example, illuminate the risk of networked infusion pumps: an attacker who introduces false dose settings not only threatens system control but also endangers patient safety and legal compliance. By sequencing threat analysis around device functions, security teams can match mitigations directly to the stakes involved.

Generic approaches to threat modeling do not suffice within the healthcare sector. Safety scenarios must reflect worst-case clinical outcomes, and privacy considerations must adhere to strict legal standards. For example, modeling an infusion pump connected directly with an EHR involves assessing the possibility that manipulated dosing information could be led to the EHR, resulting in both patient harm and a downstream cascade of inaccurate clinical documentation. Interoperability concerns are equally pressing; the same device must reliably communicate with other tools such as barcode scanners or pharmacy software, and weaknesses in one chain link can reverberate across other systems. Patient safety, data integrity, operational continuity, and compliance all intersect in such a model.

Effective threat modeling demands precise asset identification. Every element—from the devices used at bedsides, to centralized data repositories, to the sprawling wireless networks that link them, and the various user interactions—must be inventoried and assessed. For instance, a compromised ventilator could cause immediate harm, while a breach in medication management software may quietly corrupt pharmacy orders, putting dozens of patients at risk. Teams should prioritize assets using criteria such as sensitivity of data, criticality for patient care, and potential for cascading failures. This risk-driven asset prioritization ensures that resources are applied where patient outcomes are most acutely affected.

Attack trees support the technical team by allowing them to decompose high-level attacks into logical, hierarchical paths. If securing remote access to patient monitors is a goal, an attack tree could expose multiple routes: exploiting unencrypted wireless signals, leveraging hardcoded manufacturer passwords, or hijacking third-party vendor connections. By clearly visualizing steps and dependencies, attack trees let stakeholders understand which vulnerabilities, if left unchecked, create the most dangerous compromise possibilities and allow teams to formulate layered defenses.

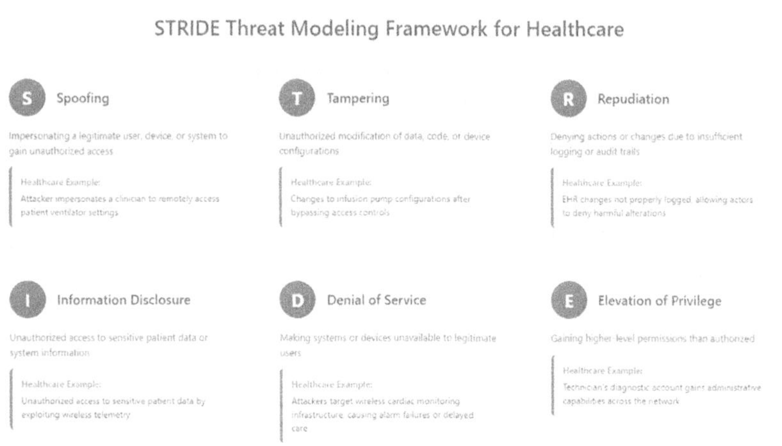

Interactive, multidisciplinary workshops should accompany every comprehensive threat modeling effort. Cybersecurity experts, clinical end users, safety officers, and IT specialists can collaborate to run realistic simulations, like a ransomware event that locks out access to imaging archives or a planned clinical downtime that intersects with life-critical surgery schedules. These sessions enable expertise sharing, help anticipate novel threat vectors, and foster the rapid solution design that healthcare requires.

As the complexity and variety of threats become visible, it is evident that a structured, quantifiable approach is necessary to make sense of risk and allocate mitigation resources. The nuanced results from robust threat modeling provide the critical foundation needed for systematic, objective risk assessment (Czekster et al., 2024; James, 2025).

Measuring and Prioritizing Cybersecurity Risks

Risk measurement stands at the heart of effective risk assessment in medical device cybersecurity. At its core, risk measurement involves systematically evaluating the likelihood and potential impact of cyber threats on devices that play a part in patient safety. The importance of this approach becomes evident when considering how a single vulnerability, if left unchecked, can jeopardize not just the integrity of devices but also the health outcomes for patients connected to critical clinical systems (*A Robust Statistical Framework for Cyber-Vulnerability Prioritisation under Partial Information in Threat Intelligence*, 2020).

A fundamental tool for risk measurement, the risk matrix, visually represents threats by mapping their estimated likelihood against their potential impact. Constructed as a grid, the X-axis shows how likely a cyber event is to occur, while the Y-axis tracks the consequences if it does. This setup allows healthcare cybersecurity teams to allocate their limited resources to the most pressing concerns. For instance, a noncritical vital signs sensor, despite being frequently scanned by attackers, sits in a lower-priority quadrant if the maximum consequence is data loss with no direct patient impact. On the other hand, an infusion pump, rarely targeted but with vulnerabilities that could permit catastrophic patient harm, lands in the matrix's top right quadrant where both likelihood and impact are severe. This matrix style prioritization guides device engineers and IT specialists to develop mitigation strategies for the highest threats first, focusing attention on vulnerabilities that could cause significant disruption or injury (AL-Dosari & Fetais, 2023).

Beyond basic matrices, more refined quantitative risk models have become central to robust cybersecurity practices. FAIR (Factor Analysis of Information Risk) adopts a probabilistic, financial view, breaking down risks into components such as threat event frequency and probable loss magnitude. In healthcare environments, FAIR can attach expected dollar values to incidents, supporting more informed decisions when considering investments in new security controls or balancing clinical process disruptions against potential losses. For example, if a vulnerability is found in the wireless module of a bedside monitor, FAIR can help model not just the technical risk but the possible cost of device downtime and the operational impact on patient monitoring. FAIR's advantage lies in enabling clear, business-oriented discussions, facilitating buy-in from IT managers, finance teams, and even device manufacturers.

CVSS (Common Vulnerability Scoring System), another widely adopted tool, provides a standardized numeric score for vulnerabilities based on characteristics such as ease of exploit, required attacker access, and the outcome of a successful attack. In healthcare technology, engineers can use CVSS to classify vulnerabilities in shared terms, which fosters swift communication with compliance officers, device vendors, or regulators. For example, if a software flaw with a CVSS score of 9.8 is discovered affecting infusion pumps, this score instantly communicates criticality to all stakeholders. The hospital's IT manager can then escalate the issue, triggering targeted patch management efforts, notifying device manufacturers for urgent firmware updates, and informing regulatory authorities as part of reporting obligations (AL-Dosari & Fetais, 2023; *A Robust*

Statistical Framework for Cyber-Vulnerability Prioritisation under Partial Information in Threat Intelligence, 2020).

Risk registers organize all this information into a single resource: each identified risk is listed alongside its description, calculated scores for likelihood and impact, mitigation actions, and a designated risk owner. In practice, cybersecurity professionals log the discovery of a vulnerability in a bedside monitor's wireless communication module, including its current patch status and the staff member responsible for mitigation. Over time, systematic risk register reviews ensure ongoing risks are not overlooked or forgotten, supporting rigorous follow-up. During audits or regulatory reviews, showing updated registers fosters transparency and demonstrates compliance to external parties.

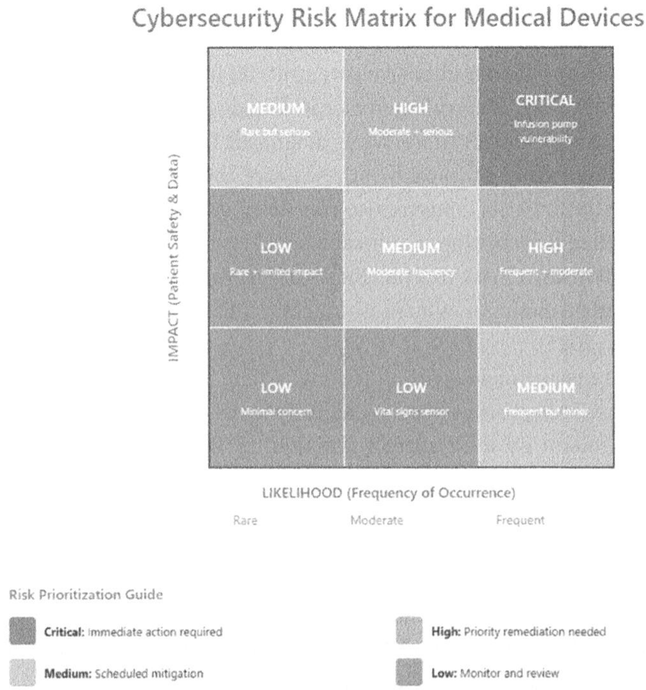

Prioritizing which risks to remediate first demands aligning risk scores with exploitability, real-world clinical impact, available remediation resources, and external compliance requirements. Risk management teams evaluate each open issue, weighing quantifiable scores against operational realities. Consider a hospital team reviewing outstanding patches for several devices: one flaw has a high CVSS score but impacts a retired laboratory device, while another with a slightly lower score affects an active life-support machine. Combining quantitative rankings with workflow constraints and clinical significance leads to decisions that might, for instance, expedite patching the life-support system while scheduling less urgent updates during routine future maintenance windows (*A Robust Statistical Framework for Cyber-Vulnerability Prioritisation under Partial Information in Threat Intelligence*, 2020).

Effective risk scoring and prioritization shape device update schedules and replacement plans that dovetail with patient care protocols. A device with a high-risk score may be targeted for upgrades or taken offline during periods of low clinical activity, safeguarding patient safety and ensuring that security measures do not disrupt ongoing medical services (AL-Dosari & Fetais, 2023).

Mapping Cyber Risks to Clinical Workflows and Patient Outcomes

The growing dependence on interconnected medical devices in healthcare brings urgency to understanding the practical impact of cyber vulnerabilities on clinical processes. When latent weaknesses turn into active threats, the resulting cyber incidents can disrupt essential workflows, affecting direct patient care and altering safety outcomes, especially in high-acuity environments. Safeguarding these environments requires deep insight into how devices are used clinically, where their vulnerabilities overlap with care delivery, and how disruptions reverberate through entire care systems (Aldosari, 2025).

Mapping clinical use provides a fundamental lens for connecting specific device functions to the critical points within a clinical workflow. For example, consider the continuous infusion pump that delivers precise medication dosages in an intensive care unit. Its uninterrupted operation is vital for managing unstable patients. Likewise, real-time monitoring devices display patient vital signs, supporting rapid decisions during emergent care, while diagnostic imaging tools inform treatment planning in acute and outpatient settings. In each of these cases, the device's cyber risk is not only about its standalone failure but how its compromise can delay care, cause diagnostic errors, or lead clinicians to act on falsified information. For instance, if a ransomware attack disables a ventilator during an emergency, the direct risk to a patient's survival becomes evident, but so does the increased stress and workload on clinicians who must urgently find alternatives while maintaining other critical tasks (Aldosari, 2025).

Using clinical use mapping, stakeholders can illuminate where device vulnerabilities intersect with key patient care actions. This process starts by identifying devices within each clinical area and documenting all tasks reliant on their correct functioning. Dependencies may include the order of device usage in a procedure, necessary coordination among devices, and the links between device-generated data and downstream care steps. After mapping, teams can analyze what happens if a device fails or is manipulated—does the workflow break, are delays introduced, or is patient safety left at risk? For example, in an operating room setting, a compromised patient monitor could result in wrong or missing data during anesthesia, potentially leading to missed critical warnings or inappropriate intervention. Mapping clarifies these points of vulnerability and reveals which devices contribute most directly to safe and continuous care.

High-impact risk areas become visible as devices are mapped to their specific clinical use cases. Critical devices, such as defibrillators or surgical anesthesia machines, play a direct, immediate role in sustaining life. Important devices, like infusion pumps in medical wards, influence ongoing clinical outcomes, while ancillary devices, such as postoperative transport monitors, support recovery but do not drive urgent interventions. Documenting device dependencies for each category enables prioritization: teams can then trace, for instance, how failure of an ICU ventilator might cascade into respiratory emergencies, or how disrupted access to portable imaging could lead to care delays, especially in trauma scenarios. This level of assessment ensures that resources and protections are directed toward assets whose compromise presents the gravest consequences for patient safety, as well as care continuity (Aldosari, 2025).

Determining device criticality is integral in ranking medical equipment by its potential impact on life safety and service interruption. Devices are categorized according to their function: those required for emergency care (such as ventilators and defibrillators), for continuous vital monitoring (e.g., bedside patient monitors), for direct life support (infusion pumps delivering lifesaving drugs), and for ancillary support functions (transport or documentation). For a practical illustration, if a large-scale ransomware attack renders both anesthesia machines and wireless temperature sensors nonoperational, the assessment framework allows the security and clinical response teams to allocate resources quickly to restore high-criticality ventilators and anesthesia machines first, while deferring less urgent devices for later action (Aldosari, 2025).

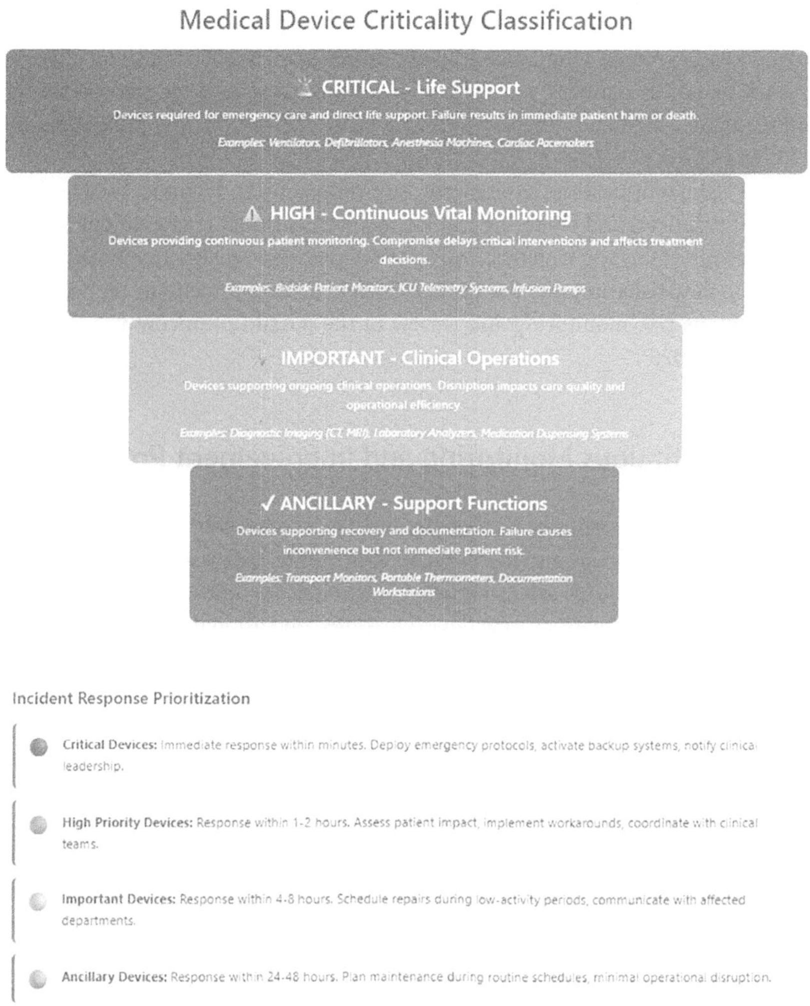

Worst-case scenario planning underpins comprehensive preparedness. Simulating plausible cyber incidents enables healthcare organizations to model and anticipate the ripple effects on patient safety, outcomes, and operational continuity. Scenario planning may begin by identifying a credible cyber threat, then mapping projected consequences through each stage of the clinical workflow. For example, a planned simulation might examine a multistage ransomware attack that disables patient monitors during a cardiac operation. The scenario details loss of visibility into vital signs, risk escalation for clinical decision makers, and delays in intervention, with compensatory response plans including manual monitoring protocols and clinical handoff procedures. Such analysis grounds the organization's risk controls and escalation pathways in real clinical contexts, making them actionable and relevant for practitioners and technical staff alike (Aldosari, 2025).

Effective medical device risk assessment depends on interdisciplinary collaboration. IT, clinical, safety, and compliance leaders bring complementary perspectives, making it possible to bridge the knowledge gap between technology vulnerabilities and practical patient care impacts. For instance, during a hospital-wide cyber risk workshop, clinicians may clarify the minute-to-minute patient safety demands attached to specific devices, while IT professionals elucidate technical risk factors and attack vectors. This collaboration ensures mapping accuracy and aligns protective strategies—such as access controls, network segmentation, and backup procedures—to clinical realities and regulatory mandates, thereby enabling faster recognition and mitigation of incidents (Aldosari, 2025).

The process of risk identification, assessment, and mitigation is dynamic. Evolving technology, changing clinical practices, and emerging threats require continual reassessment and integration into daily care delivery. Ongoing vigilance and adaptive processes ensure that health organizations respond nimbly to new risks and maintain robust patient safety protections, reinforcing the foundation for continuous risk monitoring and review in the pursuit of uncompromising healthcare resilience (Aldosari, 2025).

Establishing Continuous Monitoring and Improvement Processes

Organizations in healthcare confront a changing risk environment for medical device cybersecurity, which demands a persistent and methodical approach to risk assessment through all device life cycle stages. Static or one-time assessments are unsuitable for addressing the realities of rapidly evolving technologies and threats, as new device features, operational workflows, and threat vectors emerge. Adopting a life cycle approach enables continuous identification, reassessment, and mitigation of cyber risks, thereby reducing the chance of persistent vulnerabilities and aligning with patient safety imperatives.

Automated Monitoring for Real-Time Threat Detection

In healthcare environments with numerous interconnected devices and substantial variability in device profiles, automated risk detection solutions such as Security Information and Event Management (SIEM) platforms and intrusion detection systems (IDS) become essential. These solutions systematically collect and analyze telemetry from networked medical devices, application logs, and device communication channels. By continuously scrutinizing aggregated data, automated platforms can uncover behavioral anomalies that signal potentially unauthorized device access or anomalous traffic patterns in near real time.

For example, an SIEM system might trigger real-time alerts when a hospital infusion pump attempts to establish an unfamiliar outbound network connection, or when IDS sensors detect a rapid sequence of failed authentication attempts from an external IP address targeting diagnostic imaging equipment. Early detection is particularly critical in healthcare settings where asset quantities may run into the thousands and where device heterogeneity increases the risk of poorly understood vulnerabilities. Automation supports rapid triage and response, lessening risk exposure windows and reducing the probability of unnoticed compromise, as evidenced by industry adoption of centralized dashboards that merge risk metrics, vulnerability findings, and compliance statuses from disparate assets (*DevSecOps Principles: Tutorial and Best Practices*, 2025).

Scheduled Reviews for Structured Risk Reassessment

Periodic, structured risk reviews address the inherent limitations of point-in-time assessments by embedding reassessment routines into organizational workflows. Reviews should follow formal schedules—quarterly, semiannually, or in response to significant system or regulatory changes. This cadence ensures that risk profiles remain accurate, especially as new vulnerabilities become public, threat intelligence evolves, and clinical environments undergo transformation.

A quarterly risk review could uncover a newly released remote code execution vulnerability affecting firmware in infusion pumps deployed across several clinical departments. Such reviews compel organizations to update risk matrices and mitigation strategies, rebalancing priorities in light of new high-severity findings (Jha, 2024). Similarly, a semiannual review that systematically integrates updated threat intelligence—such as intelligence on exploitation trends or tactics used against hospital networks—forces organizations to rescore vulnerabilities and adjust resource allocations for patching and monitoring activities.

Scheduled reviews also provide a vital mechanism to demonstrate ongoing compliance with international standards and regulatory requirements. Formal documentation of these reviews supports audit readiness and ensures risk management strategies reflect the latest clinical practice, regulatory mandates, and organizational objectives.

Incident Feedback Loops for Adaptive Risk Reduction

Feedback obtained from actual cybersecurity incidents and near misses constitutes an essential dimension of a dynamic risk management strategy. Incident analysis, especially root cause investigations, enables organizations to identify—and subsequently close—unanticipated gaps in security controls and clinical workflows. During a postincident retrospective on an unauthorized device access event, analysts might discover that device onboarding procedures were inconsistently applied in certain clinical areas, resulting in vulnerable network configurations. Incorporating the findings from such investigations directly informs updates to organizational risk models, policy frameworks, and operational mitigation strategies (*DevSecOps Principles: Tutorial and Best Practices*, 2025).

The speed at which feedback is collected and acted upon determines how effectively organizations prevent recurrence. Prompt integration of incident-derived insights catalyzes improvements, enhances overall resilience, and fosters a culture of continuous learning. Tangible benefits arise from evolving security controls to better contest current adversary tactics and from aligning remediation efforts with the specific clinical realities in which incidents occurred.

Life Cycle Integration Across Development and Operational Phases

Effective cybersecurity risk assessment in healthcare persists from premarket development through to decommissioning. In early device design, risk evaluation can expose supply-chain vulnerabilities by revealing dependencies on untrusted components or insufficiently vetted software libraries. Throughout prototyping and validation, integrating threat modeling uncovers design-level exposures and enables engineers to build security requirements that counter contextual threats specific to clinical workflows.

As devices transition into production and active use, continuous risk assessment remains indispensable. Real-time monitoring facilitates rapid adaptation to new vulnerabilities arising from software updates or shifts in device usage—such as a clinical workstation transitioning to support remote consultations, which now exposes it to new attack surfaces and threat vectors (Jha, 2024). Regularly updating risk models to reflect those changes ensures that mitigation activities and monitoring efforts evolve synchronously with the device's operational context.

When decommissioning occurs, assessment remains critical to identify and neutralize residual risks, such as dormant device data or legacy connectivity channels that may persist beyond active clinical deployment. Life cycle integration therefore guarantees that risk assessment is never a one-off activity but a persistent, adaptive, and holistic practice, uniquely tailored to support the pace and complexity of contemporary healthcare delivery (*DevSecOps Principles: Tutorial and Best Practices*, 2025; Jha, 2024).

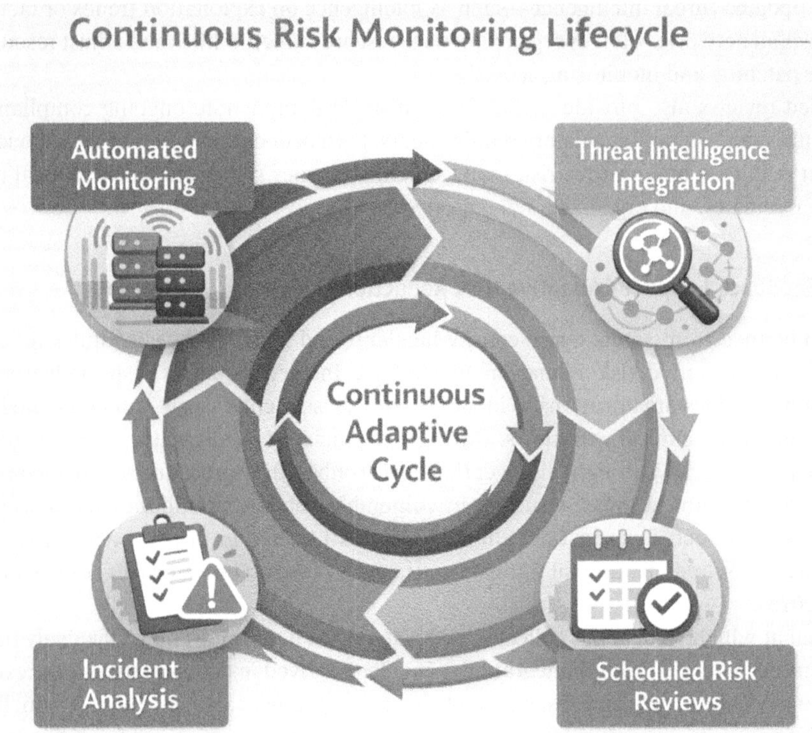

Bringing It All Together

Now that we understand the critical role of thorough risk assessment in medical device cybersecurity—from identifying vulnerabilities and applying context-specific threat modeling to quantifying risks, mapping clinical impacts, and implementing continuous monitoring—healthcare organizations and device manufacturers are better equipped to safeguard patient safety in an evolving threat landscape. Integrating these foundational practices enables targeted mitigation strategies that align with clinical workflows and regulatory requirements, fostering resilience against emerging cyber threats. Moving forward, sustained collaboration across technical, clinical, and compliance teams combined with adaptive, life cycle-based risk management will be essential to ensure medical devices remain secure, reliable, and trusted components of modern healthcare delivery.

References

A Robust Statistical Framework for Cyber-vulnerability Prioritisation Under Partial Information in Threat Intelligence. (2020). Arxiv.org. https://arxiv.org/html/2302.08348v4.

AL-Dosari, K., Fetais, N. (2023, January 1). *Risk-Management Framework and Information-security Systems for Small and Medium Enterprises (smes): A Meta-analysis Approach*. Electronics. https://doi.org/10.3390/electronics12173629.

Aldosari, B. (2025, May 6). *Cybersecurity in Healthcare: New Threat to Patient Safety*. Cureus; Springer Science and Business Media LLC. https://doi.org/10.7759/cureus.83614.

Company, T. (2025, June 25). *Web Application Security Testing: The Ultimate Guide to Protecting Web Apps*. Medium. https://medium.com/@jignect/web-application-security-testing-the-ultimate-guide-to-protecting-web-apps-279522017f32.

Czekster, R. M., Webber, T., Furstenau, L. B., Marcon, C. (2024, November 1). Dynamic Risk Assessment Approach for Analysing Cyber Security Events in Medical IoT Networks. Internet of Things; Elsevier BV. https://doi.org/10.1016/j.iot.2024.101437.

DevSecOps Principles: Tutorial and Best Practices. (2025). Drata.com. https://drata.com/grc-central/compliance-as-code/devsecops-principles.

James, P. (2025, May 15). Cyber Threat Modeling and Risk Assessment in Smart City Infrastructures. https://www.researchgate.net/publication/392408362_CYBER_THREAT_MODELING_AND_RISK_ASSESSMENT_IN_SMART_CITY_INFRASTRUCTURES.

Jha, S. (2024, July 16). *Vulnerability Management Lifecycle: The Ultimate Guide to Business Security*. Strobes Security. https://strobes.co/blog/vulnerability-management-lifecycle-the-ultimate-guide-to-business-security/.

Mastering Web Vulnerability Discovery: Manual Techniques and Powerful Tools Explained—ExamCollection. (2025). Examcollection.com. https://www.examcollection.com/blog/mastering-web-vulnerability-discovery-manual-techniques-and-powerful-tools-explained/.

7

Designing Security into the Medical Device Life Cycle

"Security is not an add-on—it must be built into every step of the medical device's journey." This principle reflects a growing recognition in healthcare technology, where protecting devices means more than fixing vulnerabilities after they arise. The complexity of medical environments and the high stakes of patient safety demand that security measures become integral from the very start of design, continuing through development and deployment. As threats evolve and regulations tighten, this approach transforms how engineers, clinicians, and regulators collaborate to safeguard both patients and data. Understanding how security intertwines with technical design, human interaction, supply management, and device resilience reveals the depth required to create trustworthy medical technology today.

Security by Design: Principles from Day One

In safety-critical healthcare environments, the consequences of inadequate security design can include patient harm, data breaches, and interruption of essential clinical workflows. Security by design asserts that protecting medical devices is not a step applied at the end of development but a foundational philosophy permeating each stage of the engineering process. Robust security must be woven into the fabric of the device life cycle, starting from conception and spanning implementation and operational deployment. The stakes differ from consumer technologies because regulatory requirements such as FDA premarket submissions, HIPAA privacy mandates, and EU MDR obligations elevate the need for explicit, proactive cybersecurity integration. Attackers increasingly target networked medical devices for extortion or as entry points into broader hospital infrastructure. Given the inherent risks to patient safety and the realities of persistent adversaries, security by design becomes more than industry best practice—it is an ethical and regulatory imperative (OWASP, n.d.; *Principles of Security—OWASP Developer Guide*, 2025).

Defense in Depth

Medical device security rests on a layered model where overlapping controls collectively diminish attack opportunities while reducing the risk that any single weakness leads to catastrophic compromise. Defense in depth applies parallel streams of protection at the physical, technical, and administrative layers. In a typical hospital, a wireless infusion pump may reside in a secured pharmacy storage area, protected by badge-controlled access and CCTV. Its firmware may validate digital signatures before executing code updates to prevent tampering, while all communications to the central pharmacy server are secured with mutually authenticated TLS sessions. Even if a network segment's firewall misconfiguration allows illicit traffic, authentication and encryption controls above the network stack reduce the likelihood of meaningful exploitation. Administrative procedures further enhance resilience—device logs may be subject to continuous review, and audit trails for medication dispensation provide accountability.

Layered defenses also directly confront the technical makeup of medical devices. Consider an MRI scanner that encrypts stored exam images locally, rejects untrusted USB peripherals at hardware input, and requires operator badge authentication for each login session. These overlapping controls, integrated at the hardware, operating system, and application levels, force adversaries to overcome multiple, independent barriers, increasing the difficulty and cost of attacks. This arrangement compensates for the possibility that one layer may be misconfigured or bypassed, supporting a more robust security posture where single points of failure rarely exist.

Least Privilege

Restricting permissions to the bare minimum required is central to medical device safety and accountability. In practice, least privilege means that device firmware grants subsystems only those capabilities needed for their functions while blocking access to unrelated modules. A ventilator's monitoring module, for example, receives and displays sensor data but cannot modify calibration parameters or initiate diagnostic firmware updates. This segregation not only guards against external adversaries but also limits accidental errors by authorized users. For hospital staff, granular access controls may authorize nurses to administer medications using smart pumps, while only clinical engineers access device settings relevant to maintenance.

Enforcement of least privilege further extends to network protocols. Medical imaging equipment can be configured to accept only queries from designated PACS servers, blocking unsolicited inbound requests from unauthorized network zones. By strictly segmenting access paths and minimizing systemic trust, the blast radius of any accidental or malicious action remains tightly controlled, directly diminishing the risk of privilege escalation or lateral movement within clinical networks (OWASP, n.d.; *Principles of Security—OWASP Developer Guide*, 2025).

Secure Defaults

Devices are often deployed by staff without advanced cybersecurity expertise, and an insecure default state can leave life-critical equipment exposed to exploitation. Shipping medical devices with secure defaults addresses this risk by proactively minimizing attack surfaces. For instance, patient monitors may ship with remote management disabled, requiring explicit enablement through authenticated, audited procedures. Password enforcement may mandate long, complex credentials for all user roles. In systems equipped for network connectivity, open ports unnecessary to the core function—such as legacy file sharing or debugging services—remain disabled, denying adversaries an easy entry. These defaults reflect a stance where access must be intentionally granted rather than merely restricted after installation, preventing common oversight errors in hectic clinical environments (*Principles of Security—OWASP Developer Guide*, 2025).

Fail-Safe Mechanisms

Given the inevitability of failures and the sophistication of modern attacks, medical devices must be designed to transition gracefully into secure states when faced with anomalous or unstable conditions. If a critical device detects attempts to compromise its operating parameters or sees signature mismatches in a firmware update, it may lock down all network-facing interfaces, enter a diagnostic-only mode, and notify responsible staff with alarms. In complex infusion systems, fail-safe routines may revert the device to a predetermined, known-safe state and require engineer intervention before reactivation. Such approaches ensure that systemic integrity and patient safety persist, even if components break down. Designing for fail-safe operation also preserves audit data and maintains forensic trails necessary for postincident analysis, reinforcing trust throughout the device life cycle (OWASP, n.d.).

Collectively, these interlocking design philosophies strengthen the overall posture of medical devices, protecting clinical workflows and patient safety. Embedding these values early in engineering fosters systems that can robustly endure attacks and unanticipated failures, setting a solid groundwork for further technical development and operational resilience.

Secure Development Life Cycles (SDLC) in Practice

Developing secure medical devices requires more than just strong technical mechanisms—it depends on rigorous, repeatable security processes embedded throughout the device's life cycle. In medical device engineering, where vulnerabilities can have life-or-death consequences, these structured approaches ensure that security is not an afterthought but a foundational discipline. Defense in depth, least privilege, secure defaults, and fail-safe mechanisms must shape every step from requirements to retirement. This is not possible unless development teams maintain continuous security competence. Ongoing training is not a luxury but a necessity. Medical device engineers face a uniquely dynamic threat landscape, with attackers leveraging novel vectors and regulatory requirements evolving in tandem. As a result, frequent security training modules are critical. Many organizations deploy scenario-based learning, such as simulated attacks on embedded insulin pumps, challenging developers to identify and patch design weaknesses. Regular workshops on emerging vulnerabilities, like supply-chain attacks against Bluetooth stacks, equip teams to recognize attack paths common in medical device ecosystems. These trainings do more than inform—they foster a workplace where any employee can spot and communicate risks, whether in system design discussions or when writing a single line of code (Prates & Pereira, 2024).

An SDLC relies on deliberate process controls to prevent missteps from reaching production. Security gates function as enforceable checkpoints within the SDLC, supporting both compliance and resilience. For example, when engineers submit code changes, a mandatory code review checks for adherence to secure defaults, such as strong encryption parameters and avoidance of hard-coded credentials. Code reviewers, using structured checklists reinforced by automation, verify that input validation routines will reject malicious payloads. At the system design level, architecture assessments evaluate how least privilege is enforced—one team might conduct a peer review focused on segmenting networking permissions so web interfaces never expose patient data services. Formal threat modeling sessions, involving engineers, product managers, and clinical staff, pinpoint exposure points in device connectivity protocols. For instance, a cross-disciplinary team might catch that telemetry data lacks adequate signing, risking patient record tampering. Engaging a wide group uncovers issues that isolated technical teams might miss, preventing the advancement of insecure solutions and dramatically enhancing overall risk management (Research Publication, 2019).

Automated security assessments further enforce quality at a speed compatible with modern healthcare release cycles. Static application security testing (SAST) scans source code for unsafe function calls, improper buffer handling, and exposure of encryption keys. A device firmware engineer might save hours when a SAST tool flags the use of insecure random number generators before human review ever begins. Dynamic application security testing (DAST), integrated into the test phase, analyzes outputs for vulnerabilities like weak session handling during simulated authentication flows. Fuzz testing attacks device interfaces with random or malformed data inputs, surfacing crash conditions or critical failures that standard unit testing overlooks. For example, fuzzing a hospital infusion pump interface might uncover that certain malformed network packets cause device reboots, potentially jeopardizing patient care if exploited. By embedding these automated tools in the CI pipeline, organizations create rapid feedback loops. Each commit triggers an array of scans, preventing the propagation of flaws downstream and giving developers immediate, actionable insights (Prates & Pereira, 2024).

The integration of DevSecOps in medical device engineering springs from the need for both security and accelerated deployment. Embedding security within DevOps practices allows teams to automate checks for regulatory alignment, such as those needed for HIPAA data privacy or EU MDR device safety. For instance, a pipeline may include automated verification that logs never include patient identifiers and that device logs are encrypted end to end. Continuous monitoring tracks device behaviors in simulated clinical environments and, if permitted, in postmarket use. Real-time detection mechanisms alert teams to deviations, like unauthorized firmware changes or unexpected data exports, so they can respond before threats escalate. Automated compliance verification ensures that every deployed build meets not only industry best practices but also regulatory mandates (Research Publication, 2019).

Practical examples abound in modern device teams. When a development team at a cardiac monitor manufacturer systematically employed threat modeling at the architecture stage, they prevented a replay attack vector before implementation commenced. Another team used automated risk detection tools to identify a broken access control issue and delivered a software patch to clinics in days instead of weeks, minimizing patient impact. Embedding these practices into daily routines means that security is a visible, shared goal, not just a box checked before release.

As automated assessments and monitoring advance, the interplay between technological solutions and those who use and maintain them grows ever more significant. Real-time tools and continuous controls support robust technical foundations, but meaningful protection ultimately draws from a culture that values user safety, trustworthy design, and constant vigilance (Prates & Pereira, 2024; Research Publication, 2019).

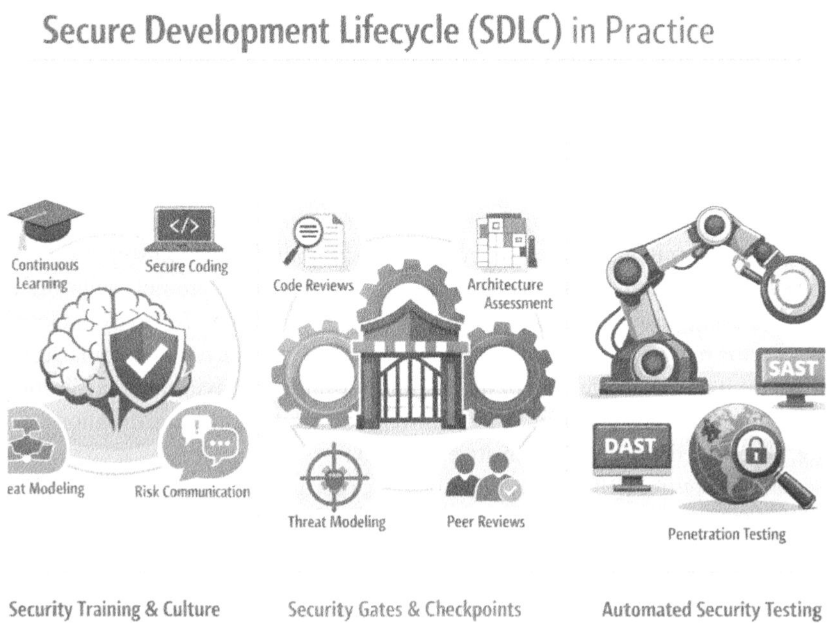

Human Factors in Secure Device Design

Secure medical device engineering starts by embedding security controls from the earliest development phases, yet the effectiveness of those controls relies on how real-world users interact with the device. Technical solutions alone cannot ensure robust security if they do not consider the cognitive, operational, and ergonomic realities of clinical practice. When security measures are designed with only theoretical users in mind, they can clash with actual workflow patterns, causing increased risk through accidental misuse or deliberate workarounds (Tzimourta, 2025). Integrating human factors throughout the device life cycle enables development teams to design controls that complement, rather than interfere with, critical care delivery.

Security and usability exist in constant tension. Consider authentication systems. Complex, multistep login protocols might theoretically provide high assurance of identity, but in a trauma bay or emergency department, such measures can slow access to lifesaving functions. Providers under pressure may resort to risky workarounds: taping login badges to monitors, sharing passwords, or propping open session timeouts. Each workaround represents a weak point in the device's security posture that was unintentionally introduced by failing to adapt the security protocol to urgent clinical conditions (Kushniruk & Kaufman, 2024). Successful device security balances the need to protect sensitive functions with the requirement for rapid, predictable access. For example, biometric logins, tap-to-authenticate badges, or layered emergency access provisions allow rapid authentication when seconds matter, while still logging access and providing audit trails for accountability.

Usability testing must extend beyond laboratory scenarios. In real clinical environments, devices are often managed by staff who operate on rotating shifts, face interruptions, and must multitask under stress. Security controls must not only work in theory but must allow for human error, interruptions, and diverse levels of user familiarity (Tzimourta, 2025). For example, medication pump interfaces can incorporate double-check systems for dose entry—such as requiring confirmation of critical values by a second clinician. These checks must present information clearly and integrate with workflow, so users do not bypass them out of frustration or time pressure. Iterative design, supported by routine feedback from clinicians, helps fine-tune these safeguards, ensuring they support rather than disrupt efficient care.

Interface clarity is a crucial defense against accidental security breaches and operational errors. Poorly differentiated alarms or cryptic error messages increase cognitive load and can lead to mistakes that undermine both workflow and security. Alarm fatigue arises when clinicians are exposed to frequent, ambiguous alerts, causing them to tune out warnings, silence alarms, or ignore prompts—thus missing critical security or safety notifications (Kushniruk & Kaufman, 2024). Concise labeling, consistent iconography, and confirmation prompts help orient users, enabling them to distinguish between types of alerts and respond properly. For example, designing medication selector menus with clear, color-coded sections and preventing similar drug names from being adjacent minimize the risk of inadvertent selection and associated medication errors. Dropdown menus must prevent misselection through separation of critical options, predictive highlighting, and immediate, context-sensitive feedback.

Role-based access control (RBAC) is a pillar of device security, yet its effectiveness depends on careful mapping to the realities of clinical roles and workflows. Devices are used by a diverse range of staff—nurses, physicians, biomedical engineers, IT support, and administrators—each requiring different access and authority levels. Secure configuration means limiting the scope of action for each role: nurses may be permitted to administer medication and adjust basic settings but not change device firmware or export data logs. The system should accommodate complex realities, such as handovers during shift changes or situations where privilege escalation is necessary but temporary. For instance, a nurse manager could be granted short-term administrative privileges during an after-hours emergency, with the system logging and automatically revoking access after resolution (Tzimourta, 2025). Implementing flexible, transparent access rules reduces resistance, while preserving auditability and restricting lateral movement within the system.

Continuous, contextual security education forms the human link in the defensive chain. Many breaches in healthcare originate with user action—lost credentials, falling victim to phishing, improper disposal of devices, or failure to recognize new threats. Security training must be relevant, ongoing, and adapted to the clinical world. Targeted awareness programs, regular hands-on drills, and integrated hints in device interfaces foster mindful behavior. Simulated phishing exercises, for instance, build vigilance against social engineering attempts. Onscreen prompts can warn users before exporting sensitive information, reminding them of policy and potential consequences (Kushniruk & Kaufman, 2024).

Embedding human factors into the heart of device security not only increases robustness but also prevents the friction that engenders risky shortcuts. Each device interface, policy, and protocol becomes a vehicle for security awareness, tailored to the user's perspective. Effective, user-centered security design underpins broader organizational defenses, ensuring that controls in supply-chain management, procurement, and system integration can build upon resilient, well-accepted device-level safeguards (Kushniruk & Kaufman, 2024; Tzimourta, 2025).

Secure Supply-Chain Management

Medical device vulnerabilities do not arise solely from user error or operational lapses in the clinical environment. Even the most securely designed and meticulously deployed devices can be compromised if their supply chains are not adequately protected. Each link in the procurement and assembly process introduces unique attack surfaces; adversaries may insert malicious firmware, clone crucial components, or intercept shipments, undermining device integrity and patient safety at a foundational level. Robust supply-chain management is not simply a compliance requirement—it is a vital layer of defense that shapes long-term device trustworthiness.

Supplier Vetting

Effective supply-chain security begins by rigorously vetting suppliers. Organizations must establish defined security benchmarks for supplier selection—and maintain formal, ongoing review throughout the contractual relationship. Objective evaluation methods should include requests for evidence of comprehensive information security programs, such as documented cybersecurity policies and procedures, workforce security training, and periodic vulnerability assessments. Many organizations require suppliers to maintain internationally recognized certifications, such as ISO/IEC 27001, which demonstrate systematic information security management. In some cases, obtaining the results of recent penetration tests and reviewing compliance with standards like Control Objectives for Information and Related Technologies (COBIT) or Telephone Consumer Protection Act (TCPA) further ensure alignment with best practices (Monette, 2022).

Beyond initial onboarding, continuous monitoring is crucial. Suppliers' security postures can shift due to business changes, ownership transitions, or even regulatory evolution. Contractual agreements should mandate recurring risk assessments, annual audit cycles, and escalation procedures for addressing identified gaps. Risk scoring frameworks—where high-impact, mission-critical suppliers receive more frequent and in-depth scrutiny—allow organizations to align resources with exposure levels. Demonstrable maturity in a supplier's security ecosystem significantly lowers the chances of introducing counterfeit components or unauthorized firmware into medical devices (DiMase et al., 2016).

Secure Logistics

Physical movement of device components represents another high-risk vector. Tampering during transit, theft, or substitution of consignments can transform a trusted shipment into a liability. Employing tamper-evident packaging—a solution where tamper seals break visibly if breached—alerts recipients to any unauthorized access. Logistical partners should be carefully chosen for established protocols involving shipment security, employee screening, and environmental controls. Carrier contracts may specify requirements for chain of custody adherence, restricted cargo access, and responsive loss reporting.

Logistics controls are enhanced by adopting traceable, serialized barcodes linked to manifests through digital management systems. GPS-based real-time tracking facilitates continuous oversight during transit; any deviation from predefined routes or unauthorized stops can trigger immediate alerts and response measures. For high-value or sensitive microelectronics integral to device safety or data handling, special handling protocols—such as segregated, monitored containers—are critical. These practical controls serve as deterrents and ensure swift identification if a shipment's integrity is compromised (Monette, 2022).

Chain of Custody

Documenting an unbroken chain of custody fortifies provenance for every device component. Each transfer—from supplier warehouses, to local distributors, through assembly lines and into finished product stock—introduces potential for misplacement, unauthorized substitution, or deliberate adulteration. Organizations achieve end-to-end traceability through comprehensive recordkeeping: transfer logs should include unique shipment identifiers, time and date stamps, recipient and sender signatures, and, ideally, digital hash verification. These records function as forensic evidence in the event of an incident, speeding investigations and guiding remediation.

Advanced implementations utilize integrated digital ledger systems or blockchain frameworks that automatically record each custody handoff. These ledgers are both immutable and transparent to authorized stakeholders, preventing retroactive data manipulation and cementing individual accountability throughout the supply journey (DiMase et al., 2016; Monette, 2022). The visibility provided by custody documentation becomes particularly critical when an adverse incident or recall questions device authenticity and safety.

Counterfeit Detection

The presence of counterfeit or unauthorized components represents a direct threat to medical device reliability. Illicit parts may fail unpredictably or harbor embedded exploits. To counter this, organizations deploy authentication technologies, such as RFID tags with unique cryptographic signatures, which verify component origin at every key stage of assembly. Similar systems leverage one-way hash functions or public key infrastructure to validate hardware authenticity before integration.

A layered inspection regime improves counterfeit avoidance. Risk-based selection for enhanced destructive testing—where samples are physically dismantled for microscopic or chemical analysis—detects intricate forgeries, while automated tools scan for irregularities in electrical signatures, weight, or appearance (DiMase et al., 2016; Monette, 2022). Cross-trained receiving and quality-control teams learn to identify subtle discrepancies, such as font mismatches or irregular barcoding, which once led to rapid quarantine and removal of a batch of counterfeit microprocessors intended for a high-dependency implant. Combining technical and personnel controls fosters a culture of vigilance and resilience against subversion (DiMase et al., 2016).

The mutual reinforcement of these practices enables medical device manufacturers to detect, document, and deter threats at the earliest stage—ensuring product integrity all the way to deployment. Yet, even with an impregnable provenance, devices must also resist attack at the hardware and firmware levels, highlighting that the journey from secure sourcing to operational defense demands constant adaptation and technical rigor.

Device Hardening and Tamper Resistance

Trust in healthcare technology hinges on absolute integrity of both the hardware and software embedded in every device. Ensuring that medical devices operate reliably means embedding strong, technically sound protections against compromise from the earliest design stages. Device hardening unites these efforts, focusing attention on direct, technical defense of devices in the field. This encompasses firmware lockdown, case integrity, interface restrictions, and environmental protection. Such layered strategies coalesce to deny attackers any easy avenue of exploitation, while allowing ongoing safe operation and maintenance as required for clinical utility.

Firmware lockdown sits at the core of technical defense, acting as a primary barrier against unauthorized code injection and system hijacking. In practice, effective firmware lockdown starts with cryptographic controls that authenticate updates before installation. For example, requiring firmware images to be signed and verified with public-key cryptography ensures that only approved vendors can release new operational code. In an infusion pump, any attempt to load unsigned firmware is blocked, preventing the installation of malicious routines that could alter pump dosages or disable alarm mechanisms. In addition, robust lockdown policies mandate that the firmware update mechanism itself be restricted. Update capabilities should operate only in controlled environments or require multifactor authentication by authorized users. Devices must block all remote or local update attempts unless all security provisions are fully satisfied. Attackers aiming to manipulate firmware using code repositories or data poisoning techniques cannot succeed if all images are cryptographically protected and update mechanisms are tightly controlled (Towards Secure MLOps: Surveying Attacks, Mitigation Strategies, and Research Challenges, 2015).

Case integrity strengthens device security through attention to form factor, mechanical robustness, and tamper evidence. Physical security begins with chassis and enclosures fabricated from strong, resilient materials that resist prying, drilling, or cutting. For implantable cardiac devices, manufacturers typically use highly robust metals for the enclosure, deterring any manipulation that could compromise battery leads or digital circuitry. Tamper-resistant screws, such as security Torx or custom hardware, thwart unauthorized disassembly by requiring proprietary tools. Tamper-evident seals—applied around enclosure seams, screw points, and access panel covers—offer visible cues if a device has been opened. For example, a blood glucose meter could feature a holographic seal that irreversibly fragments upon case separation, warning the user and clinical support team of the breach. These case-level protections are reinforced by access logging; monitoring physical access events allows security teams to detect and respond to suspected tampering in real time.

Interface restrictions reduce attack surface by ensuring exposed ports and wireless channels are limited only to those strictly needed for clinical function and authorized maintenance. Diagnostic, debug, or programming interfaces—whether USB, serial, or proprietary—represent common targets for attackers seeking to bypass system protections or extract sensitive data. Removing or physically disabling unnecessary ports is a foundational best practice. In patient monitoring devices, unused USB ports can be covered with tamper-evident plugs or filled with secure resin. Where access must remain, multifactor authentication and strong crypto authentication for connected devices help block unauthorized use. Wireless communication modules are similarly managed, often isolated by using network segmentation or access controls that enforce session authenticity and ensure only registered endpoints can interact with the device. Custom command whitelists can be programmed into both the device and its remote management software, tightly constraining what maintenance or diagnostic functions may be invoked remotely.

Environmental protection is essential for safeguarding reliability and security. Medical devices are often used in settings where moisture, dust, temperature shifts, and electromagnetic interference threaten delicate circuitry and sensor precision. To counter these threats, designers specify enclosures that meet internationally recognized ingress protection standards, such as IP65 or IP67, which ensure resistance to liquids and particulates. An infusion pump rated to IP67 can withstand accidental liquid exposure during cleaning or transport within hospitals, maintaining both safety and operational capacity. For protection against electromagnetic attacks or incidental interference, manufacturers integrate casing-level shielding and filtering around connectors and internal circuits. MRI-compatible patient monitors are equipped with special filters and shielded cabling to prevent electromagnetic pulses from corrupting operating code or sensor readouts. These physical defenses build on system-level protections like information system partitioning, sensor capability management, and event logging, combining to safeguard availability while denying malicious efforts to cause device malfunction or interruption.

Device hardening for medical systems demands precision and technical rigor at every step. By integrating cryptographic lockdown of firmware, fortified case design, strategic interface restriction, and robust environmental protections, manufacturers ensure health technology remains trustworthy and operational within complex clinical environments. Effective implementation of these measures directly supports mission-critical system integrity and empowers continuous care delivery in the face of evolving cyber and physical threats (Towards Secure MLOps: Surveying Attacks, Mitigation Strategies, and Research Challenges, 2015).

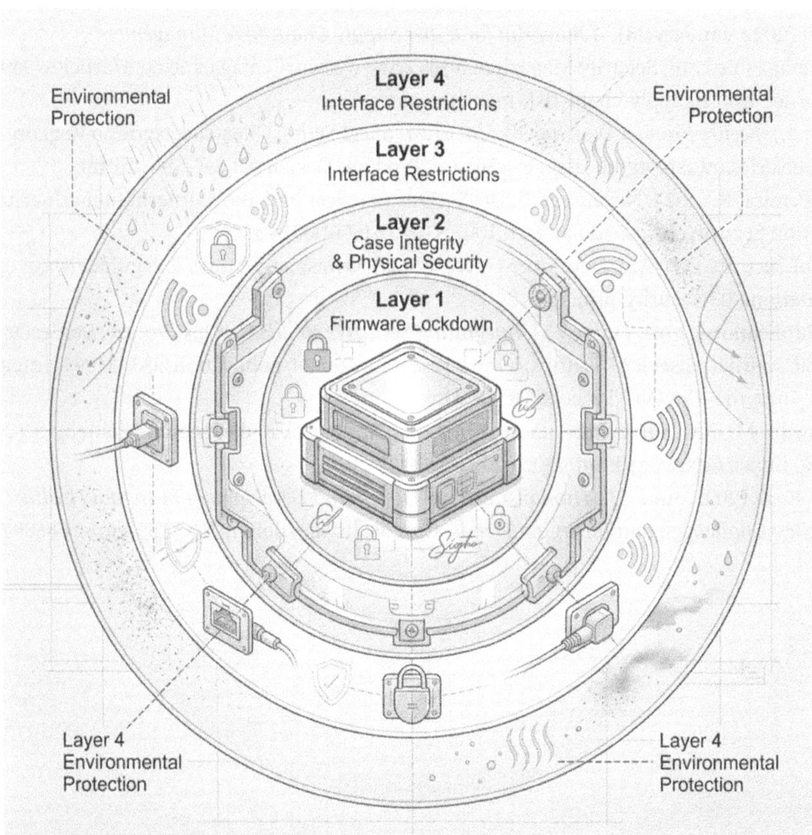

Final Thoughts

Now that we understand the essential principles and practices for embedding security throughout the medical device life cycle—from design foundations and human factors to supply-chain controls and technical hardening—we can build more resilient devices that withstand emerging threats and protect patient safety. Integrating these layered defenses early and continuously enables healthcare organizations and manufacturers to anticipate risks, respond swiftly to vulnerabilities, and maintain compliance with evolving regulations. By fostering collaboration across engineering, clinical, and regulatory teams and embracing ongoing education alongside automated security processes, the medical technology community is better positioned to deliver trusted, secure solutions that support safe and effective care in increasingly complex clinical environments.

References

DiMase, D., Collier, Z. A., Carlson, J., Gray, R. B., Linkov, I. (2016, January 21). Traceability and Risk Analysis Strategies for Addressing Counterfeit Electronics in Supply Chains for Complex Systems. Risk Analysis. https://doi.org/10.1111/risa.12536.

Kushniruk, A., Kaufman, D. (2024, August). *Human Factors and Organizational Issues in Health Informatics: Review of Recent Developments and Advances.* Yearbook of Medical Informatics; Georg Thieme Verlag KG. https://doi.org/10.1055/s-0044-1800744.

Monette, E. (2022, January 28). *A Blueprint for Cyber Supply Chain Risk Management.* Securitymagazine.com; Security Magazine. https://www.securitymagazine.com/articles/96992-a-blueprint-for-cyber-supply-chain-risk-management.

OWASP. (n.d.). *Secure Product Design - OWASP Cheat Sheet Series.* Cheatsheetseries.owasp.org. https://cheatsheetseries.owasp.org/cheatsheets/Secure_Product_Design_Cheat_Sheet.html.

Prates, L., Pereira, R. (2024, November 5). DevSecOps Practices and Tools. International Journal of Information Security. https://doi.org/10.1007/s10207-024-00914-z.

Principles of Security - OWASP Developer Guide. (2025). Owasp.org. https://devguide.owasp.org/en/02-foundations/03-security-principles/.

Research Publication. (2019, January 1). *Integrating Security into the DevOps Process (DevSecOps).* SSRN Electronic Journal; Elsevier BV. https://www.researchgate.net/publication/383334897_Integrating_Security_Into_the_DevOps_Process_DevSecOps.

Towards Secure MLOps: Surveying Attacks, Mitigation Strategies, and Research Challenges. (2015). Arxiv.org. https://arxiv.org/html/2506.02032v1.

Tzimourta, K. D. (2025, June 13). *Human-Centered Design and Development in Digital Health: Approaches, Challenges, and Emerging Trends.* Cureus. https://doi.org/10.7759/cureus.85897.

8

Secure Software Development for Embedded Medical Devices

Over 90% of medical device failures linked to software vulnerabilities result from coding errors, memory mismanagement, or insecure update processes. In clinical environments where patient safety hinges on reliable device performance, even minor software flaws can have severe consequences. Embedded medical systems operate under strict regulatory scrutiny, demanding precise control over every line of code and every data exchange. Ensuring these devices function securely requires more than conventional programming practices; it calls for specialized approaches tailored to resource-constrained hardware, real-time constraints, and evolving threat landscapes. This chapter delves into the foundational principles and practical methods essential for developing secure embedded software that withstands the unique challenges of medical technology.

Secure Coding Standards: Building in Quality

Safe language subsets serve as a decisive foundation for secure embedded C and C++ development, especially in medical devices where predictable behavior underpins both safety and compliance mandates. Certain hazardous language features and APIs, such as the use of `gets()`, `strcpy()`, and unrestricted pointer arithmetic, must be categorically prohibited in regulated codebases, as they provide no protection against memory overflows and buffer mismanagement. Functions that accept unchecked input or fail to verify buffer lengths often lead directly to exploitable vulnerabilities, such as buffer overflows or memory corruption. Relying on dynamically allocated memory (`malloc()`, `free()`) can also prove perilous in embedded environments. Heap fragmentation and unpredictable failure modes undermine device reliability and real-time guarantees essential in medical contexts, influencing both patient outcomes and compliance (Best Practices for Developing Secure Embedded Systems, 2025).

Minimizing encounters with undefined or implementation-dependent constructs—like uninitialized variables, unspecified evaluation order, or direct hardware register manipulation without clear abstraction—further enhances reliability across compilers and architectures. Unanticipated divergence in program behavior, whether due to compiler interpretation or hardware idiosyncrasies, introduces insidious, hard-to-diagnose faults in life-critical systems. Thus, only a tightly controlled subset of C/C++ should be permitted, accompanied by compiler flags and static analysis that treat any instance of undefined behavior with maximum severity.

Input validation is perhaps the most frequently targeted and most essential defense in embedded device code. Any data arriving from external interfaces, whether through UART, Ethernet, or user touchscreen, should be treated as untrusted by default. Effective input validation entails a three-pronged approach: type checking (ensuring inputs are of the correct base type), format verification (conforming to defined syntax or value patterns), and bounds enforcement (input remains within acceptable limits). For example, when controlling intravenous infusion rates, any command received from a remote console must be strictly checked for permitted dosage values and rate limits and enforced against a whitelist of valid commands. This prevents an attacker or a malfunction from issuing errant, unsafe dosing instructions. Sanitizing all inputs at the earliest processing stage restricts the scope of a successful attack and immediately raises the bar for any adversary seeking to exploit deeper vulnerabilities within device logic (Code with C, 2023).

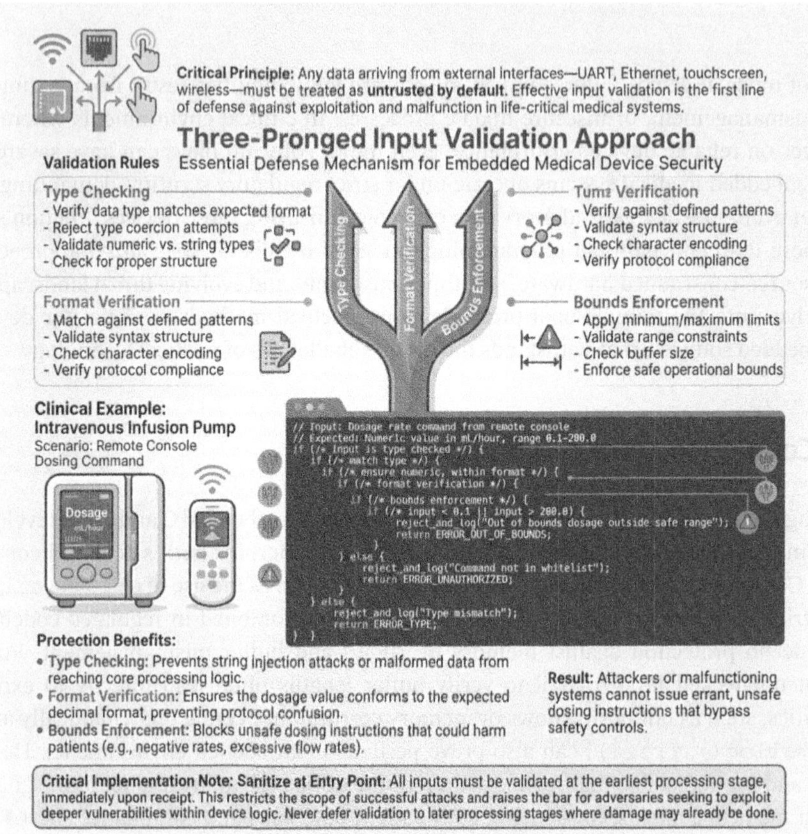

Error handling protocols must ensure that abnormal operating conditions do not lead to increased rather than decreased risk. In regulated medical environments, error routines are specified to revert devices to the safest possible state, often involving cessation of therapy, locking actuators, or entering a controlled stand-by mode. For instance, a defibrillator that encounters a sensor malfunction should abandon therapy delivery and inform the user rather than attempt unclear corrective actions. All errors are logged with precise timestamps, execution pathways, and anonymized error codes, providing forensic value for postincident review and regulatory reporting while safeguarding sensitive patient information. Error messages that disclose internal device structure or logic should never be exposed externally, neutralizing a frequent vector for adversary reconnaissance (Best Practices for Developing Secure Embedded Systems, 2025).

Robust code review routines further reinforce the security and reliability posture of embedded medical devices. Manual peer reviews enable trusted experts to spot contextual, system-aware vulnerabilities that may slip past automated tools. Human insights reveal logic flaws, incomplete requirements conformance, or inappropriate error recovery paths that jeopardize device safety or violate compliance expectations. Automated static analyzers, on the other hand, systematically identify buffer overflows, unchecked return codes, null dereferences, and the use of forbidden APIs at scale and with repeatable accuracy. By requiring checklists—verifying all buffer boundaries, confirming input validation on every exposed API, and banning legacy or deprecated calls such as `scanf()` and `sprintf()`—development teams formalize their defense against classes of vulnerabilities common in medical contexts. Real-world cases, such as a buffer overflow blocked during code review in a pulse oximeter's wireless communication stack, illustrate how layered review strategies halt defect propagation before clinical deployment (Code with C, 2023).

Committing to these secure coding practices reduces exploitable errors, enforces uniformity across geographically distributed teams, and sustains regulatory confidence in long-lived medical products. Effective use of safe language subsets, thorough input validation, safe error handling, and structured code review lay the foundation for robust memory safety and advanced fault mitigation. These foundational techniques create the conditions necessary for subsequent discussions on implementing hardened memory management strategies, supporting resilience even as threat landscapes continue to evolve (Best Practices for Developing Secure Embedded Systems, 2025; Code with C, 2023).

Memory Safety in Embedded Systems

Secure coding standards provide the foundation for reliable embedded medical devices, but memory safety practices must bring these principles into applied engineering. Practical measures begin with robust stack and heap protection strategies. Compiler features like stack canaries embed guard values next to return addresses on the stack. When a buffer overrun attempts to modify critical control data, the overwritten canary triggers immediate runtime detection, halting potential exploitation before malicious payload execution. Medical device development teams can enable stack canary options in toolchains such as GNU compiler collection (GCC) or Clang to systematically harden process memory against common attack vectors that seek overflow flaws (Szekeres et al., 2013).

Address Space Layout Randomization (ASLR) introduces additional resilience by shuffling process memory segments at each application launch. Predictable memory arrangement—often a prerequisite for buffer overflow attacks—disappears with ASLR, raising the bar for adversaries who try to exploit stack or heap addresses. Where real-time operating systems and device hardware permit, engineers should activate ASLR, recognizing that some medical-grade platforms may impose technical limitations or require vendor-specific implementations. Even partial adoption of ASLR enhances defense because it disrupts the memory mapping assumptions behind code injection and reuse techniques (Szekeres et al., 2013).

Memory tagging and runtime sanitizers provide further safeguards during development and testing. Memory sanitization tools, such as AddressSanitizer or dedicated hardware tag-based approaches, inject metadata or surveillance logic to capture out-of-bounds accesses or type confusion exploits missed by static analysis. While resource-constrained devices may struggle with production deployment, integrating these tools into continuous integration pipelines can reveal subtle stack or heap violations before field release. For instance, memory sanitization detects if a device telemetry handler reads past buffer boundaries, which could otherwise corrupt patient records or induce device instability. Such early discovery and remediation of defects align with both patient safety goals and legal mandates under medical device registration frameworks (Szekeres et al., 2013).

Defensive programming practices remain indispensable regardless of technological countermeasures. Medical device engineers should consistently define and enforce buffer size limits, replacing unchecked memory functions with bounded variants. Each buffer operation—whether copying sensor values, processing network packets, or handling patient monitor outputs—must check the intended access against object boundaries explicitly. Using fixed-size buffer allocations wherever feasible, particularly for real-time functions like EKG waveform storage, improves predictability, simplifies timing analysis, and closes off exploits depending on memory fragmentation or dynamic heap growth. A real-world case highlights the danger of neglecting these basics: a telemetry handler lacking explicit buffer length validation might overwrite adjacent structures, corrupting configuration data and jeopardizing device reliability. Strict enforcement of size limits, paired with tested error handling for out-of-bounds attempts, transforms this risk into a controlled and auditable failure mode (Szekeres et al., 2013).

The careful selection and vetting of libraries and external dependencies reinforce the device's memory safety perimeter. Only libraries with a proven track record, active maintenance, and timely security patches should find their place in firmware projects. Medical device manufacturers expose themselves to systemic risk by integrating orphaned or deprecated packages, since even rigorous in-house code cannot compensate for vulnerabilities lurking in unmaintained modules. Medical imaging subsystems, for example, have faced remote code execution weaknesses stemming from outdated third-party JPEG parsers. Teams that institute strict dependency audits—scanning for known Common Vulnerabilities and Exposures (CVEs) and adhering to supplier review processes—demonstrate due diligence to regulators and avoid introducing avoidable memory flaws into safety-critical pathways (Szekeres et al., 2013).

Dynamic analysis plays a crucial role beyond initial software engineering phases. Embedded-aware fuzzing tools generate unusual data patterns and sequences to systematically stress device software, surfacing dormant buffer handling weaknesses. Custom setups capable of simulating clinical sensor input or malformed wireless packets enable the discovery of vulnerabilities too rare or complex for manual review. AddressSanitizer, equipped with a configuration for memory-constrained targets, assists in catching runtime exceptions like use-after-free or off-by-one errors under realistic workloads. Meanwhile, the use of real-time heap and stack monitors allows engineers to recognize deviations from expected usage profiles. For example, in a networked infusion pump, intensively growing heap consumption traced by a monitoring tool may expose a silent memory leak before it crashes the device or interrupts therapy (Szekeres et al., 2013).

Every technical tactic described not only is essential for reliability but also substantiates claims for compliance with regulatory frameworks in healthcare technology. Many standards, including IEC 62304 and the U.S. Food and Drug Administration (FDA) guidance, call for evidence of systematic risk reduction and proactive management of software faults. These memory safety interventions lay convincing groundwork for certification and audit, supporting device longevity and clinical effectiveness.

While comprehensive memory safety strategies mitigate exploitation risk prior to deployment, they also foreshadow the next domain of security practice: mechanisms required to sustain and restore trust in running medical systems. Secure software update mechanisms, integrity checks, and operational monitoring are necessary to maintain device trustworthiness long after initial validation, reinforcing a holistic approach to memory-safe medical device design and maintenance (Szekeres et al., 2013).

Update Mechanisms Without Sacrificing Safety

Integrating secure and reliable update mechanisms into embedded medical devices is essential for sustaining system integrity and uninterrupted care in clinical environments. Cryptographic code signing stands at the core of trusted update pipelines. Before any firmware or software update reaches a device, the manufacturer digitally signs the update package using an asymmetric key pair. The device stores the associated public key in protected, hardware-backed storage elements designed to prevent unauthorized key extraction or manipulation (Brightwood, 2024; Palo Alto Networks, n.d.). Upon receiving a new update, the device computes a cryptographic hash—using algorithms such as SHA-256—across the package contents, then verifies the provided digital signature against the on-device public key. The verification process ensures both the authenticity of the source and the integrity of the package, guarding against tampering, substitution, or distribution of malicious code.

A practical example involves a ventilator that receives a firmware update. The update manager extracts the signature and the manufacturer's certificate bundled in the package. It hashes the firmware image and confirms the signature's validity against the public key retrieved from hardware-secured memory. If validation fails or the hash does not match, the device securely rejects the update, preventing execution of unauthorized or corrupted code. Lapses—such as mismanagement of signing keys, acceptance of unsigned packages, or weak signature algorithms—create opportunities for attackers to deploy backdoors, exfiltrate patient data, or disrupt critical treatment, demonstrating both patient safety and liability risks (Brightwood, 2024; Palo Alto Networks, n.d.).

Atomic switching and rollback support form the foundation for safe recovery during update failures. Updates should be staged in nonvolatile memory regions separate from the currently running system image. After validation, the device atomically switches pointers or boot partitions, ensuring that either the old or the new image is executable but never a partially programmed or corrupted state. In the event of a failed update—detected through failed health checks or error logs—the device triggers a self-contained rollback, reactivating the previous stable version to maintain continuous functionality. For instance, if an infusion pump's new software consistently fails built-in test routines following an update attempt, the system disables the faulty version and restores operation from the last known good image with minimal downtime.

However, rollback mechanisms themselves can be targeted by adversaries who seek to reinstall vulnerable legacy versions. To counter this, devices must enforce strict access controls over rollback triggers and record rollback events in append-only, signed log files, which are auditable by compliance teams. Rollback operations, like forward updates, should undergo digital signature verification, and devices should only accept rollbacks signed by authorized personnel or manufacturers, not arbitrary requests (Brightwood, 2024).

Deploying updates across diverse device fleets in dynamic healthcare settings benefits from a staged rollout approach, minimizing the risk of widespread disruption. Updates are initially pushed to a subset of devices—often those in low-risk or less critical operational settings—with sophisticated telemetry collection enabled. Health status, operational metrics, system errors, and patient safety indicators are monitored in real time or near real time. Data is aggregated and analyzed to detect early signs of instability or malfunction, such as sudden resets, communication faults, or unexpected behavior (Palo Alto Networks, n.d.). Example: A new patient monitor firmware release reaches 5 out of a 100 devices in an oncology ward. Postdeployment telemetry reports no anomalies, enabling the next phase of rollout. If health checks reveal inconsistencies, the process pauses for engineering review and possible intervention. This measured, feedback-driven deployment supports manual gating, where IT teams or clinical technology experts can halt or remediate updates if needed, preserving device availability and clinical workflow continuity (Brightwood, 2024).

Robust notification and consent systems ensure that all stakeholders are aware of and can properly respond to update events. Devices should be programmed to deliver alerts through secure channels—such as encrypted hospital IT dashboards, authenticated administrator consoles, or even on-device notifications—detailing update status: pending, in progress, completed, or reverted. When regulations or hospital policies require explicit user or administrator consent prior to applying certain updates, technical controls must block the installation process until required approvals are captured and logged (Palo Alto Networks, n.d.). For example, before updating antimicrobial stewardship software in a smart infusion system, IT administrators and lead clinicians receive a timestamped notification describing the new features, operational impacts, expected downtime, and postupdate verification steps. Consent capture forms become part of the audit trail, helping organizations demonstrate regulatory compliance.

These update strategies directly strengthen safety and data integrity by rejecting unauthorized or compromised updates, supporting seamless error recovery, and involving human stakeholders in change management. Ensuring communication channels and update-triggered data flows are secured and monitored further enhances resilience, setting the stage for equally robust security in application interfaces and data transmission pathways that follow (Brightwood, 2024; Palo Alto Networks, n.d.).

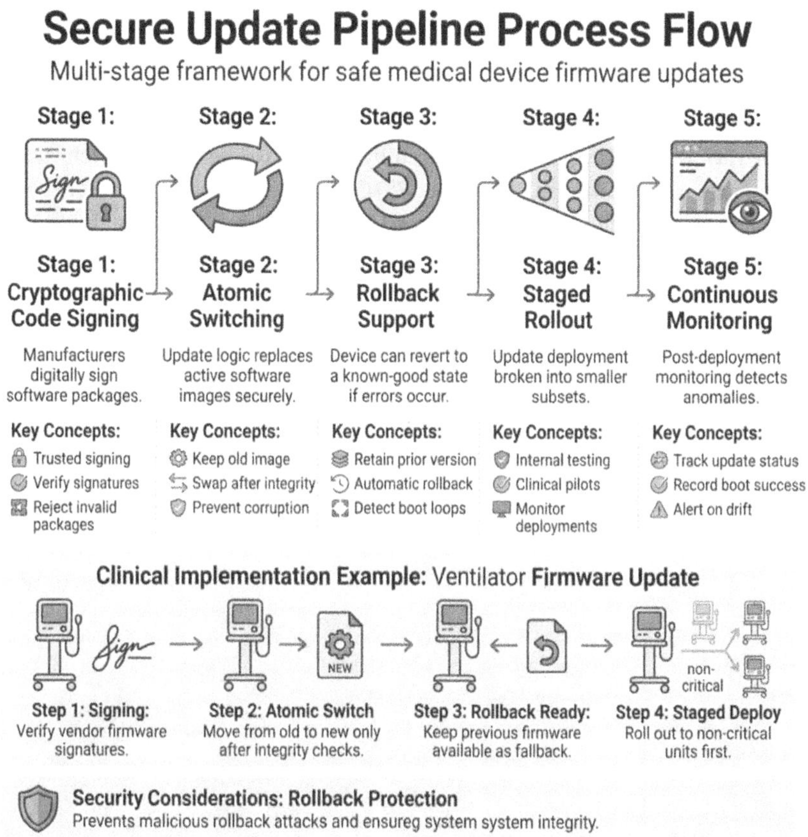

Secure APIs and Data Interfaces

Maintaining secure communication and data integrity after deploying software updates remains a top concern for embedded medical devices. Once operational in clinical environments, device APIs and data pathways represent primary attack surfaces, constantly exposed to cybersecurity threats from both internal and external sources. Attackers often target these interfaces, attempting to exploit weaknesses that could compromise patient safety or the confidentiality of protected health information (PHI). The practical implementation of robust controls at these junctions directly impacts device reliability, legal compliance, and the overall resilience of healthcare technology infrastructure (Chibogwu Igwe-Nmaju, 2024).

A sound approach begins with enforcing strict authentication and authorization mechanisms for every API and remote entry point a medical device exposes. Token-based authentication schemes, such as OAuth 2.0 and JWTs, have become industry standards because they verify session legitimacy without exposing user credentials. The stateless nature of JWTs removes the need for server-side session storage, minimizing attack vectors and system complexity (Chibogwu Igwe-Nmaju, 2024). In highly regulated scenarios, such as intra-hospital device communications, mutual transport layer security (mTLS) raises the trust bar by requiring both the client and the device server to authenticate their certificates. This dual verification model offers strong protection against impersonation attacks and ensures that only trusted devices or software endpoints can exchange sensitive telemetry or control signals.

Role-based access control adds an essential layer by mapping permissions to explicit job functions or software modules. For example, an infusion pump API should ensure that a standard clinician's security token cannot invoke firmware updates or calibration routines reserved for biomedical engineers. If a device operator attempts such an action, the API automatically rejects the command and logs it for review. Such event logs, captured by platforms like Elasticsearch, Logstash, and Kibana (ELK) stack or Amazon Web Services (AWS) CloudTrail, supply vital evidence for real-time monitoring and forensic auditing, supporting regulatory demands for accountability and enabling immediate intervention during a suspicious incident (Chibogwu Igwe-Nmaju, 2024; Mathijssen et al., 2020).

Input and output filtering comprise the next line of defense, designed to defeat a broad range of attack vectors. API endpoints must rigorously validate all inbound parameters, enforcing strict type checking and rejecting malformed or out-of-range values before processing. Adopting industry-standard data schemas, such as Health Level 7 (HL7) Fast Healthcare Interoperability Resources (FHIR) or custom JavaScript Object Notation (JSON) specifications, allows for automated compliance tests and immediate detection of protocol deviations or attempts at buffer overflows. Output data also requires careful sanitation. When a device responds to requests, it should only disclose whitelisted fields—omitting internal status codes, debug traces, and other metadata that could aid adversaries. For instance, firmware version numbers or internal logic flow details must be withheld unless explicitly required for maintenance operations. Outbound data streams carrying PHI must be encrypted using secure sockets layer (SSL)/TLS and digitally signed to guarantee integrity and confidentiality, aligning with the Health Insurance Portability and Accountability Act (HIPAA) mandates (Chibogwu Igwe-Nmaju, 2024).

Legacy compatibility presents significant risks, as embedded medical devices typically coexist with aging hospital software and communication protocols that lack modern security features. To reduce exposure, segregate legacy-facing APIs into network zones isolated from core assets using firewalls, virtual local area networks (VLANs), or dedicated gateways. Devices should employ protocol wrappers that translate legacy messages into secure formats before relaying them through primary data paths. Secure mediation tools can validate and sanitize inputs coming from outmoded protocols. A notorious example involved an MRI device that left an outdated Digital Imaging and Communications in Medicine (DICOM) service port exposed, which attackers leveraged to exfiltrate anonymized patient data. By enforcing phased deprecation plans and providing detailed migration documentation, device manufacturers help clinical users adapt safely without abrupt disruption. Maintain availability of deprecated versions for a defined period but alert users through the management portal about planned cut-off dates and security vulnerabilities inherent to legacy interfaces (Chibogwu Igwe-Nmaju, 2024).

API versioning and deprecation strategies underlie long-term device stability. Each new API version should introduce improvements in a backward-compatible manner where practical, with comprehensive release notes mapping changes, new security features, and deprecated endpoints. Without structured version management, updates can inadvertently expose devices to new vulnerabilities, as seen in a ventilator software update that removed critical input checks, allowing unauthorized commands by legacy clients. New versions should run in parallel with older ones for a transitional window, giving all stakeholders, including Electronic Health Record (EHR) vendors and hospital IT, time to migrate or refit their integration layer (Mathijssen et al., 2020). Announce all changes through centralized portals and direct communication to ensure no client system is left vulnerable due to lack of awareness or preparation.

Designing device APIs and data interfaces with these principles fortifies embedded medical systems, ensuring only trusted actions occur and sensitive data remains protected in transit and at rest. These architectural choices enable forthcoming software testing strategies and verification routines to operate effectively, building an ecosystem of trust, resilience, and proven safety for patient-critical devices (Chibogwu Igwe-Nmaju, 2024; Mathijssen et al., 2020).

Testing Techniques for Medical Device Software

API and data interface security lay the foundation for robust software assurance, but safety in medical device development demands continual examination at every level. Rigorous, multilayered testing ensures devices meet their intended security and reliability requirements—not only in code abstracts or isolated modules but also as fully realized, interconnected systems in clinical operation (*CSSLP—Certified Secure Software Lifecycle Professional Exam— | Testprep Guide*, 2024; Palo Alto Networks, n.d.).

Three-tiered testing—unit, integration, and system—serves as the backbone of practical assurance strategies. Unit testing establishes correctness at the most granular level, verifying inputs and outputs for individual functions, methods, or classes in isolation. These tests are ideally written alongside code creation, using frameworks compatible with the embedded environment. Unit tests must cover success, failure, and boundary conditions to identify regressions in frequently touched logic. Harnesses such as test-driven development (TDD) offer systematic code coverage proof, which both improves reliability and deters common programming flaws.

Integration testing moves beyond isolated components, instead focusing on the interfaces and control flows that emerge as multiple modules interact. Designing integration tests involves mapping out component dependencies—such as communication stacks interfacing with patient data storage or driver-layer logic connecting to device sensors. Automated test suites should simulate both valid and malformed interactions, illuminating vulnerabilities that arise only in context—such as mismatched data formatting, timing errors, or trust boundary violations. Medical device teams typically schedule integration test runs with every major code merge or architectural change, ensuring new updates do not introduce interoperability failures.

System-level testing evaluates the fully assembled firmware and software stack as it will function in the target medical environment. These end-to-end tests validate safety features, error handling, data integrity, and compliance with device requirements, such as emergency shutdown upon fault detection or proper encryption of patient data during transfer. System testing, executed at regular intervals and before every production release, should simulate real-world operation—including complex use scenarios, multiuser workflows, and hardware fault conditions. Effective strategies require comprehensive tracking and documentation of test coverage, with traceability back to explicit functional and nonfunctional requirements (*CSSLP—Certified Secure Software Lifecycle Professional Exam— | Testprep Guide*, 2024).

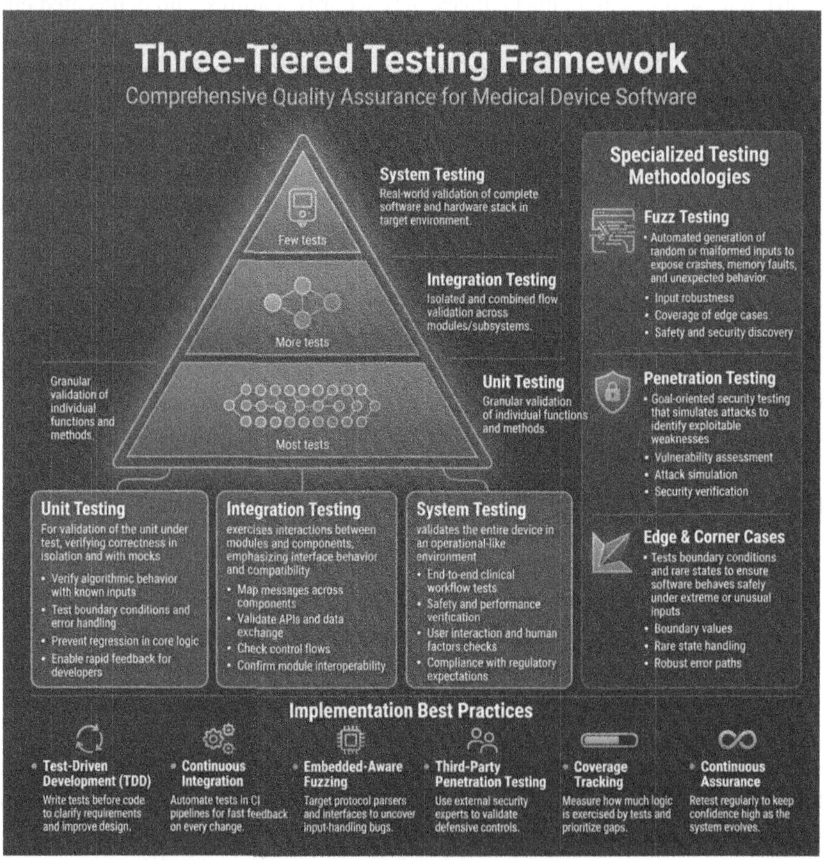

Coverage extends further through specialized edge and corner case tests, which are indispensable for uncovering faults in rarely encountered or stress-inducing operational scenarios. For example, device input parsers may behave correctly under standard operating conditions but mishandle input sizes or malformed data seen in exceptional network events. Ensuring thorough error and boundary condition handling is paramount, particularly when human safety is at stake.

Automated fuzz testing introduces persistent, unpredictable input scenarios that catch flaws traditional test cases may overlook. Fuzzers generate random, edge-case, or intentionally malformed data and feed it through device interfaces—such as USB protocol stacks or wireless communication parsers. Selecting or customizing a fuzzer for embedded systems often requires adaptation for proprietary protocols or constrained execution environments, leveraging tools like AFL or open-source variants of libFuzzer configured for memory and resource usage profiles suitable for medical devices. Targeting modules that handle the most external data—API endpoints, file parsers, or communication links—greatly increases the likelihood of uncovering input validation shortcomings or memory safety errors.

Continuous integration pipelines should automatically trigger fuzz testing on every code change, with findings forwarded as actionable events for developer triage. Persistent fuzzing, as opposed to one-off campaigns, delivers ongoing discovery and resolution of latent vulnerabilities, ensuring evolving codebases remain resilient against novel attack vectors (*CSSLP—Certified Secure Software Lifecycle Professional Exam— | Testprep Guide*, 2024; Palo Alto Networks, n.d.).

Manual penetration testing brings an adversarial perspective to software assurance, simulating the actions of a determined attacker in a controlled, ethical evaluation. Skilled testers design and execute attacks tailored to medical device risk profiles—for instance, bypassing device authentication to gain unauthorized access, attempting buffer overflow exploits in custom protocol stacks, invoking APIs with malicious payloads, or subverting the device's update and installation processes. Engagements require documented scope, signed agreements, and adherence to clinical safety protocols, with test findings cataloged for severity and mapped to risk-based remediation priorities. Clear procedures dictate prompt triage and mitigation of the highest-impact vulnerabilities, establishing controls that prevent exploitation before clinical deployment (*CSSLP—Certified Secure Software Lifecycle Professional Exam— | Testprep Guide*, 2024).

Continuous assurance extends security beyond the release date. Instrumented telemetry captures device events, anomaly signals, and operational metrics for near real-time detection of faults and attacks. Rigorous patching, system audits, and policy reviews must align with regulatory guidance as it evolves. Scheduled audits and external reviews bolster compliance and incident readiness. Telemetry analysis enables early risk detection—such as unusual access patterns or failed integrity checks—enabling preemptive action to maintain device safety and protect patient data (Palo Alto Networks, n.d.).

Concrete implementation examples reinforce best practices. Implementing TDD increases code path coverage and resilience. Integrating embedded-savvy fuzzers like AFL uncovers parser flaws before deployment. Partnering with experienced third-party penetration testers ensures compliance and deep risk discovery. Organizations that pursue these advanced and persistent testing strategies assure stakeholder confidence, support regulatory approval, and—most importantly—uphold the safety and reliability essential to modern clinical care.

Summary and Reflections

Having established a comprehensive framework for secure software development in embedded medical devices—encompassing rigorous coding standards, memory safety protocols, resilient update mechanisms, fortified APIs, and advanced testing methodologies—we are now equipped to advance toward integrating these practices into cohesive development life cycles and operational strategies. This foundation enables healthcare cybersecurity professionals, device engineers, and compliance officers to collaboratively enhance device trustworthiness throughout their life cycle, ensuring not only initial compliance but also sustained protection against evolving threats. Moving forward, applying these principles will support robust risk management, facilitate regulatory approval, and ultimately safeguard patient safety and clinical efficacy in increasingly interconnected healthcare environments.

References

Best Practices for Developing Secure Embedded Systems. (2025, March 20). Simplexity Product Development. https://www.simplexitypd.com/best-practices-for-developing-secure-embedded-systems/.

Brightwood, S. (2024). *(PDF) The Importance of Secure Firmware Updates in Maintaining System Integrity*. ResearchGate. https://www.researchgate.net/publication/384687421_The_Importance_of_Secure_Firmware_Updates_in_Maintaining_System_Integrity.

Chibogwu Igwe-Nmaju. (2024, December 1). Organizational Communication in the Age of APIs: Integrating Data Streams Across Departments for Unified Messaging and Decision- Making. International Journal of Research Publication and Reviews; Genesis Global Publication. https://www.researchgate.net/publication/392472264_Organizational_Communication_in_the_Age_of_APIs_Integrating_Data_Streams_Across_Departments_for_Unified_Messaging_and_Decision-_Making.

Code With C. (2023, October 10). Code with C. https://www.codewithc.com/secure-coding-practices-in-embedded-c-revealed/.

CSSLP - Certified Secure Software Lifecycle Professional Exam I Testprep Guide. (2024, June 21). Testprep Training Tutorials. https://www.testpreptraining.com/tutorial/certified-secure-software-lifecycle-professional-exam/.

Mathijssen, M., Overeem, M., Jansen, S. (2020, June 18). *Identification of Practices and Capabilities in API Management: A Systematic Literature Review*. ArXiv.org. https://doi.org/10.48550/arXiv.2006.10481.

Palo Alto Networks (n.d.). *What is Secure Software Development Lifecycle (Secure SDLC)?* Palo Alto Networks. https://www.paloaltonetworks.com/cyberpedia/what-is-secure-software-development-lifecycle.

Szekeres, L., Payer, M., Wei, T., Song, D. (2013, May 19). *SoK: Eternal War in Memory*. IEEE Symposium on Security and Privacy. https://doi.org/10.1109/sp.2013.13.

9

Network Security in Healthcare Environments

"Every device connected to our network is a potential doorway for attackers," remarked the lead clinical engineer during a recent security briefing at a major hospital. This statement underscores a critical challenge faced in healthcare environments: securing a vast array of medical devices and IT systems that support patient care while safeguarding sensitive data. As healthcare technology becomes increasingly interconnected, vulnerabilities multiply, requiring sophisticated strategies that address not only individual devices but also the network architecture as a whole.

This chapter examines comprehensive approaches to protecting healthcare networks through techniques such as segmentation of network zones, endpoint security measures tailored for diverse medical equipment, and robust wireless protections. It also explores the importance of rapid incident isolation to contain security breaches without disrupting essential care, alongside secure remote access solutions to accommodate the growing demand for telemedicine and distributed healthcare services. Together, these elements form an integrated framework designed to defend critical infrastructure and maintain compliance with regulatory standards in complex clinical settings.

Network Segmentation: Containing the Blast Radius

Healthcare networks demand a zoning strategy that systematically classifies assets by risk, function, and regulatory requirements, creating layered security barriers suited to a complex clinical environment. Zoned architecture begins with a comprehensive asset inventory, grouping systems into zones such as life-supporting clinical devices, general IT servers, and administrative endpoints. Each zone reflects the sensitivity of operations and the potential impact to patient care if compromised. For instance, infusion pumps and patient monitors are segmented apart from hospital workstations and printers; radiology equipment like CT or MRI machines remains in a dedicated imaging segment with tightly limited communication to picture archiving and communication systems (PACS), but no access to broader clinical or administrative networks (Toll, 2025). This architectural discipline reduces the total attack surface and contains any breaches to isolated zones. It also streamlines compliance with regulations such as the Health Insurance Portability and Accountability Act (HIPAA) and IEC 80001 by ensuring that protected health information and medical device communications are only accessible within controlled, policy-driven segments, supporting mandatory data access and audit obligations.

Implementing zoned architecture requires robust boundary enforcement mechanisms at each segment's edge. Firewalls and access control lists (ACLs) are the core technologies that make logical zoning operational. Firewalls deployed at the intersection between clinical device networks and administrative systems serve as gatekeepers, evaluating every packet for compliance with defined policies. ACLs work in tandem, explicitly listing which protocols or ports are permitted, and which devices or user groups are trustworthy. Consider a configuration where clinical laboratory information systems (LIS) only accept connections from specific analyzers and lab workstations; traffic from nonsanctioned hosts is automatically blocked at the segment boundary (Toll, 2025). Similarly, imaging workstations are allowed to reach PACS servers via designated secure protocols, while outbound internet access is blocked to prevent unauthorized data exfiltration. Carefully crafted ACLs are also crucial for highly targeted environments such as VoIP networks, enabling voice devices to contact only the call management servers and each other on voice virtual local area networks (VLANs) while denying access to administrative systems (Toll, 2025). Correct configuration is critical—overly permissive rules undermine segmentation by allowing attackers to traverse zones via misconfigured ports or trusted but unnecessary inter-segment access.

Microsegmentation advances these concepts by applying zoning at the granularity of individual devices, containers, or workloads. This identity-based technique, foundational to modern "zero trust" security, is especially important in healthcare because of the diversity and sensitivity of connected devices. In a patient monitoring network, each cardiac monitor or infusion pump can be placed in its own microsegment with a policy tailored for its role. For example, a cardiac monitor might be allowed to transmit data to a central monitoring server but cannot communicate with other monitors or access hospital database systems (Palo Alto Networks, 2025). If an attacker compromises one monitor, that access cannot be leveraged to move laterally to other monitors or steal patient data from elsewhere—the propagation of malware like ransomware or credential theft tools is stopped at the microsegment level (Toll, 2025). In clinical pharmacy systems, automated medication cabinets can only reach dedicated pharmacy management servers; segmenting these devices not only safeguards medication data but ensures the cabinet cannot become a pivot point into electronic health record (EHR) networks (Toll, 2025).

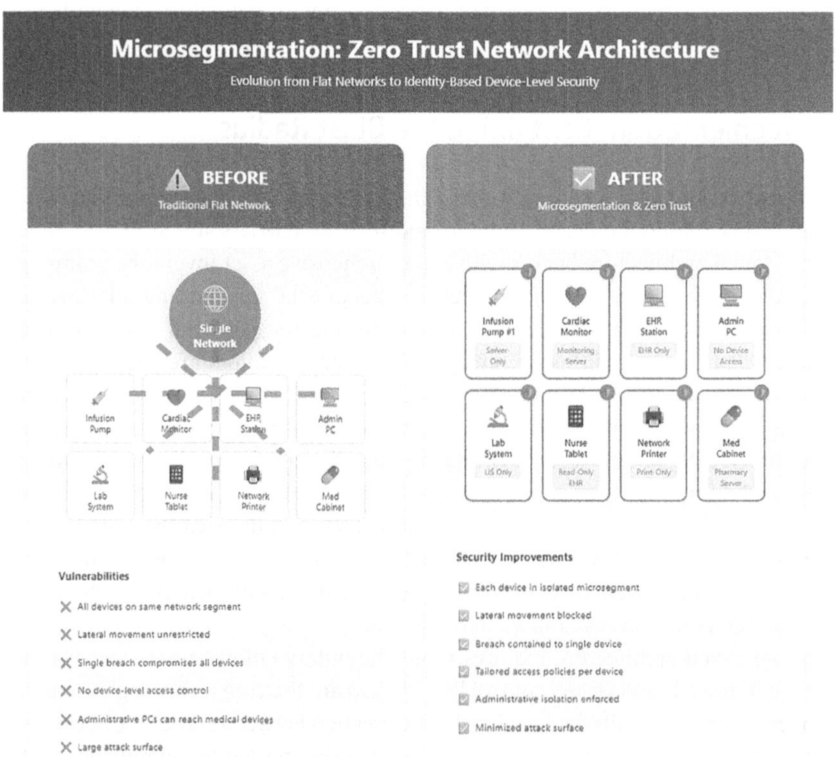

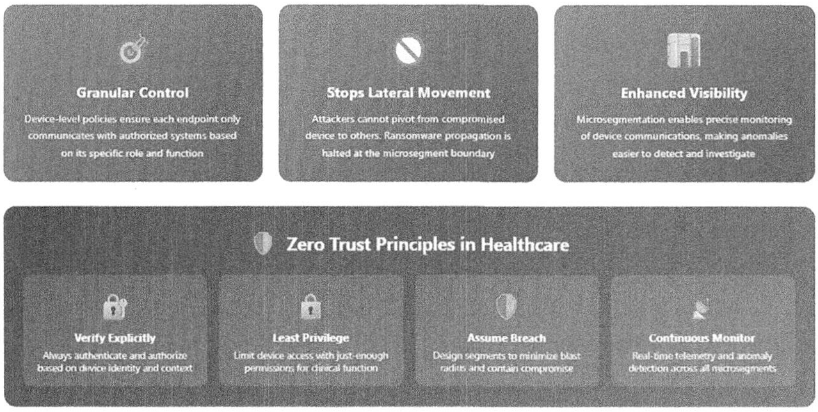

Effective segmentation is inseparable from continuous monitoring. Given the rapid pace and diversity of healthcare operations, real-time visibility into segment boundaries and device communications is essential. Monitoring platforms ingest logs and flow data from firewalls, switches, and endpoint agents, correlating activity to the specific roles and expected workflows of clinical equipment (Palo Alto Networks, 2025). For instance, if an infusion pump begins generating network requests outside its normal traffic pattern—such as attempting to access a clinical management server never used in its workflow—anomaly detection rules can trigger alerts for immediate investigation and optionally isolate the asset from the network without impacting other patient care systems (Toll, 2025). Likewise, sustained login failures on PACS servers from administrative

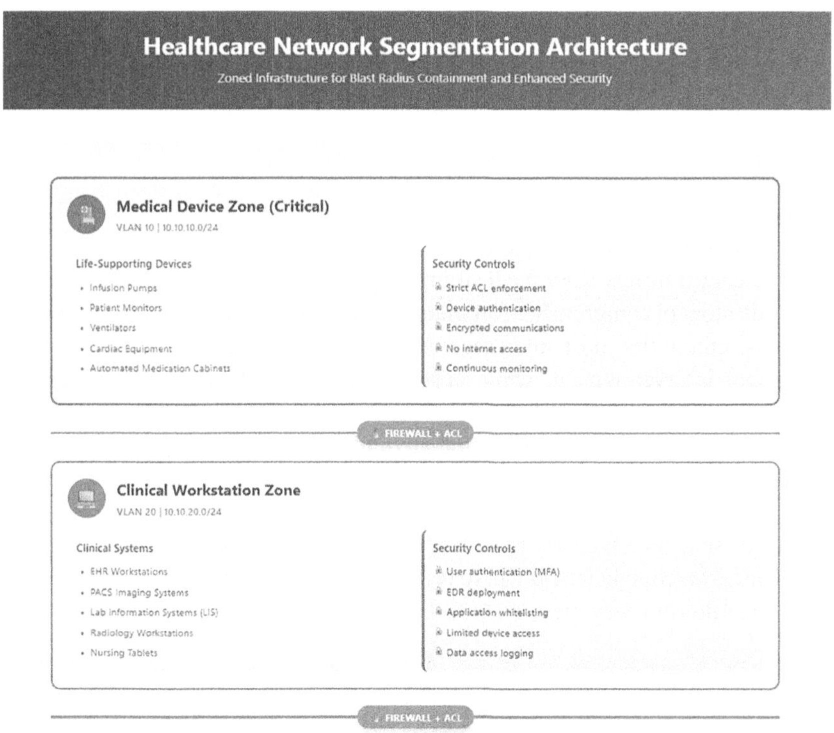

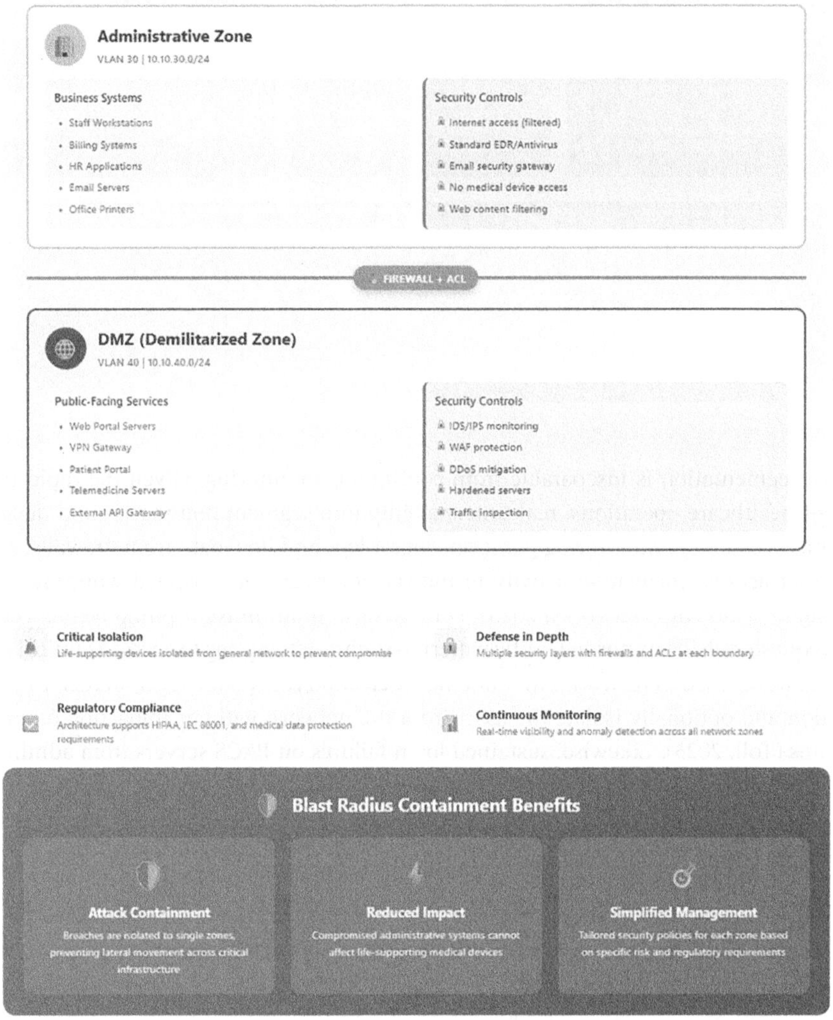

VLANs or large data transfers from medical imaging zones to unauthorized endpoints serve as high-fidelity indicators of compromise. Automated response workflows can block specific device communications, quarantine host microsegments, or escalate incidents to security operations teams for manual intervention, all while ensuring that critical medical workflows continue uninterrupted.

The fundamental principles of segmentation are under constant pressure from the variety and number of endpoints present in a healthcare setting, where legacy medical devices, IoT sensors, and virtual clinical applications require tailored policies and flexible controls. Designing, enforcing, and monitoring these segments are the precursors to effective endpoint security and the only way to maintain resilient protection in a field where regulatory mandates, patient safety, and operational uptime intersect (Palo Alto Networks, 2025; Toll, 2025).

Endpoint Security for Devices and Hosts

Safeguarding healthcare networks demands device-level protection that complements broader network segmentation. Even when segmentation is well-designed, threats can traverse zones through compromised devices, making granular endpoint security essential for minimizing risk and enforcing operational resilience. Every device or host—be it a patient monitor, medication dispenser, or clinician's tablet—requires individualized defense mechanisms to detect and neutralize attacks that evade perimeter controls (Anjan Kumar Gundaboina, 2025). In a dynamic clinical setting, the ability to isolate and address threats at the device layer preserves both system integrity and patient safety while underpinning compliance with medical data protection requirements.

Endpoint Detection and Response (EDR)

EDR serves as an advanced anchor in endpoint defense, leveraging continuous real-time telemetry collection from diverse endpoint types. Sensors embedded within the operating environments of medical devices, nursing stations, and even specialized lab equipment observe patterns such as unexpected file access, unauthorized application launches, and anomalous network connections. Through analytics-driven detection engines, EDR platforms recognize subtle threats—including zero-day exploits and customized malware—that may escape signature-based security tools. In medical environments, EDR's automated processes are vital; a typical example is a patient monitoring workstation displaying atypical memory usage and attempting outbound communication to an external IP address. The EDR solution raises an immediate alert, quarantines the endpoint autonomously, and initiates forensic data capture for further investigation (Anjan Kumar Gundaboina, 2025). This real-time responsiveness greatly reduces an attacker's dwell time, shrinking the window of opportunity for lateral movement or data exfiltration. Integration with security operations centers (SOCs) further amplifies visibility, enabling coordinated response efforts and retrospective threat-hunting using historical device telemetry.

Anti-Malware for Medical and Embedded Devices

Standard anti-malware solutions must be adapted for healthcare's critical devices where performance and uptime directly affect patient care. Imaging systems, infusion pumps, and diagnostic equipment often operate on bespoke platforms or require round-the-clock availability. Deploying anti-malware protection here means balancing proactive security with uncompromising stability. Security teams rigorously test signature updates and behavioral detection logic in laboratory replicas of clinical environments before approving them for production deployment (Anjan Kumar Gundaboina, 2025). For example, a vulnerability discovered in the software architecture of a connected infusion pump is addressed first in a simulated environment, ensuring that protective updates do not interfere with medication dosing accuracy or device reliability. The layered nature of endpoint controls ensures that if malware bypasses one checkpoint, subsequent layers—such as EDR, application whitelisting, or intrusion prevention—will intervene, reducing the risk of cascading failures.

Whitelisting and Application Control

Heightened precision is required for application management on medical endpoints. Whitelisting and application control mechanisms enforce a strict, explicit policy allowing only verified, approved programs to execute on devices. By limiting the scope of software operation, hospitals curtail exposure to both opportunistic and targeted malicious code. In practice, on a laboratory workstation integral to critical testing, a whitelisting policy blocks the launch of a ransomware payload delivered via infected USB storage. The incident concludes with patient data unharmed, and operations recover without downtime. Technically, these solutions demand frequent policy updates and close alignment with workflows to avoid inadvertently restricting necessary clinical applications. Their use also fulfills regulatory obligations in medical settings under data protection legislation, positioning organizations to demonstrate strong risk mitigation and compliance practices (Anjan Kumar Gundaboina, 2025).

Asset Inventory and Tracking

A complete, dynamic inventory of network-connected assets forms the backbone of any robust endpoint security posture. Asset management platforms consolidate information on every device, continuously updating details such as physical location, configuration, ownership, and software status. This real-time inventory empowers security teams to instantly pinpoint unauthorized or suspicious endpoints—such as a rogue wireless infusion pump that suddenly initiates unusual network scans (Anjan Kumar Gundaboina, 2025). Automated asset discovery and tracking further optimize vulnerability management by ensuring updates, patches, and security policies are deployed swiftly to all endpoints, including those temporarily brought online for specific clinical tasks. With every asset accounted for, response strategies—like containment and remediation—can be executed methodically and with minimal impact on core healthcare services.

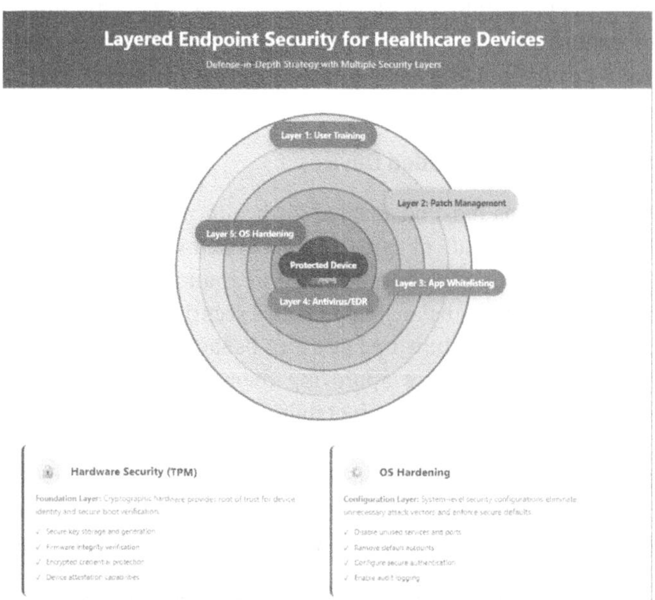

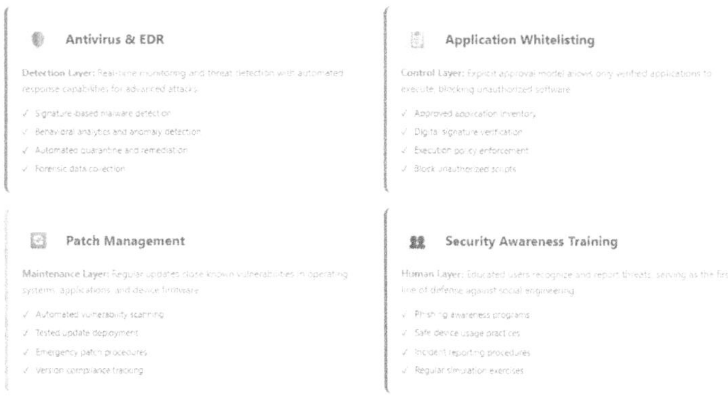

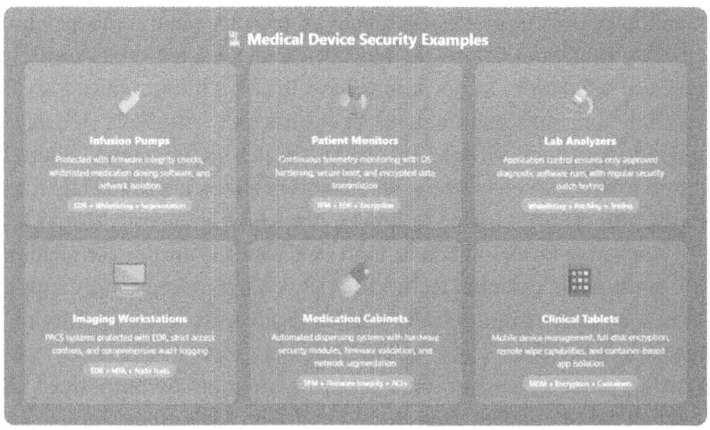

Each distinct layer—endpoint monitoring, anti-malware, whitelisting, and asset tracking—interlocks with the next, constructing a security architecture tailored to clinical priorities and operational demands. Protecting individual devices within medical networks becomes the groundwork for securing the entire ecosystem, reducing both the incidence and impact of cyber events.

A comprehensive endpoint security strategy naturally sets the stage for the unique complexities introduced when devices communicate via wireless connections, where new risks and required protections continue to evolve.

Wireless Network Security Essentials

Wireless networking in healthcare environments introduces unique threat surfaces that must be addressed through layered security controls. Strong endpoint protections can be swiftly undermined if wireless communications remain vulnerable. Enforcement of contemporary encryption standards remains foundational for the confidentiality and integrity of all patient data and machine telemetry in transit. Wi-Fi networks in healthcare settings should mandate WPA3, which applies a robust authenticated key exchange mechanism, forward secrecy, and Advanced Encryption Standard (AES) operating in Galois/Counter Mode. This creates a formidable barrier against interception and eavesdropping, especially as WPA2-PSK networks remain susceptible to dictionary and Aircrack-based attacks against preshared keys, with successful compromise enabling attackers to infiltrate traffic and laterally pivot into sensitive clinical systems (Najar & Mir, 2021). Outdated encryption implementations have regularly contributed to exposures, such as when legacy wired equivalent privacy (WEP) managed on portable diagnostic devices allowed adversaries to recover patient record transmissions simply through radio scanning and brute-forcing weak RC4 keys. Institutions must prioritize regular updates not only of access point firmware but also of every connected device, including infusion pumps and mobile workstations, to ensure that critical vulnerabilities are patched, cryptographic libraries are current, and known protocol weaknesses are eliminated.

Encryption requirements extend beyond standard Wi-Fi. Bluetooth low energy (BLE) frequently forms the backbone of patient wearable communications and asset tracking. Adoption of secure pairing methods, such as Secure Connections Only mode introduced in BLE 4.2, ensures elliptic-curve cryptography protects data as sensors pair with nursing tablets or shared infrastructure. Healthcare environments leveraging proprietary radio-frequency protocols should verify that proprietary implementations do not opt for deprecated or custom ciphers without rigorous peer-reviewed validation, as deviations have historically introduced severe flaws. Even so, encryption without ongoing maintenance is insufficient. Attackers actively exploit failed or overlooked upgrades, as demonstrated in several breach investigations where biotechnology systems were compromised not solely by the cryptosystem but by neglected bug fixes in the device operating system that attackers leveraged to subvert transmission security (Najar & Mir, 2021).

Some administrators may falsely attribute increased security to the act of hiding service set identifier (SSIDs), omitting an essential technical reality: SSID hiding simply suppresses network advertisement in beacons but cannot prevent a capable adversary from identifying and targeting the network. Specialized wireless tools routinely detect "hidden" SSIDs by passive or active monitoring, capturing probe responses, or associating attempts by authorized clients, thereby exposing the hidden network name in clear text. Reliance on SSID obfuscation leads to dangerous overconfidence; attackers have often bypassed hospitals' "hidden" networks and exploited weak or default credentials, enabling lateral movement toward EMR servers or medication dispensing units (Najar & Mir, 2021). Good wireless hygiene in healthcare must instead focus on enforcing strong authentication, minimum-privilege access control lists, and detailed audit trails for every attempted connection, whether the SSID is visible or not.

Wireless medical device firmware and software must regularly be updated to patch vulnerabilities and add critical features. Over-the-air (OTA) updates offer operational convenience but require rigorous controls to avoid introducing new security weaknesses. Encrypted transport protocols, such as TLS 1.3 with mutual certificate authentication, are mandatory for all OTA deliveries. Without cryptographic signing and validation of every update, an adversary can stage a man-in-the-middle attack, injecting trojanized firmware or ransomware-laden modules into clinical environments. Actual incidents have occurred where attackers hijacked legacy wireless infusion pump update mechanisms, granting themselves command-and-control privileges and enabling tampering with dosage administration through malicious OTA deployments. Best practice mandates that only digitally signed firmware artifacts be accepted, verified at both download and installation, with privilege escalation for updates limited to tightly managed roles. Activity logs capturing update provenance and confirmation are essential for postdeployment auditing.

Persistent and automated rogue device detection rounds out the wireless defense matrix. Hospitals must continuously monitor for unauthorized endpoints or rogue access points, which may mimic legitimate staff Wi-Fi or bridge air-gapped segments with external attacker infrastructure. Technical strategies include beacon and probe request analysis, radio fingerprinting, and MAC address anomaly detection. Onsite security teams must be equipped to immediately isolate detected rogue devices at the switch or controller level. For example, a clinical adversary establishing a rogue access point with legitimate SSID and passphrase could easily siphon telemetry from wireless medical carts and exfiltrate protected health information; however, active alerting tied to real-time radio analytics can prompt instant containment, preventing unauthorized exfiltration or lateral exploit escalation (Najar & Mir, 2021).

Elevating wireless protections in healthcare demands standard-based approaches that adapt to emerging attacks while enabling fast, coordinated incident response. Exposure from wireless attacks can rapidly trigger network segmentation or isolation procedures, so linking detection to containment tools is as vital as prevention itself. Strategic alignment of wireless security enables prompt neutralization of threats before they can propagate deeper into clinical environments (*Products—Cisco Unified Wireless Network Solution Guide*, n.d.).

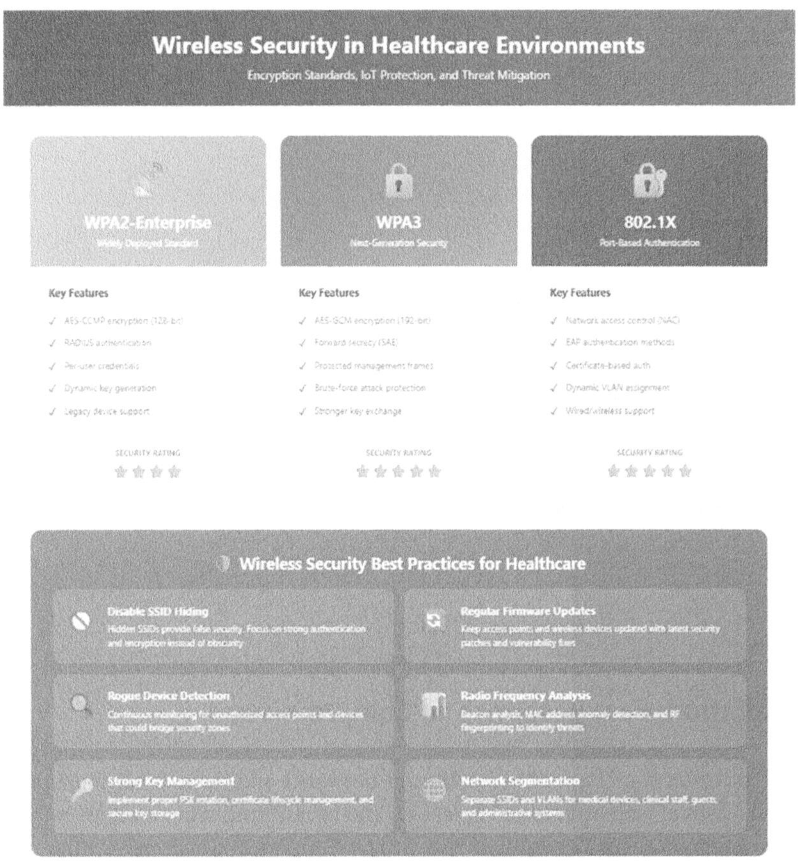

Incident Isolation: Fast Response to Compromises

Incidents that threaten the integrity of healthcare networks and devices can materialize swiftly even under the watch of well-architected detection and monitoring schemes. Preventive measures such as layered endpoint security, continuous device monitoring, and aggressive vulnerability management reduce risk, yet they cannot eliminate the possibility of a threat actor bypassing defenses. In these cases, sustaining patient safety and continuity of care requires a rapid, decisive response: the prompt isolation of affected systems. Incident isolation acts as a critical safeguard, curtailing the spread of compromise and protecting sensitive patient data from further exposure.

Isolation must occur with precision and minimal clinical disruption. Step-by-step disconnection protocols guide responders, outlining how to safely separate compromised segments or devices. Hospitals develop isolation playbooks that, for instance, prioritize disconnecting nonessential systems first, such as administrative workstations or noncritical sensors, before considering disconnection of high-dependency medical devices. Criteria for isolation may include observed malware activity, unexplained outbound traffic, evidence of unauthorized data access, or direction by a central security monitoring platform. Each protocol incorporates device criticality and redundancy plans—ensuring, for example, that when an infusion pump shows signs of malicious tampering, it is detached from the hospital's main network but left operational to maintain drug delivery while a spare monitoring system assumes oversight duties. These protocols, structured in compliance with device interoperability methods and network segmentation best practices, align with both clinical priorities and network isolation goals.

For these protocols to succeed, staff must understand and execute them without hesitation. Simulation drills, response checklists, and digital runbooks foster preparedness across all shifts and disciplines. Clinical and IT staff participate in mock incident scenarios, practicing their roles in the isolation chain to internalize actions—from alerting supervisors to physically detaching devices or toggling network ports. Readily available, intuitive documentation empowers nurses and technicians to act decisively when screens display unfamiliar warnings or endpoint agents flag a compromise. In one common scenario, a nurse equipped with protocol knowledge can promptly power down a suspicious infusion device or notify IT to trigger an electronic network isolation command, reducing the time adversaries remain active in the system. Through immersive training and clear workflows, institutions reduce reliance on memory under duress and lessen the potential for accidental disruptions to patient care.

The rapid pace and scope of modern attacks demand support from automated containment technologies. Software-defined networking (SDN) frameworks and EDR platforms translate threat intelligence into action at machine speed. Imagine, for instance, an EDR agent deployed on a radiology workstation recognizes abnormal file activity suggestive of ransomware. Within seconds it signals the network controller, which places the device into a restricted VLAN, cutting its communication with other medical equipment and sensitive databases. In the background, alerting systems notify both clinical and technical responders, while tailored access controls preserve the device's local functionality if needed for ongoing procedures. These automated responses, calibrated to initiate only when policy-defined triggers are met, ensure threats are contained before they escalate or affect wider patient populations (*Projects | Apart Research*, 2025).

Despite sophisticated automation, exceptional situations demand careful human judgment. Security tools can misinterpret complex clinical workflows, prompting unnecessary or even harmful isolation if left unchecked. Clinical engineers and security officers keep oversight of the isolation process, ready to intervene when urgent patient needs or nonstandard device use arises. Open communication channels between IT, biomedical engineering, and care teams are preserved, often through secure messaging platforms or integrated incident dashboards, allowing quick escalation and collaborative risk assessments. For example, a cardiac monitor used in a critical care unit might show signs of anomalous traffic due to a needed software update. Technicians, informed by direct communication with nurses familiar with the patient's trajectory, determine whether a controlled reboot or continued monitoring outweighs immediate disconnection.

Hospitals also invest in fallback and fail-safe systems to preserve care during isolations. Secondary monitoring tablets, backup ventilators, or redundant infusion pumps, set up in anticipation of device interrupts, can be swiftly activated. If electronic record-keeping systems must be brought offline to prevent further harm, clinicians switch to preprinted chart templates, ensuring the traceability and integrity of patient data. Fail-secure device modes allow vital equipment to operate in a minimal-risk state, delivering core medical functions even if isolated from central management. When a networked drug-dosing system is disrupted, pharmacy teams and clinicians revert to manual dosing protocols, guided by established treatment algorithms, until automated systems are restored.

Secure Remote Access and Telemedicine Channels

Healthcare networks face growing threats from remote access vulnerabilities as providers expand telemedicine, remote diagnostics, and distributed care. As clinical staff, patients, and vendors increasingly connect from outside the traditional secure perimeter, network security must adapt with precise technical safeguards. A robust remote access solution forms a critical element in ensuring medical devices continue to operate reliably, patient data remains protected, and organizations respond rapidly to real-time threats.

Virtual Private Networks (VPNs)

VPNs facilitate secure connectivity between remote users and healthcare networks by creating encrypted communication tunnels over public or untrusted networks. This design ensures that sensitive health data, session commands, and device telemetry are unavailable to eavesdroppers, even when users connect via home Wi-Fi or mobile carriers. VPN gateways typically require authentication before establishing access, with traffic subsequently routed through encrypted tunnels that meet regulatory obligations for privacy. However, VPNs come with operational constraints. High volumes of telemedicine consultations or remote radiology sessions can saturate available bandwidth, leading to latency that disrupts clinical workflows. Endpoint configuration also poses a risk; a compromised staff laptop can introduce malware if VPN policy does not enforce strong baseline security such as up-to-date antivirus controls and restricted split-tunneling. These inherent limitations demand continuous monitoring of VPN usage, session audits, and traffic analysis to identify anomalies suggesting compromised credentials or unauthorized access attempts (Zhou et al., 2019).

Access Logging and Audit Trails

Recording granular activity during remote sessions forms a backbone of security and compliance in healthcare IT. Detailed audit logs capture each remote login, privilege escalation, command execution, configuration change, and file access event. Logs support both proactive and reactive security needs. In a real-world scenario, when an attacker successfully steals valid remote credentials, comprehensive audit trails allow incident responders to identify the time, systems, and actions involved—tracing everything from unauthorized access to potential medical record theft. In managed telemedicine infrastructure, audit logs might highlight repeated failed login attempts or an unexpected file export, triggering a rapid incident response and system isolation. Audit logs also support routine compliance reporting against frameworks such as HIPAA, where documentation of data access and administrative actions is required. For privacy-sensitive workflows, like remote radiology or remote consultation review, audit records help confirm that users only access records necessary for care and that disclosures are justified. Automated audit review tools increasingly flag abnormal activity in real time, enabling organizations to contain incidents before they threaten sensitive data or the integrity of connected medical devices (Zhou et al., 2019).

Multi-Factor Authentication (MFA)

MFA strengthens remote user identity validation by requiring two or more verification mechanisms. In healthcare environments, this often involves combinations of hardware tokens, time-based one-time passwords (OTPs) delivered to mobile devices, biometric identifiers, and traditional passwords. OTPs generated on a provider's phone, combined with digital certificates or smart cards, drastically reduce the risk posed by password theft or phishing. Even if a password is stolen or leaked, the absence of a valid second factor prevents unauthorized remote access. MFA can be tailored to device capabilities: for telemedicine staff using tablets, fingerprint or facial recognition adds both security and ease of use; for vendor engineers working on medical device software, USB hardware tokens or digital signatures stored on secure elements support robust static authentication controls (Suleski et al., 2023). MFA deployment must balance security with usability, ensuring that authentication processes remain swift during urgent care moments. Effective MFA also includes ongoing key and token life cycle management, with policies for prompt revocation if devices are lost or access is suspended (Suleski et al., 2023).

Session Management and Timeouts

Strong session management ensures remote access is strictly limited in both duration and scope. Automatic timeouts for inactivity are vital to prevent unauthorized session persistence, such as when telemedicine physicians leave workstations unlocked or remote device management consoles are left unattended. Forced reauthentication after periods of inactivity reduces the risk of adversaries using dormant sessions to infiltrate healthcare systems. Configurable timeout policies aligned with clinical workflows can prevent session lockouts during lengthy, legitimate consultations. In a remote diagnostics context, automatic logoff after a set window ensures that vendor access ceases when troubleshooting is complete, eliminating lingering backdoors into critical device networks. For added protection, session identifiers can be dynamically regenerated following reauthentication events or privilege escalation. This practice further complicates session hijacking and credential reuse attempts.

Each of these technical controls operates as a complementary component of a larger cybersecurity strategy. VPNs provide fundamental encryption and perimeter extension, access logging supplies both accountability and early warning, MFA fortifies user validation against evolving social engineering and credential-targeted threats, and disciplined session management ensures access remains purposeful and temporary. Together, they help healthcare organizations minimize the risk of data breaches and device tampering across their expanding remote access landscape, while supporting uninterrupted patient care delivery and regulatory compliance requirements (Suleski et al., 2023; Zhou et al., 2019).

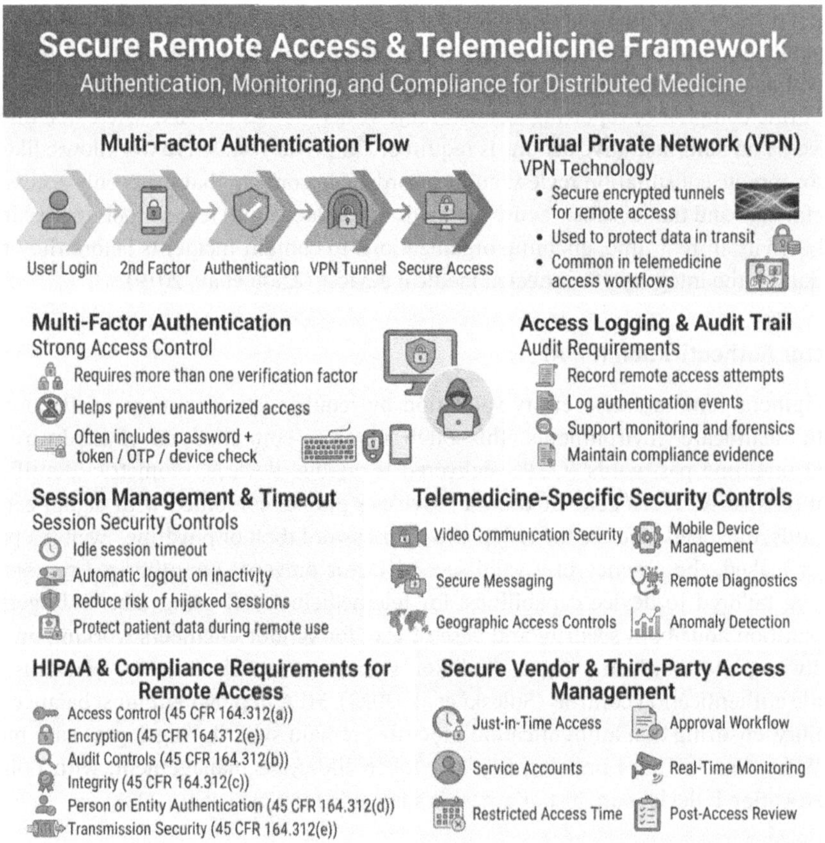

Bringing It All Together

Now that we understand the critical strategies of network segmentation, endpoint security, wireless protection, incident isolation, and secure remote access, healthcare cybersecurity professionals and related stakeholders are better equipped to build resilient defenses that protect sensitive devices and patient data. Implementing these integrated controls not only reduces risk but also supports compliance and operational continuity in complex clinical environments. Moving forward, continuous adaptation and coordination across technical teams, device designers, and regulatory bodies will be essential to address emerging threats and evolving healthcare delivery models. By fostering proactive monitoring, rigorous access management, and rapid incident response capabilities, organizations can confidently safeguard their networks while enabling innovative care solutions such as telemedicine and remote diagnostics, ensuring both patient safety and robust cybersecurity posture into the future.

References

Anjan Kumar Gundaboina. (2025, July 1). *Endpoint Detection and Response (EDR) in Healthcare: Mitigating Threats on Critical Devices*. Journal of Technology and Systems; CARI Journals Limited. https://urfjournals.org/open-access/endpoint-detection-and-response-edr-in-healthcare-mitigating-threats-on-critical-devices.pdf.

Najar, Z. A., Mir, R. N. (2021). Wi-Fi: WPA2 Security Vulnerability and Solutions. Wireless Engineering and Technology. https://doi.org/10.4236/wet.2021.122002.

Palo Alto Networks. (2025). *What is Microsegmentation?* Palo Alto Networks. https://www.paloaltonetworks.com/cyberpedia/what-is-microsegmentation.

Products—Cisco Unified Wireless Network Solution Guide. (n.d.). Cisco. https://www.cisco.com/c/en/us/products/collateral/wireless/unified-wireless-network-sg.html.

Projects | Apart Research. (2025). Apartresearch.com. https://apartresearch.com/project.

Suleski, T., Ahmed, M., Yang, W., Wang, E. (2023, May 22). *A Review of Multi-factor Authentication in the Internet of Healthcare Things*. Digital Health. https://doi.org/10.1177/20552076231177144.

Toll, W. (2025, June 23). *15+ Network Segmentation Control Examples to Strengthen Healthcare Security*. Elisity.com; Elisity. https://www.elisity.com/blog/15-network-segmentation-control-examples-to-strengthen-healthcare-security.

Zhou, L., Thieret, R., Watzlaf, V., DeAlmeida, D., Parmanto, B. (2019, June 12). A Telehealth Privacy and Security Self-Assessment Questionnaire for Telehealth Providers: Development and Validation. International Journal of Telerehabilitation. https://doi.org/10.5195/ijt.2019.6276.

10

Data Security and Patient Privacy Enhancement

"Have you ever wondered what happens the moment patient data slips through an unseen crack in the system?" The question lingered in the air during a cybersecurity briefing at a regional hospital, where the palpable weight of responsibility settled over every attendee. In healthcare, safeguarding sensitive information is not just about technology, it's about maintaining trust, protecting lives, and navigating a complex landscape of regulations and risks that evolve daily. Years of experience have shown that even the smallest oversight can ripple into far-reaching consequences.

Reflecting on these critical moments reveals how multifaceted the challenge truly is: it involves securing data at rest and in transit, ensuring that patient identities remain shielded even as information moves across devices and networks, and managing consent in ways that empower individuals without impeding care. It requires constant vigilance to detect breaches early and respond decisively before damage spreads. This delicate balance between accessibility and protection shapes the foundation of modern healthcare cybersecurity.

The insights gathered from frontline professionals, engineers, regulators, and researchers converge to form a picture of resilience built on layered defenses and proactive strategies. Understanding this interplay provides a crucial perspective for those tasked with fortifying healthcare infrastructures against ever-present threats.

Protecting Health Data at Rest

Health information stored in electronic systems presents a unique set of challenges in the healthcare environment. Patient data is both highly sensitive and of significant value, making it an attractive target for cyber threats, while patient privacy regulations such as the Health Insurance Portability and Accountability Act (HIPAA) impose high stakes for any breach. The complexity of healthcare IT environments, characterized by a diverse ecosystem of electronic health record (EHR) systems, connected medical devices, and cross-organizational data exchanges, further complicates secure storage. Protecting data at rest is therefore a critical requirement, demanding mechanisms that guard against unauthorized access, loss, or tampering, even when hardware is compromised or stolen (Infrastructure, 1997).

Encryption stands as the primary defense line for health data at rest. Two leading cryptographic approaches, full disk encryption and field-level encryption, address this need with distinct operational trade-offs. Full disk encryption secures the entire storage device, whether it is a workstation drive, server disk, or removable storage medium. It ensures that, even if the hardware is physically lost or stolen, all data remains inaccessible without the appropriate cryptographic keys. This method is widely applied to portable devices used by clinicians, such as laptops and mobile computing tools that routinely leave controlled environments (Infrastructure, 1997).

Figure 10.1: Data Protection Lifecycle

Comprehensive Security Strategies for Patient Data Across All States

Data at Rest

Information stored in databases, file systems, and backup media

Protection Methods:

- ✓ Full disk encryption (AES-256)
- ✓ Field-level encryption
- ✓ Database access controls
- ✓ Physical media security
- ✓ Audit logging

Data in Transit

Information moving between devices, networks, and cloud systems

Protection Methods:

- ✓ TLS 1.3 encryption
- ✓ VPN tunneling
- ✓ Digital certificates (PKI)
- ✓ Network segmentation
- ✓ Integrity verification

Data in Use

Information actively processed by applications and users

Protection Methods:

- ✓ Role-based access control
- ✓ Secure memory handling
- ✓ Session management
- ✓ Real-time monitoring
- ✓ Consent enforcement

⚠ Common Attack Vectors Across Data States

Physical Theft
Lost laptops, stolen backup tapes, compromised portable devices

Eavesdropping
Network traffic interception, man-in-the-middle attacks

SQL Injection
Database exploitation through web application vulnerabilities

Insider Threats
Unauthorized access by employees or contractors with credentials

Data Tampering
Modification of patient records during transmission or storage

Improper Disposal
Incomplete data destruction allowing recovery of sensitive information

Patient Data Journey Through Healthcare Systems

Point of Care (Capture) → Encrypted Transit (TLS/VPN) → EHR Database (At Rest) → Cloud Backup (Encrypted) → Analytics/Research (De-Identified)

Protection Layer	Technology	Healthcare Use Case	Implementation Difficulty
Full Disk Encryption	AES-256, BitLocker, FileVault	Clinician laptops, mobile workstations	Low
Field-Level Encryption	Database TDE, Column encryption	SSN, addresses in EHR databases	Medium
TLS 1.3	SSL/TLS protocols	Patient portals, telemedicine sessions	Low
VPN Tunneling	IPsec, OpenVPN	Remote physician access to hospital networks	Medium
Digital Signatures	PKI, X.509 certificates	Lab reports, prescription orders	High
Secure Erasure	Cryptographic erasure, DoD 5220.22-M	Decommissioned servers, backup tapes	Low

HIPAA Compliant | GDPR Ready | NIST 800-53 | ISO 27001 | HITRUST CSF

Note: Multi-layered protection across all data states is essential for comprehensive healthcare cybersecurity. Each layer addresses specific vulnerabilities and provides defense-in-depth against evolving threats.

Field-level encryption, by contrast, targets the specific data elements most likely to compromise patient identity or privacy, such as social security numbers, addresses, or diagnostic details within database structures. Selecting certain fields for encryption minimizes unnecessary computational overhead and helps maintain usability within clinical workflows. For instance, a hospital may encrypt social security number fields in the EHR database, rendering those values unreadable even to system administrators, while nonsensitive scheduling data remains accessible for regular operations. The balance between comprehensive encryption and day-to-day usability is crucial; healthcare organizations must ensure that robust data protection does not impede clinical decision-making or slow response during emergencies. Effective implementation involves workflows and key management policies capable of quickly decrypting essential information for authorized users during critical care events, yet able to prevent unauthorized access during routine operations (*HIPAA Data Security and Retention Requirements*, 2025; Infrastructure, 1997).

Healthcare information systems are often backed by intricate database architectures vulnerable to specialized risks. SQL injection attacks, for example, exploit web-based front ends of EHR systems by leveraging poorly sanitized user input to issue unauthorized queries that extract, alter, or delete health data. Unauthorized access also remains a persistent threat, particularly where insufficient authentication or role enforcement allows privileged operations. Insider risk complicates matters further; employees or contractors with legitimate credentials may intentionally or unintentionally access and misuse data. These risks become more pronounced due to the heterogeneous and interoperable nature of healthcare IT, where many database variants and legacy applications coexist, sometimes without unified security oversight (Infrastructure, 1997).

Mitigating database-specific threats requires a multilayered defense. Role-based access control must strictly limit data access according to job responsibilities. Clinical staff may gain broad access for patient care, while administrative users are segregated to billing and scheduling records. Security patching routines must be automated and closely monitored, with vulnerability assessments conducted routinely to ensure identified issues are resolved across all database platforms. Audit logging tracks access, configuration changes, and privilege grants to support rapid incident response and retrospective analysis. For example, a major health network may institute quarterly reviews of user roles, privilege grants, and incorporate just-in-time approvals for administrative tasks, reducing the operational window for potential misuse (*HIPAA Data Security and Retention Requirements*, 2025).

Physical protection remains vital alongside digital controls, particularly for backup media and portable storage devices such as magnetic tapes, external hard drives, and USB keys. Best practices dictate that these media be stored within locked, environmentally controlled facilities with restricted personnel access. Media handling requires defined procedures controlling issuance, movement, and storage, especially during transport to offsite disaster recovery locations. Chain-of-custody documentation ensures that every interaction with a piece of media is logged and accountable, which is critical when media is shipped for secure archival or destruction. For example, a regional healthcare provider may implement badge-controlled storage vault access, sign-out logs, and GPS tracking during media transport to third-party vaults (*HIPAA Data Security and Retention Requirements*, 2025).

Ensuring data is truly irretrievable after media retirement is nonnegotiable in regulated healthcare settings. Cryptographic erasure, a technique where the destruction of encryption keys renders encrypted data undecipherable, offers efficiency at scale for modern storage arrays. Alternatively, data overwriting procedures repeatedly write patterns of random information over the physical sectors of a device, inhibiting advanced recovery attempts. For physical media no longer suited for secure overwriting or reuse, hardware destruction—such as shredding, crushing, or degaussing—is necessary to destroy residual magnetic or electronic traces. Data retention and deletion workflows must always align with applicable regulatory timelines, such as the six-year audit log retention mandated by HIPAA, to avoid premature data loss or exposure risks. For instance, a community clinic may securely destroy offsite tape backups after statutory retention periods by contracting certified shredding services and maintaining permanent certificates of destruction for compliance audits (*HIPAA Data Security and Retention Requirements*, 2025).

A comprehensive, multilayered approach to securing data at rest forms a foundation on which further protections—such as those ensuring data confidentiality and integrity during network transmission—can build in pursuit of holistic patient privacy and system resilience.

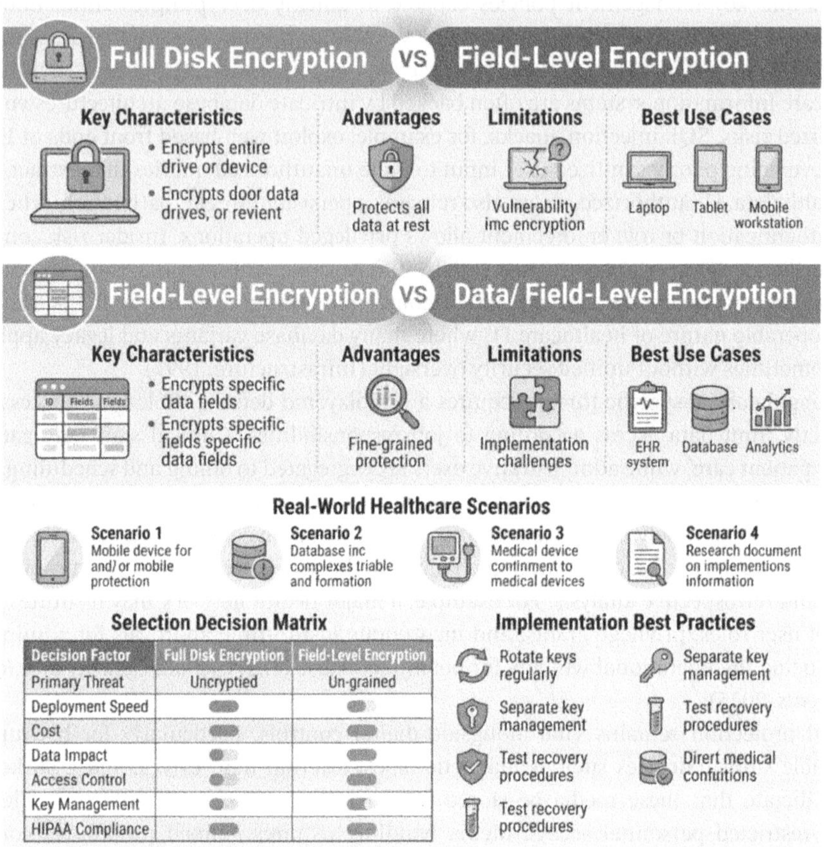

Healthcare Recommendation: Deploy full disk encryption on all portable devices immediately. Implement field-level encryption for the most sensitive database fields (SSN, payment cards, genetic data) as a phased rollout. Many organizations benefit from *both* strategies working in tandem for defense-in-depth.

Data-In-Transit Security: From Patient to Cloud

When health information leaves secure storage and moves between a patient, a healthcare provider, and the cloud, new risks emerge. These moments of data exchange are critical junctures for cyber attackers seeking to intercept or manipulate clinical details, treatment notes, or sensor data.

Even brief lapses in safeguarding transmission routes can expose sensitive details to eavesdropping or unauthorized alteration. Every phase of transit, from wearable device syncing to cloud record updates, must be protected with purpose-built security strategies and controls (Tariq et al., 2023).

Transport Layer Security (TLS) is an essential mechanism for establishing encrypted channels, especially for web-based patient portals, health information exchanges, and electronic prescriptions. TLS encrypts all data between endpoints, making intercepted health records unreadable without the correct cryptographic keys. This encryption thwarts most forms of passive interception—even if an attacker accesses the communication line, the data's confidentiality remains intact. Furthermore, message authentication features embedded within TLS ensure that any tampering or message corruption is detectable, maintaining data integrity during transmission (*Cyber Security Glossary: Cyber Security Terms Listed from a to Z*, 2024). Healthcare environments deploying mobile applications, telemedicine platforms, or internal APIs benefit from TLS by ensuring that clinical instructions, images, and lab results cannot be read or altered by outsiders.

Comparing TLS to Virtual Private Networks (VPNs) reveals different operational trade-offs for healthcare organizations. VPNs encapsulate all user traffic within an encrypted "tunnel" over public or insecure networks. When a remote physician connects from home to the hospital's EHR system, a VPN creates trusted connectivity despite an untrusted local network. This is especially useful for multisite hospitals or remote consultations involving home health devices. However, the complexity of deploying and managing VPNs, especially for always-on medical device telemetry or high-volume cross-site transfers, can introduce latency and maintenance burdens. VPN misconfigurations, such as improper endpoint authentication or incomplete segmentation, can actually increase exposure if not carefully managed (Tariq et al., 2023). In contrast, TLS is often more granular, protecting only defined application data streams, resulting in lower latency and simplified troubleshooting for targeted workflows.

Verification protocols play a central role within both frameworks. Digital certificate validation confirms that endpoints exchanging medical records are legitimate entities, not impostors executing man-in-the-middle or spoofing attacks. Public Key Infrastructure (PKI) manages these trust chains: certificate authorities issue digital credentials to devices or servers, providing strong evidence that an endpoint is trustworthy. In a typical telemedicine session, when a provider's browser connects to the hospital, certificate hierarchies ensure the session's authenticity. Certificate revocation—continuous checking of credential validity—guards against using compromised or expired certificates. Without strict validation controls, attackers could pose as legitimate portals to harvest patient data (*Cyber Security Glossary: Cyber Security Terms Listed from a to Z*, 2024). Prolonged medical device sessions or cloud-based analytics streams rely on constant monitoring of channel integrity, using automated alerts to signal certificate expirations or changes in trust relationships.

Eavesdropping—silently intercepting health information in transit—is a perennial threat. Encryption of all network packets is the front line of defense. Even with encrypted tunnels and secured sessions, sophisticated attackers might leverage unrestricted internal network segments or outdated wireless technologies to collect and analyze traffic (Tariq et al., 2023). Segmenting medical device networks from general hospital IT resources minimizes the ability for unauthorized insiders or malware to observe sensitive flows. Layering intrusion detection systems identifies abnormal data movement, such as spikes in outgoing health records or anomalous access attempts during off-hours. Prompt updates to cryptographic standards are vital; incidents have shown how outdated encryption, such as deprecated cipher suites or weak key lengths, enabled unauthorized disclosure or regulatory penalties following breaches (*Cyber Security Glossary: Cyber Security Terms Listed from a to Z*, 2024).

Data integrity remains a bedrock principle for medical workflows. Hashing algorithms generate a unique digital fingerprint for every message or transaction. If even a single bit changes during transit, recalculating the hash immediately reveals a mismatch. Digital signatures take this a step further, allowing the sender to "sign" the data before transmission, enabling the recipient to verify both its origin and unchanged state. Critical scenarios, such as automated infusion pump instructions or lab report deliveries, depend on robust integrity checks. If a hash comparison or signature verification fails, systems can block the transaction, preventing harmful errors or manipulated results from reaching patients or clinicians. These methods create audit trails and foster regulatory compliance, all while supporting patient trust in digital care processes.

Healthcare networks transmitting data across borders, between wearable devices, and into multitenant cloud platforms demand security models that span the full distance from sender to recipient. Even as these technical protections mature, privacy concerns move beyond simple encryption. Future strategies must also control how data sets can be relinked or reidentified once in the cloud—ensuring that a patient's identity remains shielded even after successful transport and storage have occurred (*Cyber Security Glossary: Cyber Security Terms Listed from a to Z*, 2024; Tariq et al., 2023).

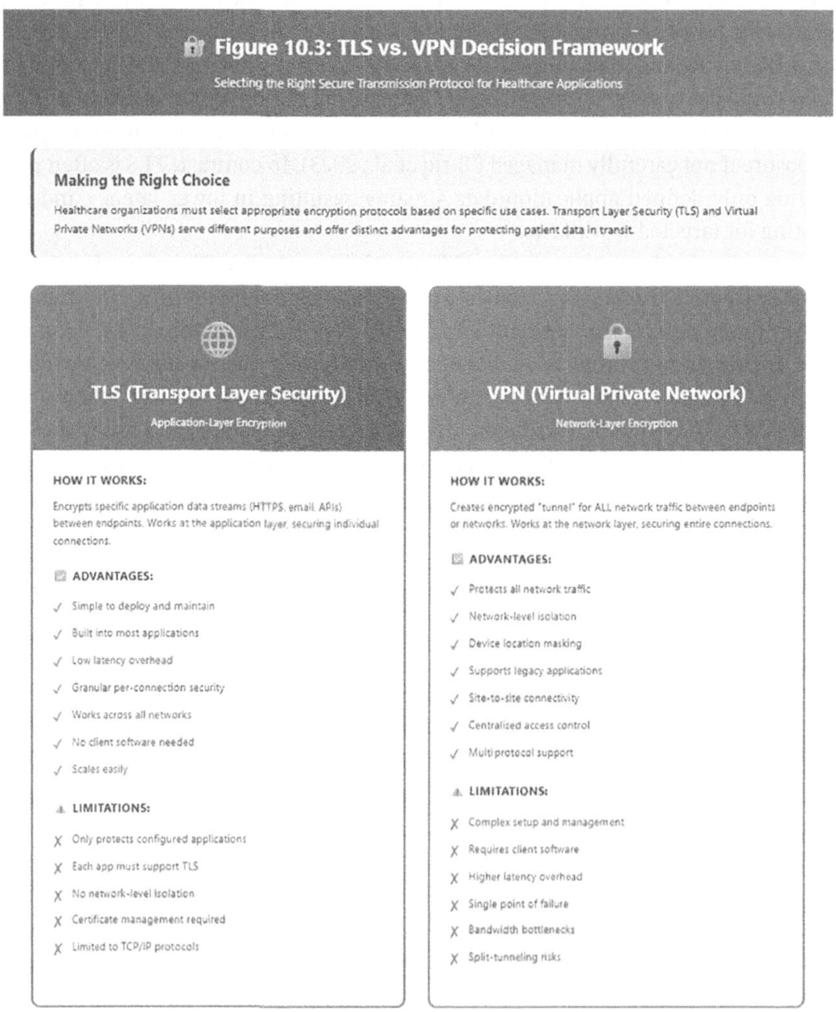

Quick Decision Tree

? What type of access do you need to secure?

Specific Applications	Entire Network
Patient portal, telemedicine, web APIs	Remote access to all hospital systems
→ Use TLS	→ Use VPN

? Who are the primary users?

Patients & Public	Staff & Clinicians
Cannot require VPN client installation	Can deploy managed software
→ Use TLS	→ VPN Acceptable

? Do you need to support legacy applications?

Yes - Legacy Apps	No - Modern Apps
Old imaging systems, billing software	Cloud EHR, modern web services
→ Use VPN	→ Use TLS

Healthcare-Specific Use Cases & Recommendations

	Recommended	Rationale
Patient Portal Access	TLS	Patients access from home networks via web browser. HTTPS (TLS) provides seamless encryption without client software.
Remote Physician EHR Access	VPN	Full access to hospital network required. VPN ensures all traffic encrypted and provides network isolation.
Telemedicine Video Sessions	TLS	WebRTC with TLS encryption built-in. Low latency essential for real-time video, VPN overhead undesirable.
Medical Device Telemetry	TLS	High-volume continuous data streams. TLS protocols optimized for device communication, lower overhead than VPN.
Site-to-Site Hospital Network	VPN	Connect entire branch clinic networks to main hospital. VPN creates persistent secure tunnel for all services.
E-Prescription Transmission	TLS	Pharmacy partners use HTTPS APIs. TLS sufficient for point-to-point prescription messaging (NCPDP SCRIPT).
Legacy Imaging System Access	VPN	Old PACS systems don't support modern TLS. VPN wraps legacy protocols (DICOM) in secure tunnel.
Cloud EHR API Integration	TLS	Modern cloud services use HTTPS/REST APIs. TLS 1.3 provides strong encryption with certificate validation.
Health Information Exchange (HIE)	Both	TLS for external organization connections. VPN for trusted partner networks. Depends on relationship and volume.

Technical Specifications Comparison

Specification	TLS 1.3	VPN (IPsec/OpenVPN)
Encryption Standards	AES-256-GCM, ChaCha20-Poly1305	AES-256, 3DES, Blowfish
Key Exchange	ECDHE (Elliptic Curve Diffie-Hellman)	IKEv2, Diffie-Hellman
Authentication	X.509 certificates, PKI	Preshared keys, certificates, tokens
Latency Impact	~10-30ms handshake overhead	~50-100ms connection + per-packet overhead
Throughput Impact	Minimal (~5% overhead)	Moderate (~10-20% overhead)
OSI Layer	Layer 6 (Presentation) / Layer 7 (Application)	Layer 3 (Network) / Layer 4 (Transport)
Certificate Management	Required per service/domain	Per VPN gateway/client
Client Requirements	Built into browsers/apps	Dedicated VPN client software
Firewall Compatibility	Excellent (standard HTTPS port 443)	Variable (may require port forwarding)
HIPAA Compliance	Yes (with proper implementation)	Yes (with proper implementation)

Implementation Quickstart

Implementing TLS

1. Obtain SSL/TLS certificates from trusted CA (Let's Encrypt, DigiCert)
2. Configure web servers to enforce TLS 1.3 only (disable older versions)
3. Enable HSTS (HTTP Strict Transport Security) headers
4. Implement certificate pinning for mobile apps
5. Set up automated certificate renewal
6. Configure strong cipher suites (ECDHE-ECDSA-AES256-GCM-SHA384)
7. Enable OCSP stapling for certificate validation
8. Monitor certificate expiration dates

Implementing VPN

1. Select VPN protocol (IPsec for site-to-site, OpenVPN for remote access)
2. Deploy VPN gateway hardware/software
3. Configure strong authentication (certificates + MFA)
4. Define network segmentation and access policies
5. Deploy VPN client software to endpoints
6. Implement split-tunneling policies (if needed)
7. Set up redundant VPN gateways for high availability
8. Configure intrusion detection on VPN traffic

Best Practice: Many healthcare organizations deploy *both* TLS and VPN in complementary roles. Use TLS for patient-facing applications and public APIs, while VPN secures internal staff remote access and site-to-site connectivity. Always enforce the latest protocol versions (TLS 1.3, IKEv2) and disable legacy versions for maximum security.

Anonymization and Pseudonymization of Patient Data

Data protection in healthcare applications reaches far beyond securing information during transmission. Once data reaches storage or analysis environments, the risk of exposure intensifies, as the value and utility of health data often depend on its use for research, quality improvement, or device optimization. Maintaining privacy while preserving clinical and analytical relevance requires advanced anonymization and pseudonymization techniques. These techniques serve as the cornerstone of responsible data stewardship and form a practical basis for regulatory compliance, such as under HIPAA and General Data Protection Regulation (GDPR) requirements (*18 HIPAA Identifiers for PHI De-Identification*, 2025; *Advanced Techniques for De-Identifying PII and Healthcare Data—Skyflow*, 2024).

Robust de-identification calls for systematically removing direct and indirect identifiers from patient records. Direct identifiers—like full names, social security numbers, or complete addresses—pose an immediate risk and must be masked or replaced entirely. Techniques in anonymization go further, eliminating or irreversibly altering attributes so individuals cannot be reidentified, even indirectly. For instance, dates of birth might be shifted or generalized by retaining only the year of birth, and zip codes can be truncated to retain only the first three digits or grouped into larger regions. Aggregating geolocation data and limiting the granularity of time stamps represent further practical steps. These approaches allow sharing data for epidemiological research or device calibration, where individual-level detail is unnecessary and increases privacy risk. Additionally, masking visual identifiers in patient medical images—blurring or cropping facial features, removing metadata, and employing secure storage—ensures images remain useful for clinical or educational settings without enabling direct identification (*18 HIPAA Identifiers for PHI De-Identification*, 2025).

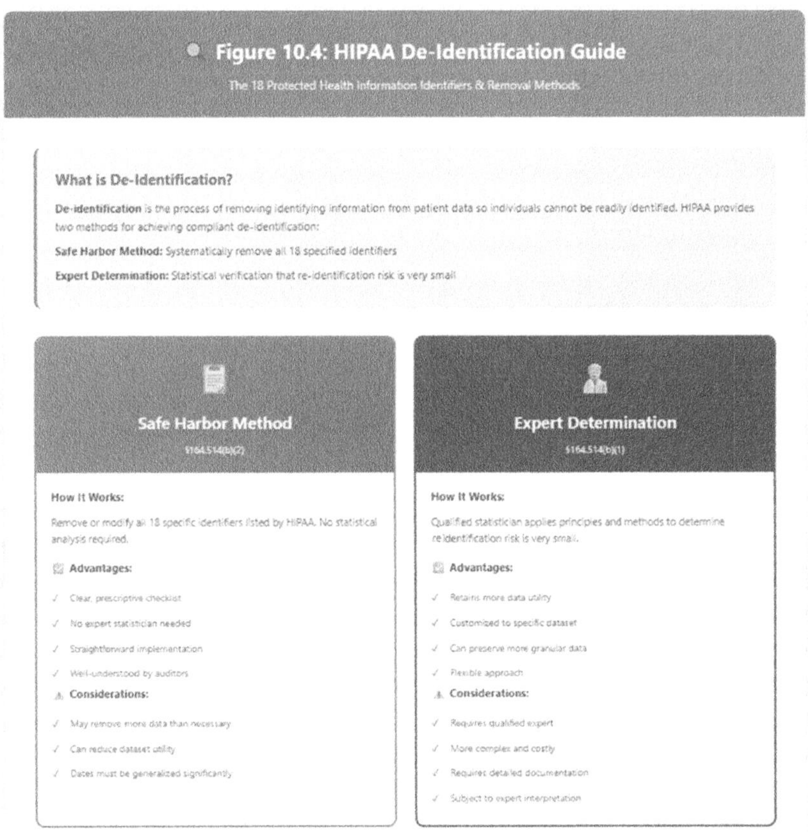

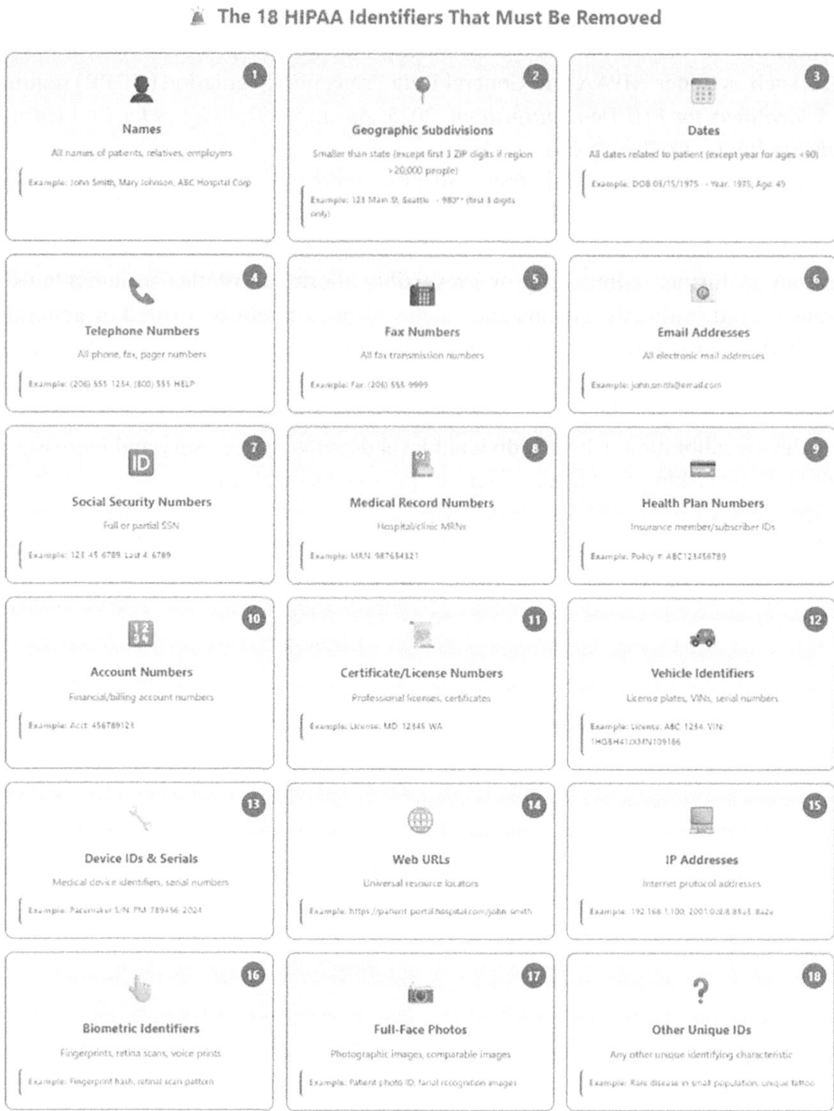

Statistical de-identification methods enhance dataset utility while minimizing reidentification risk. Controlled noise infusion subtly alters values within allowable ranges to disrupt unique patterns, mitigating the risk of relinking datasets to specific individuals. For example, small shifts in laboratory results or aggregation of rare diagnostic codes into broader categories help retain data utility for population-level studies without singling out individuals. Effectiveness evaluation of these modifications requires adversarial testing—intentional attempts to reidentify data using sophisticated algorithms or cross-linking with external data sources. Risk-scoring frameworks quantitatively measure the likelihood of reidentification, guiding the choice and calibration of anonymization algorithms. This iterative evaluation secures both compliance and analytical integrity (*Advanced Techniques for De-Identifying PII and Healthcare Data—Skyflow*, 2024).

Pseudonymization provides a valuable alternative when maintaining the ability to relink data is essential. Assigning unique pseudonyms in place of identifying information—such as replacing medical record numbers with randomly generated codes—enables authorized reconnection with patient identities under strictly controlled circumstances. Secure pseudonym key management is fundamental. Cryptographic storage protects mapping files, and access restrictions ensure that only designated individuals or systems can retrieve identifying information. For longitudinal studies, clinical trial management, or follow-up care, technical controls, such as multifactor authentication and role-based access, work hand-in-hand with policy measures to define who can perform reidentification and under what circumstances. Logging every access and administrative action in immutable audit trails reinforces both security and regulatory expectations (*18 HIPAA Identifiers for PHI De-Identification*, 2025).

Risks of unintended reidentification persist, especially when de-identified datasets are combined with auxiliary data sources or when pseudonym generation lacks sufficient randomness. Mosaic attacks, where adversaries piece together seemingly innocuous data to reveal identities, demonstrate the necessity of ongoing defense. Regular risk assessments, adaptive updates to anonymization algorithms, and continuous monitoring for emerging threats form the backbone of resilient privacy programs. Inadequate risk controls or outdated de-identification logic can expose organizations to severe consequences, including regulatory penalties and lasting reputational damage (*Advanced Techniques for De-Identifying PII and Healthcare Data—Skyflow*, 2024).

Comprehensive regulatory frameworks require more than technical diligence. HIPAA outlines two recognized de-identification approaches: Safe Harbor and Expert Determination. Safe Harbor mandates systematic scrubbing of a fixed set of 18 identifiers and retention of documentation detailing methodical removal, review checkpoints, and responsible personnel. Expert Determination involves professional assessment and documentation that the residual risk of reidentification is very small, based on current techniques and external risk factors. Transparent process documentation and immutable audit trails are essential, enabling organizations to demonstrate compliance during audits and investigations. Failing to meet these documentation standards can result in financial penalties, enforced corrective measures, and public disclosure of violations (*18 HIPAA Identifiers for PHI De-Identification*, 2025).

Ultimately, privacy protection must further incorporate patient engagement and transparent consent mechanisms. Patients should understand not only how their data is de-identified but also when and how reidentification might occur for care or legal purposes. When combined with robust technical and administrative controls, this transparency solidifies trust, enhances regulatory defensibility, and fosters a culture of accountability in healthcare data stewardship (*Advanced Techniques for De-Identifying PII and Healthcare Data—Skyflow*, 2024).

☑ Safe Harbor Compliance Checklist

- ☑ Remove all patient, relative, and employer names (Identifier #1)
- ☑ Generalize geographic data to first 3 ZIP digits (if population >20,000) (Identifier #2)
- ☑ Remove or generalize all dates to year only (except ages ≥90 aggregated) (Identifier #3)
- ☑ Remove all telephone, fax, and email contact information (Identifiers #4, #5, #6)
- ☑ Remove SSN, MRN, health plan numbers, account numbers (Identifiers #7, #8, #9, #10)
- ☑ Remove certificate/license numbers and vehicle identifiers (Identifiers #11, #12)
- ☑ Remove device serial numbers and identifiers (Identifier #13)
- ☑ Remove URLs and IP addresses (Identifiers #14, #15)
- ☑ Remove biometric identifiers and full-face photographs (Identifiers #16, #17)
- ☑ Assess and remove any other unique identifying characteristics (Identifier #18)
- ☑ Document de-identification methodology and personnel responsible
- ☑ Conduct adversarial testing to verify reidentification is not feasible
- ☑ Maintain audit trail of all de-identification procedures
- ☑ Obtain legal/compliance review before data release or sharing

Critical Reminder: De-identified data under HIPAA Safe Harbor is *not* considered Protected Health Information (PHI) and is not subject to HIPAA restrictions. However, organizations should still implement appropriate data use agreements and access controls to prevent misuse. Reidentification attempts are prohibited under many state and federal laws.

Patient Consent Management in a Digital Era

Direct patient engagement remains central to achieving meaningful privacy protections and accountability in managing health data. Digital consent management in healthcare enables individuals to directly oversee how their personal information is collected, shared, and utilized by clinicians, researchers, and third parties. Digital consent, in this context, refers to a technologically mediated process that facilitates patient authorization for specific uses of their data. Modern interfaces for consent aim to support informed and autonomous decision-making by structuring information transparently and intuitively. These platforms deploy plain language throughout all materials, making information accessible to patients of varying health literacy levels and abilities, in line with accessibility mandates. Accessibility principles require not only simple vocabulary, but also adaptable text sizes, color-contrast enhancements, and compatibility with assistive technologies to optimize usability for all users (Kostick-Quenet et al., 2025).

Concrete design elements differentiate robust consent management platforms. Step-by-step guides parse complex information into logical, sequential prompts, preventing cognitive overload. Tooltips and clickable explanatory icons offer real-time definitions and clarifications of medical or technical terms. Interactive sample data flow diagrams can visually map where data originates, how it travels through the system, and which external entities it may reach. Scenario-based examples pose realistic data use opportunities and potential outcomes, giving patients a clear sense of the direct and indirect implications of their consent choices. These features collectively reduce information asymmetry and foster a more meaningful consent process aligned with regulatory expectations for patient comprehension and voluntariness (Kostick-Quenet et al., 2025).

A central pillar within digital consent is the implementation of granular, modular permissions frameworks. Rather than adopting broad, bundled consent that covers multiple types of data and uses, modular systems empower patients to specify which categories of information they authorize and for what explicit purposes. For example, a patient might elect to share blood pressure records with a primary physician, permit the use of de-identified imaging data for clinical trials, or restrict genetic data from being used in insurance risk assessment. Some platforms permit consent at the data-element level—such as demographics, prescription history, or lab results—allowing for tailored authorization profiles. Modular control structures accommodate evolving legal requirements, patient circumstances, and research needs, satisfying requirements for explicit and ongoing consent as articulated in progressive privacy laws (Kostick-Quenet et al., 2025).

Dynamic permissioning not only addresses compliance obligations under GDPR, HIPAA, and similar frameworks, but also cultivates ongoing patient trust. As legal and ethical norms shift or data subjects reconsider their preferences, patients must be able to adjust their authorizations in real time. This operational flexibility reassures individuals that their choices are honored, promoting transparency and reinforcing loyalty to healthcare providers or research organizations. A modular approach also clarifies for each stakeholder—clinician, data controller, or third-party recipient—precisely which permissions exist for each individual dataset, reducing ambiguity and the risk of unauthorized use (Filkins et al., 2016).

Robust consent management depends on maintaining secure, auditable logs of every consent decision, along with any subsequent modifications or revocations. Healthcare IT environments must generate tamper-evident audit trails that record each patient's consent event with time stamps, user identification, and cryptographic integrity protections that prevent unauthorized alteration or deletion. Logs integrate with EHR systems and access control engines, ensuring that consent-based permissions are enforced immediately wherever the data is accessed or processed. Regulatory requirements often mandate such logging to provide traceable evidence during privacy audits or investigations, supporting organizational accountability and transparency (Filkins et al., 2016).

Immediate and reliable revocation rights are essential for sustaining patient autonomy. Digital consent interfaces typically offer intuitive dashboards where patients can withdraw previously granted permissions with a few clicks. Backend systems must synchronize these changes across all platforms, locking down access to revoked data and notifying affected parties—such as clinicians or research partners—of updated consent status. Revocation protocols require backend policies to identify all systems and data flows affected and to enforce withdrawal uniformly, ensuring no further access or processing occurs without renewed permission. Failure to swiftly update permissions risks violating privacy regulations and undermining patient trust if revoked data remains accessible (Filkins et al., 2016; Kostick-Quenet et al., 2025).

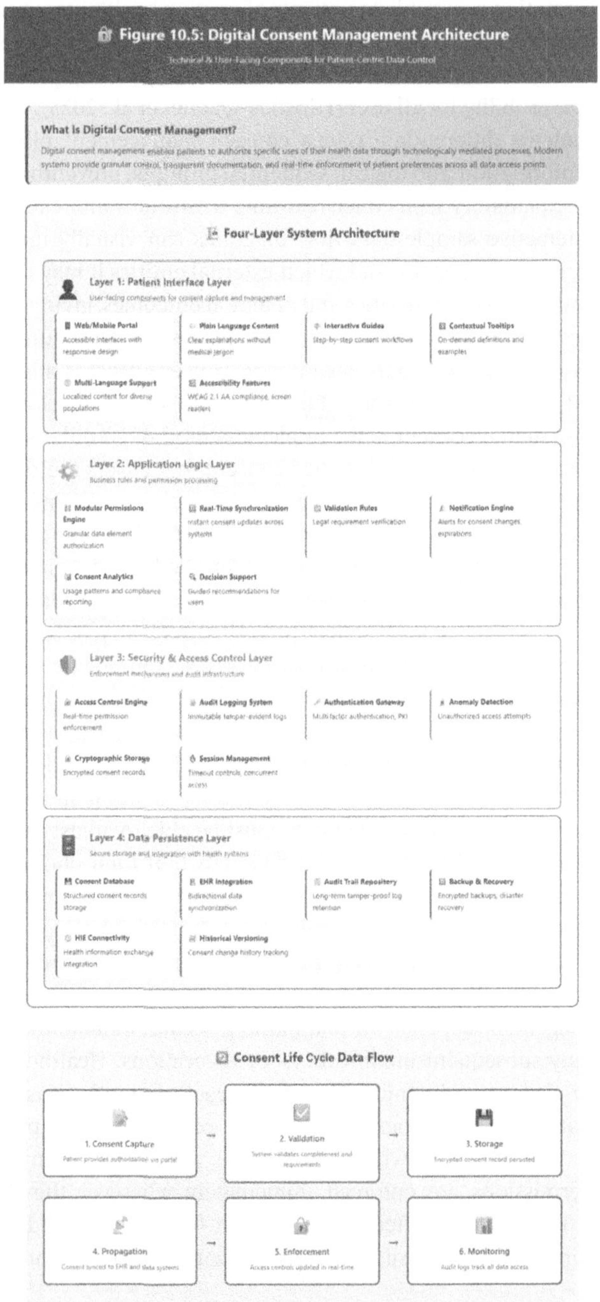

Together, these features—plain language interfaces, modular consent structures, real-time logging, and enforced revocation—work as an integrated network, elevating transparency in health data management. They reduce the ambiguity around data use, deter unauthorized disclosures, and engender trust among patients, providers, and regulators. Where these mechanisms falter, the resulting gaps can lead to misunderstandings, privacy incidents, or regulatory breaches, presenting an ongoing impetus for robust internal controls and continuous improvement. This interdependence illustrates that technical and procedural safeguards must progress in concert to ensure both effective compliance and meaningful patient empowerment.

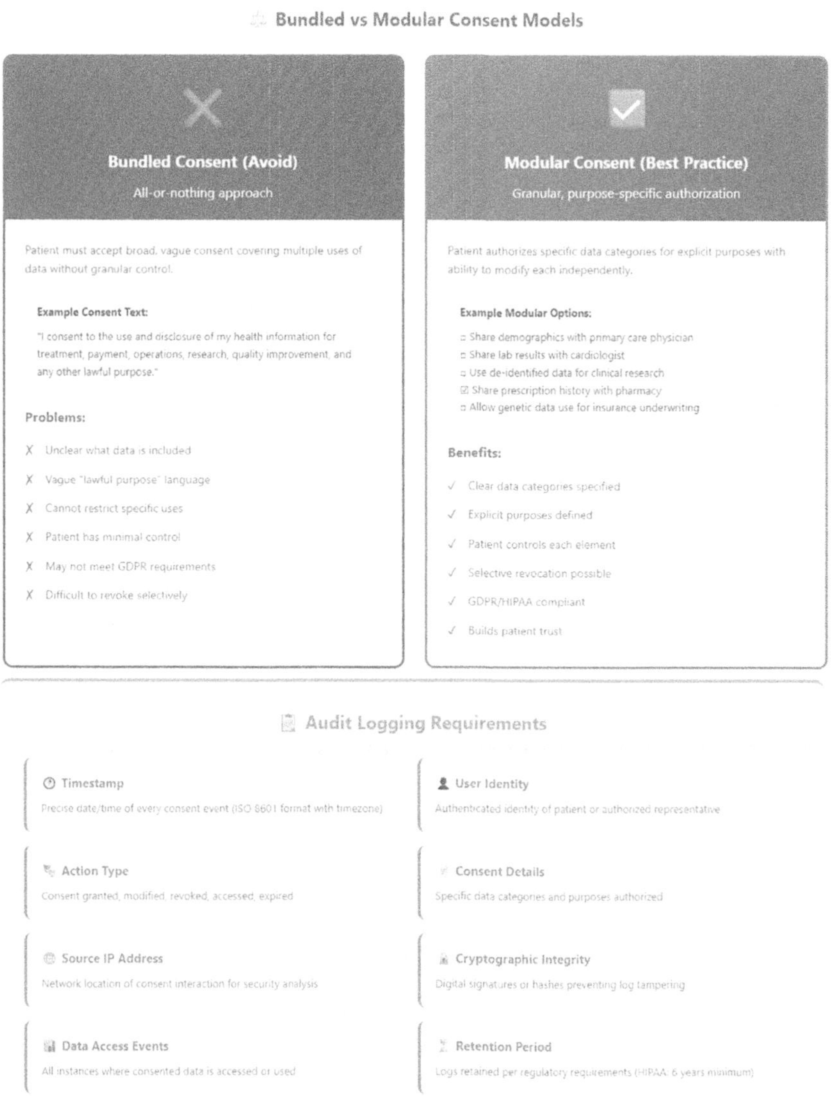

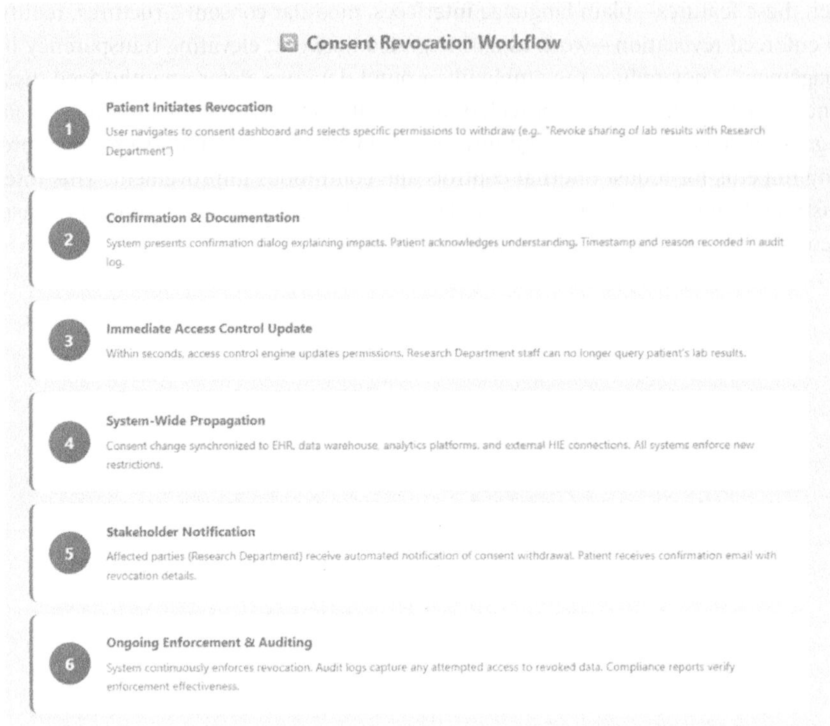

Critical Success Factors: Effective consent management requires seamless integration across all four architectural layers. Patient-facing interfaces must be intuitive and accessible, while backend systems ensure real-time enforcement and comprehensive audit trails. Regular testing of revocation workflows and consent synchronization is essential to maintain system integrity and patient trust.

Detecting and Responding to Privacy Breaches

Privacy principles form the foundation for patient control over health data, but these protections demand real-time vigilance when a breach unfolds. Robust incident detection and agile response mechanisms transform theoretical patient rights into tangible safeguards whenever clinical data is at risk. Immediate awareness of a breach, coupled with decisive action, upholds both institutional trust and legal duties while directly reinforcing patient confidence in digital healthcare frameworks.

Rapid detection is essential. Modern intrusion detection systems (IDS) and anomaly detection technologies continuously monitor networks and endpoints for abnormal patterns indicative of unauthorized access or manipulation. These automated tools, often powered by artificial intelligence, scan real-time traffic, analyze user behaviors, and correlate diverse data streams to flag suspicious activities within moments of occurrence. Automated alerting mechanisms then escalate these findings to security operations teams, activating standardized workflows through integrated incident response platforms. Quick escalation eliminates bottlenecks and ensures that threat intelligence, forensic analysis, and operational decisions happen in parallel. For instance, when an anomaly is detected—such as atypical access to electronic protected health information (ePHI) late at night—an automated workflow triggers real-time team notifications, initiates access control lockdowns, and launches immediate incident validation (The 7 Metrics That Matter in Healthcare Risk Today, 2025). Following verification, assessment of the incident clarifies scope and severity, guiding the organization through evidence preservation and rapid short-term containment.

When a breach is verified, compliance with legal requirements is nonnegotiable. HIPAA regulations and other privacy statutes establish explicit obligations for healthcare entities to report certain incidents. HIPAA requires covered entities to notify the U.S. Department of Health and Human Services (HHS) and all affected individuals without unreasonable delay and no later than 60 days following discovery of a breach. Incidents that affect more than 500 residents of a state or jurisdiction require additional notification to prominent media outlets (The 7 Metrics That Matter in Healthcare Risk Today, 2025). State laws and international frameworks, such as the GDPR or local data protection boards, may impose shorter timeframes, such as 72 hours for notifying supervisory authorities. For every scenario, organizations must clearly define reporting chains, including IT, legal, compliance, executive management, and, when necessary, risk and communications teams.

Clear documentation, timely escalation up the leadership chain, and early external counsel engagement bolster an effective legal response. As an example, consider a ransomware incident that encrypts patient files. The technical incident response team isolates the impacted systems, IT security evaluates forensic evidence, and legal collaborates with executive leadership to promptly determine notification scope. Within hours, formal notifications are drafted and reviewed by compliance and communications teams, then submitted to HHS, law enforcement agencies when appropriate, and relevant state authorities, with careful records kept for postincident review (Ultimate Guide to Healthcare Incident Response Testing, 2025).

Maintaining transparent, effective communication with patients proves just as critical. Immediate, honest notification is necessary to sustain trust and enable patients to act to protect themselves. Communications should plainly state the nature of the breach, the types of data involved, potential consequences, and steps being taken to address the incident. Practical guidance, such as credit monitoring instructions, password reset recommendations, or direct hotlines, minimizes patient anxiety and clarifies the institution's commitment to resolution. Tailoring messages to the needs of specific populations—for example, providing information in multiple languages or formats that are accessible to those with disabilities—ensures inclusivity. A message template might include, "We recently identified unauthorized access to certain patient records. We are actively investigating the incident, have contained the threat, and are working with authorities. Your social security number and treatment information may have been involved. Please review the guidance enclosed to protect your information, and contact our hotline if you have concerns." Such templates build trust while providing actionable support.

Remediation focuses on containing damage and preventing recurrence. Containment begins by quickly securing affected systems and cutting off unauthorized access. Root cause analysis investigates precisely how attackers gained entry—was it phishing, unpatched software, or credential compromise? Findings inform an action plan that includes patching vulnerabilities, updating policies, strengthening monitoring, and conducting targeted staff retraining. For example, after a phishing campaign exposes sensitive scheduling data, the hospital might deploy stricter email filtering, require multifactor authentication, and integrate more sophisticated endpoint monitoring. Supplementary training emphasizes current threats and testing with simulated attacks, tracked for engagement and effectiveness (Ultimate Guide to Healthcare Incident Response Testing, 2025). Remediation documentation, lessons learned, and updated policy guidance are shared across teams, fostering an environment of continuous improvement in both technical and organizational defenses.

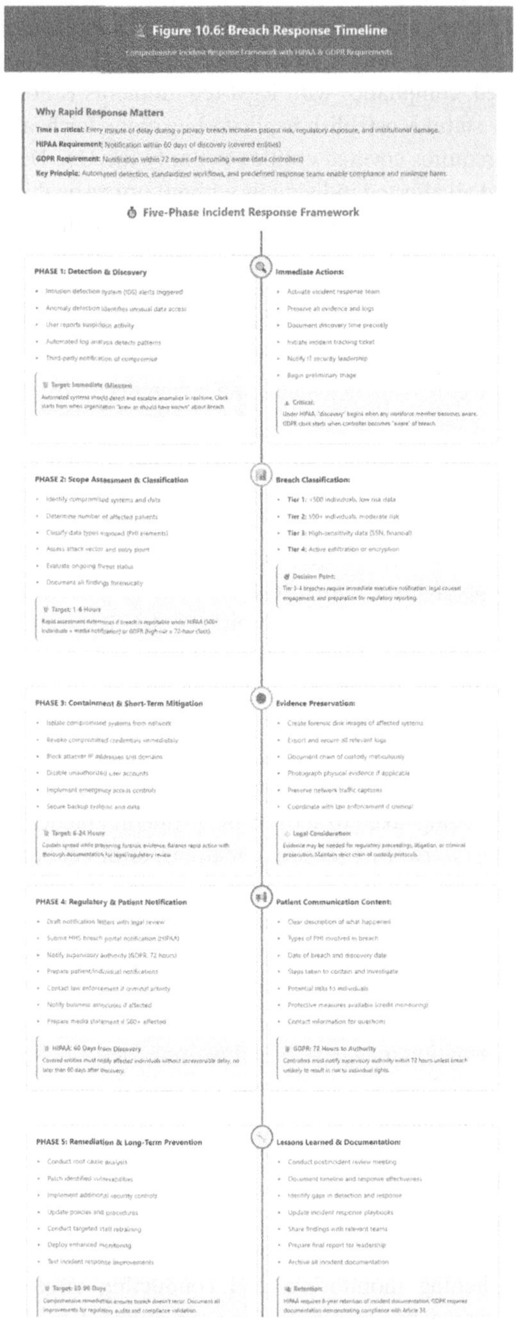

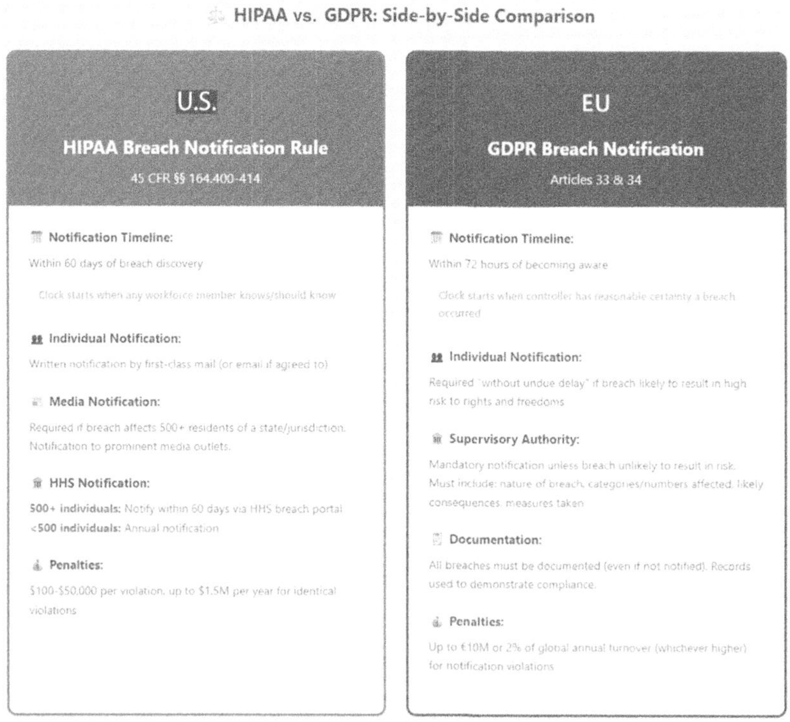

Notification Requirements Matrix

Scenario	Affected Individuals	HHS/Portal	Media	Deadline
Small breach (<500)	☑ Written notice	Annual submission	Not required	**60 days**
Large breach (500+ individuals)	☑ Written notice	☑ Immediate portal	☑ Prominent outlets	**60 days**
Breach with unknown patients	☑ Substitute notice (web + media)	☑ Portal notification	☑ If 500+	**60 days**
Urgent threat (imminent misuse)	☑ Expedited notice	☑ Portal notification	☑ If 500+	**<60 days**
GDPR-covered breach (EU patients)	☑ If high risk	☑ Supervisory authority	Not required	**72 hours**
Business associate breach	BA notifies covered entity	Covered entity notifies	☑ If 500+	**60 days from CE discovery**

Each layer in this workflow combines to fortify patient data protection and institutional readiness. Automated monitoring, prompt escalation, legally sound reporting, transparent patient outreach, and thorough remediation directly translate privacy commitments into measurable, enforceable action, safeguarding health data even in the aftermath of a breach (The 7 Metrics That Matter in Healthcare Risk Today, 2025).

☑ Comprehensive Response Checklist

Immediate Actions (0-24 Hours):

- ☑ Activate incident response team
- ☑ Document discovery time precisely
- ☑ Preserve all logs and evidence
- ☑ Contain the breach (isolate systems)
- ☑ Notify IT security leadership
- ☑ Engage legal counsel
- ☑ Begin preliminary scope assessment

Short-Term Actions (1-7 Days):

- ☑ Complete detailed forensic analysis
- ☑ Identify all affected individuals
- ☑ Classify data types exposed
- ☑ Determine notification requirements
- ☑ Draft notification letters
- ☑ Notify law enforcement if criminal
- ☑ Prepare HHS/supervisory authority report

Mid-Term Actions (7-60 Days):

- ☑ Submit HHS breach portal notification
- ☑ Mail individual notification letters
- ☑ Issue media notifications (if 500+)
- ☑ Set up patient support hotline
- ☑ Offer credit monitoring services
- ☑ Respond to patient inquiries
- ☑ Coordinate with business associates

Long-Term Actions (60+ Days):

- ☑ Conduct root cause analysis
- ☑ Implement security improvements
- ☑ Update policies and procedures
- ☑ Conduct staff retraining
- ☑ Test enhanced controls
- ☑ Document lessons learned
- ☑ Archive all incident records (6 years)

Real-World Response Example: Ransomware Attack

Scenario: Mid-Sized Hospital Network Ransomware Incident

Initial Event: Friday, 11:45 PM - Automated monitoring detects encryption of file shares across multiple servers. Network operations center receives IDS alerts indicating widespread file modifications.

Affected Data: Electronic health records for approximately 12,500 patients including names, dates of birth, medical record numbers, diagnoses, treatment plans, and limited Social Security numbers (~500 patients).

T+0 (11:45 PM Fri)	T+2h (1:45 AM Sat)	T+12h (11:45 AM Sat)	T+48h (Mon 11:45 AM)
Automated detection, incident response team activated, systems isolated	Forensic analysis begun, executives notified, legal counsel engaged	Scope confirmed (12,500 patients), threat contained, evidence preserved	Notification letters drafted, HHS report prepared, patient hotline configured
T+5 days	**T+10 days**	**T+30 days**	**T+90 days**
HHS breach portal submission completed (>500 affected)	Individual notification letters mailed, media statement released	Credit monitoring offered to 500 with SSN exposure, systems restored	Security upgrades deployed, staff retrained, post-incident review completed

Outcome: By following structured incident response procedures, the hospital met all HIPAA notification requirements well within the 60-day deadline, minimized patient harm through rapid credit monitoring deployment, and implemented enhanced security controls to prevent recurrence. Total cost: $2.8M (forensics, notification, credit monitoring, legal, remediation).

Critical Success Factors: Automated detection systems, pre-defined response teams with clear roles, standardized notification templates, and regular incident response drills are essential for meeting tight regulatory deadlines. Organizations should conduct annual tabletop exercises simulating breach scenarios to ensure all stakeholders understand their responsibilities. **Remember:** Under HIPAA, the clock starts when *any* workforce member becomes aware—not just leadership.

Bringing It All Together

Having examined the comprehensive strategies and technologies essential to protecting patient data—from securing information at rest and in transit, through effective anonymization and consent management, to vigilant breach detection and response—we are now equipped to advance healthcare cybersecurity with greater confidence. By integrating these layered defenses and fostering transparent patient engagement, professionals across healthcare and medical technology domains can build more resilient systems that uphold privacy without compromising care quality. Moving forward, embracing continuous innovation, rigorous compliance, and collaborative risk management will be vital to addressing evolving threats and regulatory demands, ensuring that patient trust remains at the heart of digital health transformation.

References

18 HIPAA Identifiers for PHI De-Identification. (2025, June 15). Censinet.com; Censinet. https://www.censinet.com/perspectives/18-hipaa-identifiers-for-phi-de-identification.

Advanced Techniques for De-Identifying PII and Healthcare Data—Skyflow. (2024). Skyflow.com. https://www.skyflow.com/post/de-identifying-pii-and-healthcare-data.

Cyber Security Glossary: Cyber Security Terms Listed From A to Z. (2024, March 2). Global Cyber Security Network. https://globalcybersecuritynetwork.com/blog/cyber-security-glossary-cyber-security-terms-listed-from-a-to-z/.

Filkins, B. L., Kim, J. Y., Roberts, B., Armstrong, W., Miller, M. A., Hultner, M. L., Castillo, A. P., Ducom, J.-C., Topol, E. J., Steinhubl, S. R. (2016, March 15). Privacy and Security in the Era of Digital Health: What Should Translational Researchers Know and do About It? American Journal of Translational Research. https://pmc.ncbi.nlm.nih.gov/articles/PMC4859641/.

HIPAA Data Security and Retention Requirements. (2025, June). Bytebase. https://www.bytebase.com/blog/hipaa-data-security-and-retention-requirements/.

Infrastructure, N. R. C. (US) C. on M. P. and S. in H. C. A. of the N. I. (1997). *Technical Approaches to Protecting Electronic Health Information*. Www.ncbi.nlm.nih.gov; National Academies Press (US). https://www.ncbi.nlm.nih.gov/books/NBK233433/.

Kostick-Quenet, K. M., Compagnucci, M. C., Aboy, M., Minssen, T. (2025, January). Patient-centric Federated Learning: Automating Meaningful Consent to Health Data Sharing with Smart Contracts. Journal of Law and the Biosciences. https://doi.org/10.1093/jlb/lsaf003.

Tariq, U., Ahmed, I., Bashir, A. K., Shaukat, K. (2023, April 19). *A Critical Cybersecurity Analysis and Future Research Directions for the Internet of Things: A Comprehensive Review*. Sensors; MDPI. https://doi.org/10.3390/s23084117.

The 7 Metrics That Matter in Healthcare Risk Today. (2025, July 25). Censinet.com; Censinet. https://www.censinet.com/perspectives/7-metrics-that-matter-healthcare-risk-today.

Ultimate Guide to Healthcare Incident Response Testing. (2025, July 14). Censinet.com; Censinet. https://www.censinet.com/perspectives/ultimate-guide-to-healthcare-incident-response-testing.

11

Clinical Workflow Security: Bridging IT and Patient Care

Hospitals are among the most targeted institutions for cyberattacks, with nearly 70% of healthcare organizations reporting security breaches in recent years. The complexity of clinical workflows, combined with the rapid use of connected medical devices and frequent human interaction, creates an environment rich with potential vulnerabilities. As care teams focus on urgent patient needs, cybersecurity often becomes a secondary concern, leaving critical systems and sensitive data exposed. This chapter examines how daily hospital procedures intersect with cybersecurity challenges and the crucial role that integrated security measures play in protecting both patients and information systems.

Understanding Clinical Workflows and Their Security Impact

Busy clinical environments generate unique cybersecurity risks through the intricate interaction of workflow, human behavior, and networked technology. Admission processes often involve multiple systems, such as electronic health records, administrative portals, and internal messaging platforms. Each touchpoint offers a potential entry point for cyber threats, particularly as staff juggle time demands and competing priorities. During patient admission, rushed staff may bypass verification steps, reuse passwords, or enter data on shared terminals. An unmonitored sign-in station in a waiting room, for instance, could invite unauthorized access if not properly configured to enforce timeouts or unique user authentication. This is compounded by the presence of floating personnel—covering different units or shifts—who may not receive full security training or may rely on colleagues' credentials to speed up admissions during peak periods (Brass et al., 2023).

Patient handoffs present additional vulnerabilities. Exchanging information between clinicians requires secure communication channels and clear demarcation of responsibility. Inadequate protocols may result in staff discussing sensitive details over unsecured messaging apps or through conversations in public spaces. If devices used for handoffs—such as tablets or handheld scanners—are left unlocked or set down unattended, unauthorized individuals could gain access to confidential data with minimal effort. Research highlights that missing or incomplete updates during handovers can leave clinicians unprepared, amplifying the risk of inadvertent data leaks or erroneous entries (Fausett & Keebler, 2022). The use of shadow IT, where staff opt for familiar but unsanctioned software to exchange information for efficiency's sake, can further increase the attack surface and introduce unmonitored vulnerabilities.

Medical device usage is another significant intersection of workflow and cybersecurity risk. Devices ranging from drug-infusing pumps to implantable monitors connect directly to critical networks and patient data repositories. Security practices for such devices often depend on the awareness and diligence of a wide cast of clinical staff, many with varying levels of technical acumen. Simple actions, such as forgetting to log off after using a device, failing to apply software updates, or leaving service technician ports unlocked, can lead to exploitation by unauthorized individuals or malware. This risk rises during staffing shortages or emergencies, when protocol adherence may give way to expediency. For example, a nurse may leave a networked glucometer logged in to avoid delays while tending to several patients at once, inadvertently making the device—and the network it connects to—vulnerable to exploitation (Brass et al., 2023).

Underlying these workflow-based exposures are the real and pressing pressures experienced by frontline staff. High patient volumes, acute care needs, and limited resources can create an environment where security feels secondary to immediate clinical goals. Password sharing, hasty data entry, and reliance on remembered logins are common workarounds to cumbersome authentication or device access procedures. These shortcuts are not simply signs of negligence; rather, they reflect a mismatch between rigid security mandates and the adaptive ways clinicians manage real-world pressures (Fausett & Keebler, 2022). Time constraints and fatigue escalate the likelihood that smart devices may be misconfigured, that incorrect patient records are amended, or that necessary updates are ignored.

Human error, magnified by convoluted processes, is frequently cited as a root cause in cyber incidents. One plausible scenario is a clinician entering patient data under the wrong profile in the heat of a rushed admission, a mistake perpetuated when other staff access these flawed records downstream. Research links such workflow-induced errors to breakdowns in security posture—either by causing improper access permissions to persist or by enabling exploitation of outdated information systems (Brass et al., 2023; Fausett & Keebler, 2022). Fatigue also plays a contributory role, eroding compliance with and attentiveness to warning prompts or best practices.

Workflow-centric security policies that incorporate protective measures into daily tasks deliver better compliance and reduce user pushback. Embedding two-step authentication directly into the patient check-in process or integrating device log-out protocols with routine equipment handover ensures that security commands are not ignored because they are essential to completing the task itself. For instance, requiring stepwise device sign-out—where each transaction is authenticated—prevents ambiguous device ownership and accounts for user accountability, all without slowing clinical tempo (Brass et al., 2023). Ongoing partnership between cybersecurity teams and healthcare practitioners remains vital; policies benefit from the lived experience of clinicians, allowing for solutions that safeguard data and devices without becoming unmanageable burdens (Fausett & Keebler, 2022).

Ultimately, robust access control emerges as a linchpin for reducing exposure. Overly broad user permissions or inconsistent application of protocols creates unnecessary risk. Mapping roles to job functions and carefully tailoring access based on actual responsibilities allow staff to operate efficiently but securely. Authentication methods adapted to distinct clinical roles—such as biometric logins for medication cabinets accessed only by pharmacy staff or context-aware sign-ins that adjust for emergency scenarios—directly address the intersection of workflow and vulnerability. These approaches lay the groundwork for advanced access management strategies that keep pace with the shifting realities of modern healthcare delivery (Brass et al., 2023; Fausett & Keebler, 2022).

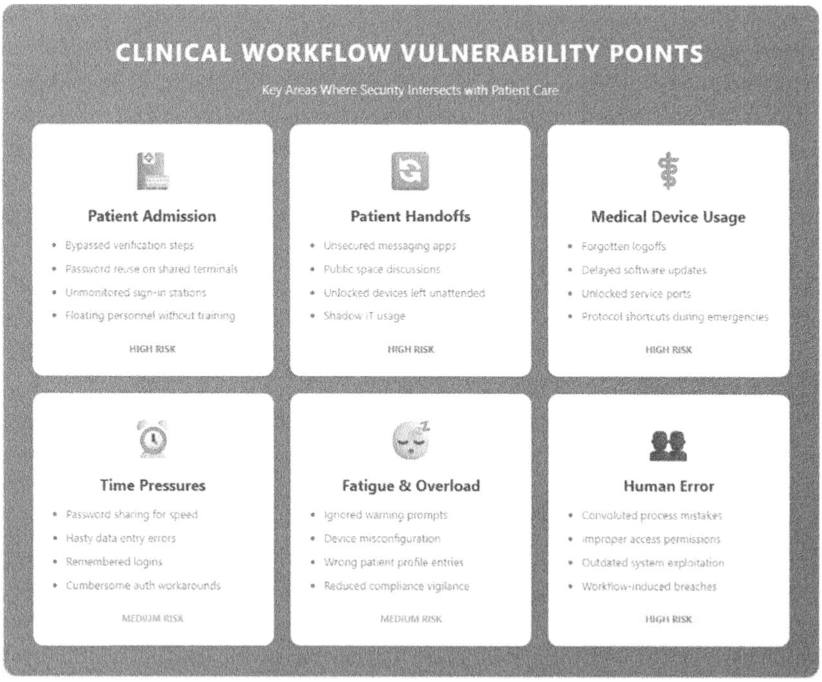

Implementing Role-Based Access Controls in Healthcare

Mapping access privileges directly to well-defined job functions provides much-needed clarity in healthcare environments where the risk of data exposure remains high. Defining role categories such as registered nurse, physician, radiology technician, or billing administrator allows each group to receive only the permissions needed for their routine clinical duties. For instance, a nurse may be granted access to view and update patient vitals and administer medications, while a radiologist has the ability to access diagnostic imaging records but has no editing rights to billing systems. This specificity reduces the chance of users entering areas outside their sphere of responsibility, a common source of accidentally exposed or misused data (Stapleton, 2025). As job titles or responsibilities change, it becomes easy to adjust access, either expanding or reducing permissions as dictated by the new role, thus eliminating lingering access linked to previous positions.

Clear mapping of access also provides a strong foundation for transparency and accountability during audits or security investigations. When each user's access matches a documented role, security teams can quickly determine whether someone's behavior or system interactions align with their approved set of tasks. During a breach investigation, if a device log shows that a billing administrator viewed sensitive clinical images, the system's audit trail will show a clear deviation from approved roles, immediately flagging inappropriate activity. This direct alignment between access rights and job function not only simplifies audit processes but also supports compliance with regulations demanding strict segmentation of privileges, such as the Health Insurance Portability and Accountability Act (HIPAA) and the EU Medical Device Regulation (Gagliardi, 2024; Stapleton, 2025).

A robust authentication workflow further protects sensitive information while adapting to the fast pace of healthcare delivery. In traditional settings, passwords, key cards, or physical badges provide basic authentication, but these can be easily compromised, misplaced, or shared. Multimodal approaches such as biometric verification—using fingerprint scanning or facial recognition—layer additional security without significant disruption to procedures. For instance, in a trauma bay, badge swipe combined with fingerprint scanning ensures that only authorized staff can access medication dispensing machines, blocking attempts with stolen or borrowed credentials. Similarly, multifactor authentication for remote access, such as combining a password with a one-time SMS code, sharply limits the impact of lost credentials and thwarts phishing attacks. By calibrating these controls so that they integrate smoothly into clinical workflows—such as fast-location biometric readers at the point of device use—hospitals support critical activity while holding the line on data protection (Stapleton, 2025).

Continuous and systematic access reviews are essential to enforce the principle of least privilege and maintain the integrity of hospital security systems. Scheduling periodic audits—monthly or quarterly—ensures that permissions stay tightly aligned with personnel changes. Automated tools and dashboards help administrators quickly scan for accounts that have accumulated excessive access rights or retain permissions long after a role transfer or offboarding. For example, a system may automatically flag a user account belonging to a former contractor or highlight a dormant account with elevated privileges, enabling rapid removal or review (Gagliardi, 2024). Compliance teams benefit from consolidated reports and event logs detailing who accessed what and when—all features supported by modern role-based access control (RBAC) tools (Stapleton, 2025). These audits make it practical to enforce compliance with legal and industry standards, closing gaps before they can be exploited.

To address privilege escalation—a scenario where users gradually amass permissions they no longer need—organizations must implement real-time monitoring solutions. Technical controls for alerting can be configured to detect sudden changes in access patterns, such as a nurse attempting to access administrative billing tools, triggering an automated notification to the security team. Role-based rules allow RBAC systems to automatically update permissions when employees change departments, so unused privileges are promptly revoked. Automated offboarding processes guarantee that all temporary or contract users lose their access as soon as their engagement ends, reducing the window for possible misuse (Gagliardi, 2024).

Supporting these measures, comprehensive documentation and staff training programs instill respect for access protocols. Training tailored to each department clarifies how access rights benefit both patient privacy and institutional security. Staff are shown how to request permissions through approved workflows and recognize suspicious system behavior, which empowers them to prevent insider threats and careless breaches (Gagliardi, 2024). Real-world scenarios drive home the risk: for example, showing how generic or shared accounts can bypass audit controls, making incident investigations nearly impossible. As a result, security culture grows stronger alongside technical improvements.

Integrating these approaches into daily operations forms the backbone for resilient access control in healthcare settings. The process of mapping permissions, enforcing strong authentication, and scheduling regular audits demonstrates the necessity of robust frameworks. Just as these systems restrict the flow of sensitive data, organizations must also set clear expectations for device use. For example, a hospital may establish that only specific staff roles may connect to certain medical devices, restricting access based on device criticality and patient safety requirements. Such policies—when enforced through the technical controls already in place—close the loop between user access and secure device management, rounding out a comprehensive approach to clinical cybersecurity.

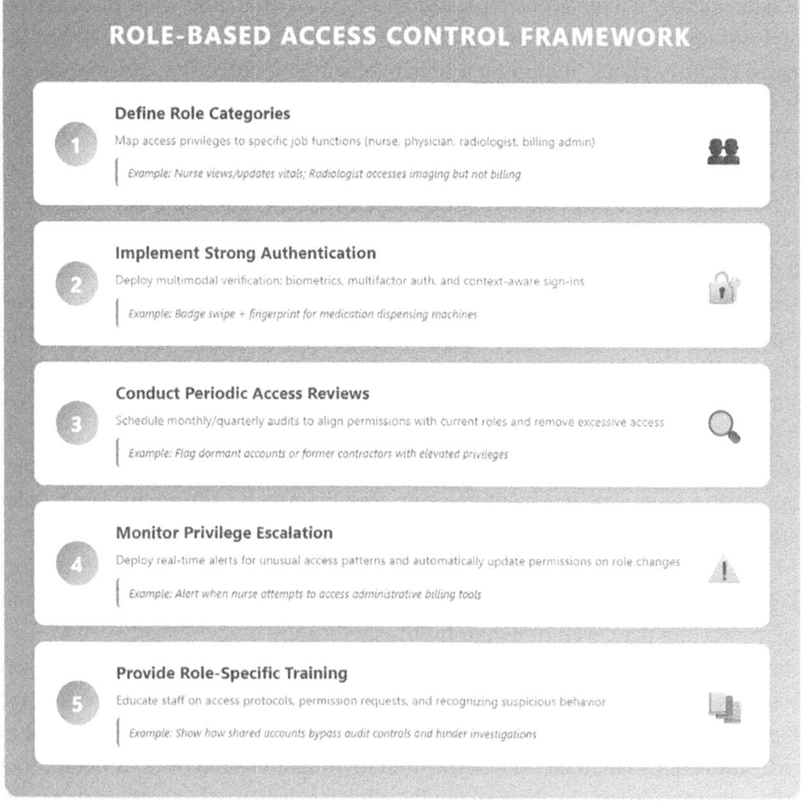

Developing Device Utilization Policies for Cyber Safety

Limiting user access on medical devices forms only one part of a comprehensive defense strategy. Access management must work in tandem with granular, scenario-specific device utilization policies that clearly define what actions are acceptable. Without specifying permissible functions and strictly enforcing them, clinical environments remain vulnerable to both human error and deliberate misuse, even where user privileges have been restricted. For example, clinical staff typically require access to patient monitoring systems for recording vitals and administering care, but should be explicitly prohibited from downloading external applications, connecting peripheral devices, or executing administrative commands. Such restrictions are often codified within device governance policies, mandating that installations and configuration changes are performed only by authorized IT professionals or biomedical engineers following an approved protocol (*Staffing for Doctors | Hire an Experienced Medical Virtual Assistant*, 2025). This reduces the risk of inadvertent exposure to malware and ensures that only vetted software operates within the patient care ecosystem, aligning technical controls directly with clinical workflow.

The design and enforcement of device utilization policies must also consider the risks posed by shared or loaner medical devices. In many hospitals, infusion pumps, imaging workstations, and physiological monitors circulate frequently between departments and patient rooms. Clear identification and tracking—via check-in and check-out records, user authentication logs, and assigned device ownership—are essential for both operational continuity and incident response. When a clinician signs out a loaner electrocardiogram (ECG) machine, their identity, department, and intended duration of use must be recorded. Before redeployment, policies should require a comprehensive security check, including a review of software integrity, anti-malware scans, and a full audit of configuration settings (*Staffing for Doctors | Hire an Experienced Medical Virtual Assistant*, 2025). Each device must be cleaned, reset to its baseline configuration, and, if necessary, reimaged to its approved state, minimizing the persistence of unapproved code or residual patient data. These practices safeguard against malware that could slip through from one user to another, as well as augment investigative capabilities in the event of a breach, allowing organizations to reconstruct device history and user actions as part of the forensic process (Hub, 2025).

Automated lockout and session timeout functionalities constitute a foundational technical control for safeguarding medical devices at the point of use. The rationale centers on minimizing the window of opportunity for unauthorized access, particularly in busy clinical spaces where users might step away unexpectedly or rotate between patient care tasks. When idle, devices such as medication dispensing carts or radiology workstations should automatically lock after a brief, configurable period—often 5–15 minutes depending on workflow sensitivity. For example, imaging workstations storing diagnostic scans enforce auto-lock features, ensuring that if a user leaves their station unattended, patient health information remains protected from view, whether by accidental passersby or deliberate intruders (*Staffing for Doctors | Hire an Experienced Medical Virtual Assistant*, 2025). Technical controls like these enforce policy compliance where human vigilance alone might waver, protecting data even in dynamic clinical settings.

To maintain robust resilience, device use policies require dynamic updating in response to security incidents and the continual emergence of new threats. Incident-driven reviews serve to close loopholes discovered through real-world breaches or near misses, translating practical lessons rapidly into policy language. When incident investigations uncover vulnerabilities, such as persistent malware introduced via physical media, effective device policies adapt immediately—for instance, by instituting an embargo on USB port usage, updating endpoint protection requirements, or enhancing device auditing following root-cause analysis (Hub, 2025). This iterative process of policy refinement ensures that each new threat informs a more sophisticated, targeted defense, reinforcing a proactive rather than reactive posture.

Medical device life cycle management integrates these policy interventions from the earliest deployment through end-of-life. Security auditing must be continuous, with periodic checks of user access logs, device configuration baselines, and patch management records. Device inventories are regularly compared against access records to spot anomalies and address orphaned assets, further reducing risk (Hub, 2025). At every step, documentation creates an evidence trail that supports compliance and enforcement, while also facilitating postincident forensics.

The success of these technical and administrative controls ultimately relies on staff engagement and understanding. Clinicians and technical personnel must not only be aware of device policies but also comprehend the rationale behind each safeguard, from lockouts to restricted application use. Compliance depends on their consistent, correct behavior as much as it does on embedded technology or automated enforcement. Technical defenses only fulfill their function when end users contribute to the secure operation of clinical workflows. This interdependence between policy, technology, and human action underscores the need for targeted, ongoing workforce training and frequent simulation exercises, setting the stage for a deeper exploration of workforce preparedness in cyber safety.

Training Healthcare Staff in Cyber Hygiene Practices

Healthcare organizations must align cybersecurity education with the distinct tasks, device interactions, and risk exposure that define clinical, technical, and support roles. Standardized, undifferentiated training fails to account for the unique workflows and technology access patterns found across hospital departments, creating measurable vulnerabilities where role-specific knowledge lapses are exploited. A nurse responsible for medication administration at shared workstations faces different threats than a biomedical engineer updating ventilator firmware or an IT specialist overseeing network segmentation, necessitating targeted pedagogical approaches. Customizing modules by experience level is equally indispensable: new staff frequently require foundational content on device usage policies, password management, access protocols, and lockout procedures, while experienced personnel must confront phishing attack variations, social engineering scenarios, and the layered impact of clinical workflow deviations on organizational security controls (*Cybersecurity Training | Cybersecurity Awareness Training*, 2025; Prümmer et al., 2024).

Appropriate assignment of foundational modules to new hires focuses on clearly articulating approved device use, the rationale behind tracking of shared endpoints, and how technical controls like workstation lockouts operationalize compliance policies throughout clinical routines

(*Cybersecurity Training | Cybersecurity Awareness Training*, 2025). This base layer ensures universal understanding before advancing to higher-risk activities or nuanced compliance obligations. For example, onboarding clinical staff might receive a digital training segment that visually tracks the login/logout sequence on medical carts, demonstrates session timeouts, and spells out the steps for reporting an unattended or compromised device. In contrast, an experienced ICU nurse benefits more from engaging in interactive, scenario-driven exercises such as responding to a phishing email masquerading as an urgent systems patch notification or handling real-world breach drills requiring rapid device isolation and secure patient data handoff. These advanced modules are constructed around the nuanced realities of clinical operations, ensuring that training always maps to likely, high-impact risk events within each staff profile (Prümmer et al., 2024).

IT administrators receive additional modules covering the technical back end of approved usage policies, including configuring group policy lockout mechanisms, monitoring device authentication patterns, and supporting audit trails that enable compliance officers to detect anomalies in device access or credential misuse. Compliance teams are provided tailored, metrics-driven content focused on interpreting audit data, investigating incident logs, and coordinating with technical and clinical stakeholders to evolve device governance in response to emerging threats. This role-based differentiation operationalizes the abstract recommendation to "clearly define and enforce approved device use," directly connecting it to daily behaviors, governance processes, and technical systems, thus supporting sustainable policy implementation (*Cybersecurity Training | Cybersecurity Awareness Training*, 2025; Prümmer et al., 2024).

For implementation, follow the method called Creating Role-Based Healthcare Cybersecurity Training Materials:

1) Define each staff role's daily workflow, device interactions, and access privileges.
   Example: Nurses interact with electronic health record terminals throughout their shifts; they access patient charts, dispense medications, and sometimes print reports from shared workstations. Nurses need training that focuses on session management and secure handling of temporary device logins. This step lays the foundation for customized content precisely matched to job reality.
2) Enumerate the primary cybersecurity threats encountered by each role, linked to their workflow.
   Example: Nurses experience frequent phishing attempts via hospital email, risk of exposure from unattended terminals, and the potential for misdirected printouts of electronic protected health information (ePHI). Effective training builds from actual incident data to address these specific risks, avoiding attention drift that plagues generic modules.
3) Develop modular training content tailored to address identified risks for each group.
   Example: Nurses participate in short video walkthroughs demonstrating how to recognize suspicious login screens, use quick logoff shortcuts, and respond if they notice their session left open on a shared device. Biomedical engineers complete simulated firmware update protocols that prompt secure USB handling, digital signature checks, and reporting requirements for suspicious device behavior. Each example immerses staff in realistic, task-specific security decision points to reinforce correct responses.
4) Assign modules based on experience and responsibility level, using baseline content for new hires and layered, scenario-based modules for advanced staff.
   Example: Entry-level medical assistants complete a mandatory basic cybersecurity course introducing password policies and device usage tracking. Senior clinical personnel unlock elective advanced modules simulating multistage spear phishing, device quarantine during

ransomware exposure, and auditing privileges during emergent clinical crises. This graduated approach ensures training maintains relevance throughout the employee life cycle.

5) Integrate simulation exercises that mirror clinical reality and are deliberately aligned with role expectations.
Example: Phishing simulation emails target nurses during shift changes, testing their ability to discern urgent, but fraudulent, IT requests when busy. Device loss drills assign support staff the scenario of locating, remotely locking, and reporting a missing barcode scanner with ePHI exposure potential. Social engineering role-playing exercises are reserved for clinical leadership and IT teams, simulating adversaries seeking unauthorized access under the pretext of equipment repair. Each drill conditions staff to respond reflexively, converting theoretical training into practical, time-sensitive action (*Cybersecurity Training | Cybersecurity Awareness Training*, 2025; Prümmer et al., 2024).

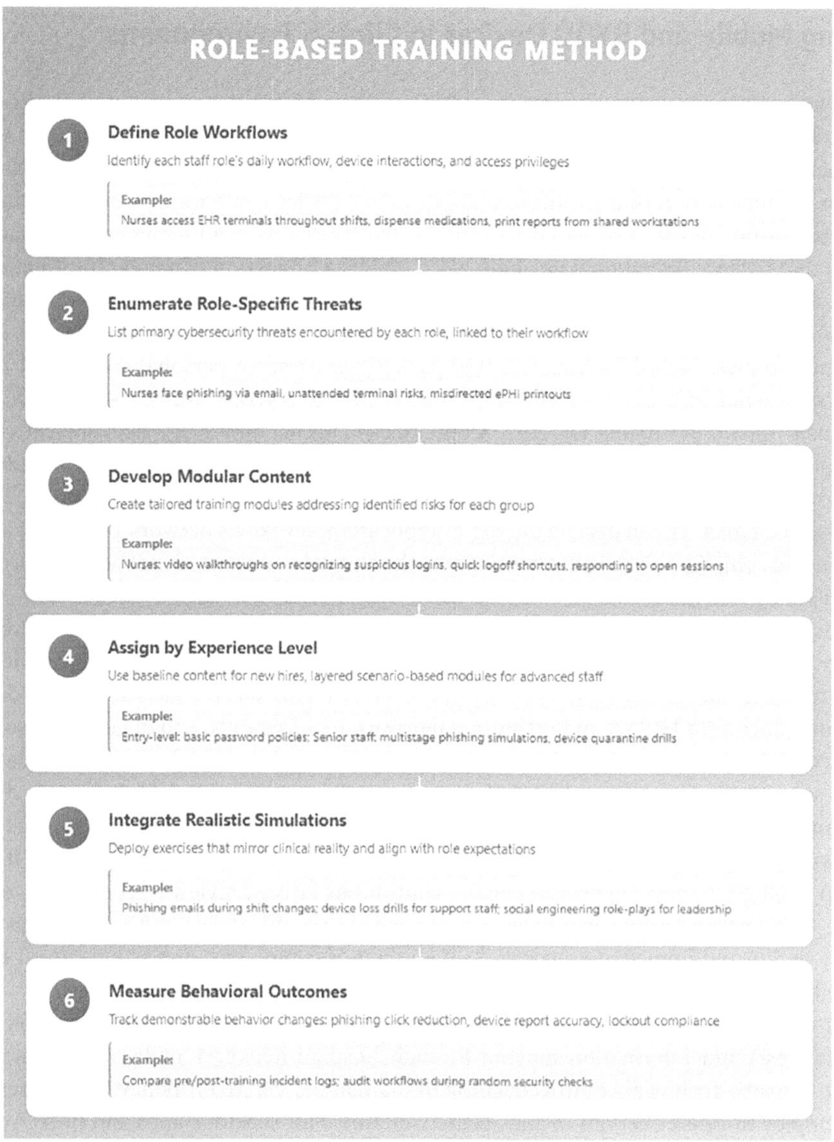

After module completion, collect outcome-based metrics centered on demonstrable behavior change and security efficacy. Use technical means to track reduction in reported phishing clicks among clinical staff by comparing pretraining and posttraining incident logs. Monitor accuracy and timeliness of lost device reports submitted through the help desk system, correlating detection improvements with exercise participation. Audit a sample of user workflows for proper application of lockout policies during random, unscheduled security checks. These focused measurements reveal tangible gaps or progress, providing actionable insights for iterative revisions of training content, resource allocation, and policy alignment (*Cybersecurity Training | Cybersecurity Awareness Training*, 2025; Prümmer et al., 2024).

Finally, reinforce that the next set of training will address mobile and bring your own device (BYOD) security with dedicated modules comprising device management, app vetting, and handling of mobile data transfers, ensuring comprehensive coverage of all vectors in future cycles.

Securing Mobile and BYOD Devices in Clinical Environments

Centralized mobile device management (MDM) systems have become essential for healthcare organizations challenged by the fluctuating, complex presence of both personal and hospital-issued devices across clinical operations. Security risks arise not only from lost or misused equipment but also from the installation of unauthorized applications, device configuration drift, and improper data transmissions during fast-paced workflows. Healthcare IT teams must implement robust solutions that balance operational flexibility with steadfast protection of patient information.

MDM offers IT administrators visibility and granular control into the status, configuration, and activity of every registered device. For example, a lost clinician's smartphone containing PHI can be located, remotely locked, or wiped, either partially to preserve personal content on a BYOD device or fully when institutional ownership is clear (*A Modern MDM Strategy: What Mobile Device Management Should Be Capable Of*, 2025; Stone, 2024). This not only averts data breaches but also meets regulatory expectations for timely data protection action. Similarly, real-time monitoring of endpoints allows for immediate identification when an unapproved app appears on a clinician's hospital-issued tablet. IT can disable the rogue application, revoke its network privileges, and trigger automated compliance reporting, minimizing the clinical and reputational fallout of accidental or willful policy violations.

A proactive application vetting and whitelisting process is fundamental in this environment. New apps proposed for workflow improvement—a secure note-taking tool, for example—must undergo technical and policy reviews before approval. IT teams assess code security, vendor reliability, compliance with HIPAA and institutional privacy requirements, and ongoing support. Only after successful vetting is an app added to the hospital's whitelist; MDM restricts installations so that any attempt to sideload external or consumer-grade apps fails by default (Stone, 2024). By maintaining this restrictive environment, the risk of malware or data-leaking software entering the clinical workflow drops significantly. Still, static whitelists offer only temporary safety: if a malicious app manages to pass superficial checks, continuous layered reviews—combining technical scanning, user feedback, and vendor alerts—are needed to swiftly detect and remove new threats. Quarterly audits and immediate revocation of any out-of-compliance app reduce dwell time for vulnerabilities and provide IT managers and compliance officers with critical agility.

Robust encryption sits at the core of safe clinical workflows, both for data stored on mobile devices (at rest) and information moving through hospital networks or between remote endpoints (in transit). Technical controls begin with mandating, via MDM policy, that all registered devices employ full-disk encryption and secure containers for sensitive apps and files (*A Modern MDM Strategy: What Mobile Device Management Should Be Capable Of*, 2025). Integration with hospital directory services and multifactor authentication helps enforce that only authorized devices—those meeting encryption and compliance standards—can connect to sensitive resources or upload patient data. A noncompliant or jailbroken phone, for instance, will be denied network authentication and access to clinical databases until issues are remediated.

For data in transit, secure network protocols such as enforced Virtual Private Networks (VPNs), secure sockets layer/transport layer security (SSL/TLS) encryption for application traffic, and dedicated enterprise file transfer tools safeguard PHI moving between hospital, remote, and cloud destinations (Stone, 2024). These encrypted pathways ensure that even if data packets are intercepted, they remain unreadable to outside attackers. Continuous network-level monitoring detects and flags abnormal data flows, such as a sudden spike in outbound files from a single device, which may indicate attempted exfiltration or a malware infection. Two-step authentication for uploading datasets—such as one-time codes delivered via a secondary app or token—further limits accidental or deliberate leaks to only those with proven identity and need-to-know status.

Technology alone cannot close every gap: staff training in authorized device usage and data transfer protocols remains critical. For instance, a nurse emailing ECG images from a personal account, rather than through the approved medical records app, can quickly expose confidential data. The corrective response should not just be postevent notification, but real-time blocking by MDM of unauthorized email clients, linked with prompts that direct users to safe, institution-sanctioned workflows. Best practice dictates continuous reinforcement of do's and don'ts, built directly into workflow tools and endpoint alerts, so technical controls and end-user behavior stay aligned.

Consider the requirements faced by clinical engineers charged with integrating new devices—scanning for persistent malware, reviewing firmware for update vulnerabilities, and ensuring that wireless connectivity abides by encryption and authentication mandates are now standard steps prior to device deployment. Meanwhile, compliance officers must monitor both logs and active device states, ready to demonstrate regulatory fidelity to auditors with on-demand reports, all made possible by the centralized management and layered policy enforcement that MDM and modern endpoint controls provide (*A Modern MDM Strategy: What Mobile Device Management Should Be Capable Of*, 2025; Stone, 2024). In this way, technical, operational, and policy measures unite to defend both the integrity of patient care and the confidentiality of the data that fuels it.

Final Thoughts

Having examined the intricate relationship between clinical workflows and cybersecurity, it is clear that effective protection of healthcare environments depends on integrating security measures seamlessly into everyday processes. By aligning access controls, device policies, and training programs with the real-world demands and behaviors of clinical staff, organizations can reduce vulnerabilities without hindering care delivery. Moving forward, healthcare institutions must prioritize adaptive, role-based strategies that evolve alongside emerging threats and technological advances, fostering collaboration among clinicians, engineers, and security professionals. This approach will enable resilient systems capable of safeguarding patient data and medical devices while supporting the dynamic nature of modern clinical operations.

References

A Modern MDM Strategy: What Mobile Device Management Should Be Capable Of. (2025). Airdroid.com. https://www.airdroid.com/mdm/mobile-device-management-should-be-able-to/.

Brass, I., Straw, I., Mkwashi, A., Charles, I., Soares Mesquita, A., Steer, C. (2023). *HEALTH AND WELLBEING Report: Emerging Digital Technologies in Patient Care Report Emerging Digital Technologies in Patient Care: Dealing with connected, intelligent medical device vulnerabilities and failures in the healthcare sector PETRAS in partnership with BSI HEALTH AND WELLBEING Report: Emerging Digital Technologies in Patient Care Acknowledgements.* https://doi.org/10.5281/zenodo.8011139.

Cybersecurity Training | Cybersecurity Awareness Training. (2025). SparxITSolutions. https://www.sparxitsolutions.com/cybersecurity/training.

Fausett, C. M., Keebler, J. R. (2022, September). *Shift Happens: Human Factors Considerations for Handoffs in Cyber-Security Operations Centers.* Proceedings of the Human Factors and Ergonomics Society Annual Meeting. https://doi.org/10.1177/1071181322661131.

Gagliardi, T. (2024). *How to Perform User Access Reviews—Drata.* Drata.com. https://drata.com/blog/user-access-review.

Hub, C. (2025, April 20). *Compliance Updates 2025: Strategic Guidance for Organizations.* Compliance Hub Wiki. https://www.compliancehub.wiki/healthcare-cybersecurity-in-2025-new-regulations-transforming-the-industry/.

Prümmer, J., van Steen, T., van den Berg, B. (2024, January). *A Systematic Review of Current Cybersecurity Training Methods.* Computers Security. https://doi.org/10.1016/j.cose.2023.103585.

Staffing For Doctors | Hire An Experienced Medical Virtual Assistant. (2025). Staffingfordoctors.com. https://staffingfordoctors.com/hipaa-compliance-policies-and-procedures.

Stapleton, S. (2025, June 27). *Role-Based Access Control (RBAC)—A Comprehensive Guide*. Pathlock. https://pathlock.com/blog/role-based-access-control-rbac/.

Stone, M. (2024, January 6). *Mobile Device Management: Securing the modern workplace*. Levelblue.com. https://levelblue.com/blogs/security-essentials/mobile-device-management-securing-the-modern-workplace.

12

The Human Firewall: Addressing Insider Threats

Insider threats pose a unique challenge in healthcare, where trusted individuals with legitimate access can either intentionally or unintentionally compromise sensitive patient information and critical systems. Studies show that a significant portion of cybersecurity incidents in healthcare stems from internal actors, highlighting the vulnerability embedded within organizational operations. Unlike external attacks, insider risks are difficult to detect because they often involve authorized actions performed under normal workflows, making traditional defenses less effective. Understanding the factors that lead to these threats and developing tailored strategies is essential for protecting patient data, ensuring system integrity, and maintaining compliance with regulatory standards. This chapter examines the complexities surrounding insider threats in healthcare environments and the approaches necessary to identify, prevent, and respond to these multifaceted risks.

Profiling the Insider Threat Landscape in Healthcare

Insider threats in healthcare organizations are uniquely complex because they involve both individuals who intentionally harm the institution and those who introduce risk through mistakes or neglect. Deliberate insider threats arise when someone with legitimate access abuses that privilege, driven by motives such as financial gain, personal grievances, or even broader ideological objectives. For example, a clinician might exfiltrate patient health records to sell to criminal groups because the black market highly values such data (Sari et al., 2022). This direct intent to harm places deliberate threats in a category that requires a strong forensic understanding and response capability. In contrast, unintentional insiders pose risks through negligence, such as a nurse failing to log out of a terminal or accidentally emailing sensitive data to the wrong recipient. These actions may lack malicious intent but can cause significant security events (Sari et al., 2022).

Understanding the mindset and motivations behind both types of insiders is critical because organizations must tailor their prevention and response strategies accordingly. Malicious insiders are often careful to disguise harmful acts within normal workflows, exploiting authorized access to compromise confidentiality or integrity. Responding to such threats demands robust access management and in-depth monitoring. Meanwhile, unintentional threats often stem from workflow pressures, low digital literacy, or an underestimation of security requirements. In one plausible scenario, a busy clinician might share login credentials out of necessity or convenience, introducing an exploitable weakness into the system (Sari et al., 2022).

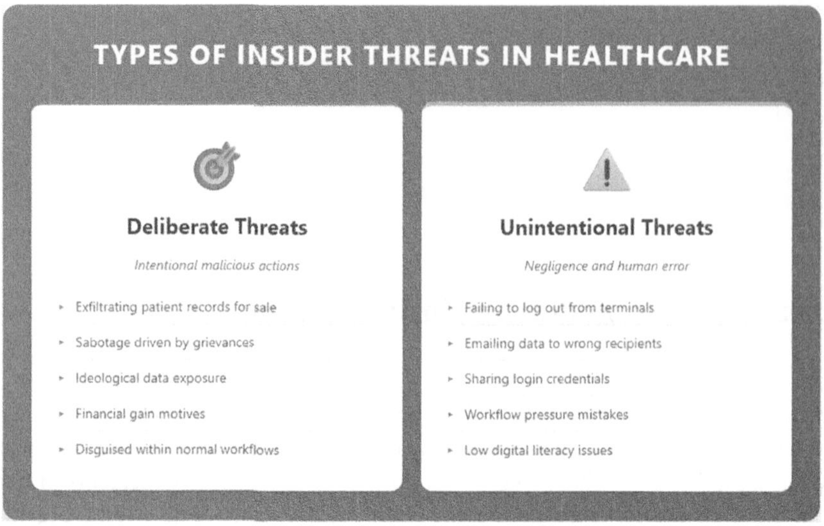

Healthcare environments have a wide variety of personnel with direct and indirect system access. While clinicians such as doctors and nurses represent key user groups, the scope of insider risks extends to IT staff, administrative employees, temporary contract workers, vendors, and even cleaning personnel. Each of these roles brings a different level of access and method of interfacing with health information systems. Doctors and nurses typically interact with electronic health records (EHRs) multiple times each day. IT staff administer the back-end infrastructure, sometimes holding privileged access across the network. Administrative staff may handle patient data, insurance documentation, and communications containing sensitive details.

Nonclinical groups, often overlooked, create distinct risk scenarios. For example, cleaning staff might be present in offices or server rooms where unmonitored computers or paper records are temporarily left accessible. A contractor tasked with maintaining medical devices could exploit trusted access to install unauthorized tools or collect confidential configuration files. In a real-world illustration, a cleaning contractor could inadvertently view patient information displayed on an unattended workstation screen, while an IT vendor might use default credentials to access and reconfigure network-connected diagnostic equipment. This broad array of access types dramatically expands the attack surface, making healthcare organizations particularly challenged in securing every entry point (Sari et al., 2022).

Motivational triggers driving these insider threats are just as diverse. Financial pressure is a recurring theme, given the high value of health information on the black market. A technician struggling with debt might consider offers to leak device firmware or patient information in exchange for payment. Dissatisfaction with hospital administration or unresolved workplace conflict could push an otherwise loyal employee to commit acts of sabotage or data destruction. In rare situations, political or ideological imperatives influence behavior, such as an employee exposing patient records to discredit the organization or advance an activist agenda. Stress and burnout are common personal triggers, increasing the likelihood of accidental breaches due to fatigue or cognitive overload. Even momentary lapses in focus can result in significant exposure if sensitive documents are left unattended or if a rushed staff member misroutes an email containing confidential data (Sari et al., 2022).

Detecting these varied forms of insider threats is complicated by the fact that insiders operate under the guise of legitimate users. Unlike external hackers, insiders exploit valid credentials, established workflows, and trusted relationships. Conventional perimeter defense tools are often ineffective because insiders act within their approved boundaries. For instance, an employee who methodically collects additional patient records over several months may not trip automated alerts, since each access event appears routine on its own. The nuance of insider threat detection lies in identifying subtle behavioral patterns, such as unusual working hours, accessing data unrelated to job functions, or consistent failure to follow documented procedures.

Continuous monitoring and behavioral analytics are necessary to reveal hidden anomalies that distinguish harmful insider activities from benign workflow variations. Implementing technical controls like multifactor authentication and privilege segregation helps reduce risk, but no single tool can fully replace close vigilance and user behavior analysis. For example, a hospital might notice an increase in system access after hours, leading to the discovery of an employee exfiltrating files under the cover of normal backup tasks. Recognizing these patterns requires a combination of automated systems, staff awareness, and proactive investigation (Sari et al., 2022).

Given the complexity of insider threats in healthcare, organizations need detection and mitigation strategies designed specifically for these environments, considering their unique risks, multifaceted access controls, and the central role of human behavior. This need sets the stage for exploring next-generation approaches tailored to the medical context.

Detection Strategies for Insider Threats

Healthcare organizations face a complex challenge in detecting and reducing insider threats, given the dynamic, high-stakes environments in which they operate. The protection of patient data, the integrity of systems, and the safety of care delivery all depend on proactive strategies that extend well beyond traditional security measures. Addressing insider threats in healthcare requires leveraging a combination of technical controls, policy frameworks, and supportive human factors.

This section explores practical, actionable methods that help security and compliance professionals anticipate and mitigate the risks posed by trusted insiders.

Behavioral Analytics in Clinical Environments

One of the most powerful tools available comes in the form of behavioral analytics. By deploying advanced analytics and machine learning, organizations can establish baselines for what constitutes normal user activity. These systems collect and analyze vast volumes of log data—tracking details such as login timings, access frequency, device usage, and file interactions. When a deviation occurs, like a radiologist accessing hundreds of patient records late at night when they typically only interact with a handful each shift, the system raises an alert for further investigation (*Account Obfuscation | Anti-Forensics*, 2025).

Healthcare presents unique monitoring challenges because authorized personnel often require broad access to perform their roles effectively. An emergency department nurse, for example, may need to access multiple patient charts during a crisis, but if a machine-learning algorithm detects access rates drastically higher than even crisis-mode patterns, it can flag this as anomalous. Behavioral analytics allows for real-time and context-aware risk identification that adapts as patterns shift in response to changing workflows, holidays, or unusual staffing scenarios. This powerful detection capability not only recognizes potential malicious activity but also helps organizations distinguish between legitimate operational surges and threats that warrant immediate review (*Account Obfuscation | Anti-Forensics*, 2025).

Least Privileged Monitoring and Periodic Access Reviews

The principle of least privilege stands as a foundational tenet in healthcare cybersecurity. Every user—whether a device engineer, IT administrator, or temporary contract nurse—should only have access rights strictly needed to perform assigned responsibilities. Least privilege monitoring enforces this principle by systematically reviewing, restricting, and auditing privileged accounts. Continuous oversight helps ensure access is appropriate for clinical and operational functions without introducing unnecessary risk (*Account Misuse | Anti-Forensics*, 2025).

Automated tools and access management platforms support hospitals in conducting regular privilege audits. These reviews can highlight when a staff member retains elevated rights after a role change or when temporary staff possess permissions beyond their assignment duration. For example, if access reviews reveal a device engineer with writable access to medical device configuration after project completion, removal of unneeded privileges is immediately triggered. Monitoring also tracks the use of administrative accounts, identifying if device or system accounts are accessed from unusual locations or in ways inconsistent with expected job functions. Prompt investigation of such discrepancies helps spot early-stage insider activities, preventing escalation or system misuse (*Account Misuse | Anti-Forensics*, 2025).

Practical Application in Healthcare

A nurse returning from leave who retains elevated emergency access, despite resuming a standard role, poses a risk. Through automated periodic reviews, healthcare IT staff can detect this and adjust access, ensuring that only essential privileges are active at any time. Privileged access management (PAM) systems and just-in-time access controls also limit prolonged, broad access, minimizing the window for exploitation (*Account Misuse | Anti-Forensics*, 2025).

Whistleblower Channels and Anonymous Reporting

Beyond technology, human vigilance is crucial for detecting and addressing insider risks. Establishing secure, anonymous reporting mechanisms gives all staff a voice, regardless of their position or level of authority. For a whistleblower channel to be effective, it must guarantee confidentiality and offer protection from retaliation. Communicative efforts across leadership and human resources, as well as visible policies around anonymity, are critical for building trust and encouraging participation (*Account Misuse | Anti-Forensics*, 2025).

Consider a hospital where an employee observes unusual database access patterns by a colleague. Using an internal reporting platform, the individual can safely notify the cybersecurity or compliance team, prompting a targeted review. Such a tip-off may lead to an internal audit that uncovers the early stages of a data exfiltration attempt, ultimately preventing a serious breach. These human-driven reports complement technical detection by providing contextual awareness or alerting teams to threats that automated systems might miss.

Postaccess Audit Trails and Forensic Investigation

When suspicious or unauthorized activities are detected, robust audit trails provide essential evidence. Tamper-proof logging solutions track every access, modification, and transfer within systems holding protected health information. These logs not only underpin forensic investigations but also fulfill regulatory obligations by demonstrating due diligence (*Account Misuse | Anti-Forensics*, 2025).

Suppose an audit reveals an abnormal surge in data transfers to a storage device by a clinician who has recently resigned. By reviewing the audit trail, the organization pinpoints the timeframe, systems, and specific files involved in the transfer and determines whether proper protocols were followed during offboarding. The evidence supports internal reviews, potential disciplinary action, and external reporting if required by law.

Organizational efforts to counter insider threats must balance automation with culture—demonstrating that success comes from both rigorous technological controls and a commitment to empowering individuals at every level. This holistic approach bridges technical solutions and ongoing investment in culture and leadership, forming strong, resilient defenses against even the most complex insider risks.

Building a Culture of Trust and Vigilance

Robust technical strategies, including real-time monitoring and anomaly detection, are the bedrock of insider threat identification in healthcare organizations. These tools flag suspicious behavior—such as unusual access to patient records or uncharacteristic data transfers. Yet, reliance on detection tools alone can leave critical gaps. No monitoring algorithm can fully compensate for weaknesses in policy or organizational culture. A truly secure environment takes shape by integrating technical vigilance with deliberate risk-based policy and a company-wide commitment to security values, especially due to healthcare's complex ecosystem of sensitive data and staff diversity (*Mastering Cybersecurity Policy: Analysis and Implementation*, n.d.).

A comprehensive defense framework begins with policy. Policies crafted from meticulous risk assessments, tailored to reflect the organization's context, act as a map for action and decision making. Defining roles and responsibilities for every staff level reduces ambiguity: for example, when a cardiac nurse, IT technician, or facilities manager knows exactly what constitutes policy-compliant actions and what escalation routes to follow in the face of anomalies (*Mastering Cybersecurity Policy: Analysis and Implementation*, n.d.). Effective policies must also fit the organization's risk appetite; a research hospital might accept more risk to allow innovative device integration, while a community clinic may focus on strict record lockdown with minimal data sharing.

But policy itself would remain abstract without leadership buy-in and visible executive commitment. When boards, chiefs, or directors publicly prioritize security, expectations are set across the hierarchy. Setting up regular executive briefings on insider threat trends, requiring department heads to share how they have implemented new policies, or ordering quarterly reviews of response plans demonstrates that security is a lived value, not a tick-box requirement. For instance, an IT manager might brief the CEO on how a simulated phishing attempt played out, offering both transparency and a model for other teams. Clearly communicated executive mandates create accountability, freeing managers to request necessary resources for ongoing policy refinement, staff training, and security improvements (*Mastering Cybersecurity Policy: Analysis and Implementation*, n.d.).

Another essential organizational pillar is open communication. Employees at every level need to feel safe admitting mistakes and voicing concerns, especially since punishments for honest errors can lead to reluctance in sharing potentially vital information. Open-door policies could mean providing an anonymous reporting channel for staff to escalate suspicious activity or security failures without fear of reprisal. Alternatively, hosting regular security Q&A forums where anyone can present concerns or ask questions in a supportive environment helps catch overlooked vulnerabilities. In some healthcare settings, debriefing sessions follow minor security lapses; staff are encouraged to explain precisely what happened and what they learned, cultivating a learning culture rather than a punitive one. These open sharing opportunities reinforce that early identification and escalation of issues—no matter how small—protect the organization as a whole (*Video Modules Podcasts Managing Human Risk Requires More than Just Awareness Training*, 2019).

Simulation exercises and security drills are practical methods to breathe life into policy. Simulations such as tabletop exercises mimic real-world breach scenarios, like an unauthorized access incident involving sensitive imaging data. Staff from diverse departments are tasked with responding—logging actions, alerting leaders, and containing the event. The debriefing after these exercises reveals both process and performance strengths and uncovers hidden vulnerabilities, leading to meaningful improvements. In a hospital context, these might take place every few months, with records kept of challenges encountered and corrective actions taken. Well-executed drills familiarize employees with their responsibilities, ensuring responses to actual incidents are swift, coordinated, and effective (*Mastering Cybersecurity Policy: Analysis and Implementation*, n.d.).

Beyond policy enforcement and scenario readiness, motivational tools harness the power of positive reinforcement. Recognizing and rewarding employees who demonstrate vigilance inspires others to follow suit. For example, a formal recognition program might spotlight a staff member who intervened after noticing unusual login attempts or who reported a potential data mishandling incident. Some organizations display leaderboards showing departments with top security behaviors, creating friendly competition. Even modest rewards, like a thank-you note or a small bonus for consistent incident reporting, can reinforce the message that security is everyone's job. These incentives keep staff engaged, raise morale, and embed the ethos of constant attentiveness through collective effort (*Mastering Cybersecurity Policy: Analysis and Implementation*, n.d.).

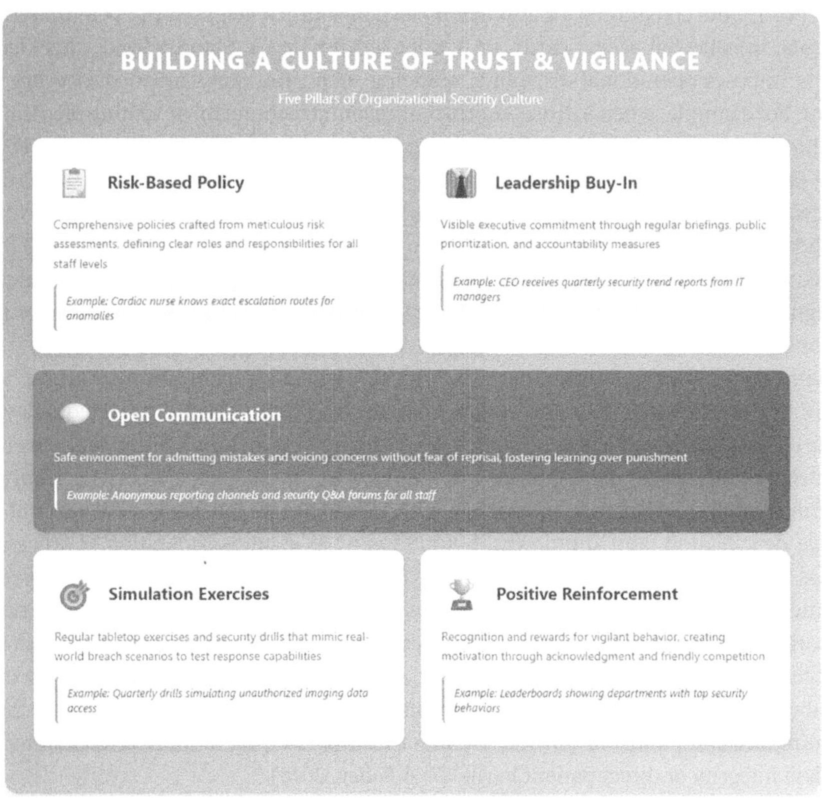

Cultivating a security-aware workforce, supported by comprehensive policy, leadership engagement, open communication, practical drills, and positive incentives, creates multilayered protection far stronger than any single technical control. These human-centric practices foster an alert, adaptive culture better positioned to spot and mitigate insider risks before they result in harm. As healthcare organizations fine-tune this approach, they must not lose sight of ongoing technical and procedural rigor, particularly when managing system access and the secure removal of credentials for departing personnel—a focus to be explored in depth in the next section (*Mastering Cybersecurity Policy: Analysis and Implementation*, n.d.; *Video Modules Podcasts Managing Human Risk Requires More than Just Awareness Training*, 2019).

Access Reviews and Termination Protocols

When managing sensitive healthcare environments, strict user access control forms the foundation of protecting both patient safety and data privacy. Hospitals, clinics, and research labs grant a myriad of users—ranging from surgeons and IT staff to contract technicians and temporary researchers—differing levels of access to medical devices, EHRs, and critical administrative consoles. Any lapse in managing these permissions, particularly during the offboarding of staff, poses substantial risks to network security and regulatory compliance. Meticulous offboarding processes help ensure that no departing staff member or vendor inadvertently retains digital or physical entry points, closing common avenues for insider threats and accidental data exposure (*UVA Policy*, 2016).

Human error, time constraints, or ambiguous responsibilities heighten the potential for forgotten accounts, uncollected access cards, or overlooked privileges. Standardized, checklist-driven offboarding imposes operational discipline, reducing such gaps. Steps must cover every possible touchpoint. For example, when a nurse, contract ultrasound technician, or visiting medical student departs, the checklist should include deactivating EHR credentials, retrieving hospital ID badges, collecting keys to medication storage, reclaiming organization-owned mobile devices, disabling VPN access, and reviewing permissions on software medical devices and administrative portals. Every offboarding action requires documentation and sign-off by responsible parties. This structured process instills both accountability and traceability, supporting compliance with internal and federal regulations, and establishing an audit-ready trail to withstand external reviews (*Leaving USC—Employee Gateway*, 2025).

Even with rigorous procedures, timing is critical. Immediate revocation of access upon an employee's exit ensures that data and systems remain uncompromised. Protracted delays—even those as brief as a few hours—permit malicious actors or well-meaning but misinformed former staff to retrieve confidential data, tamper with equipment, or create backdoors for future access. Automation tools play a vital role, integrating with HR platforms to trigger swift account suspension, badge deactivation, and shutdown of cloud logins as soon as separation is processed. For instance, as soon as a departing IT administrator's termination is entered into the human resources system, enterprise identity management software can lock all network accounts, disable two-factor authentication tokens, deprovision privileged access, and alert system owners to confirm receipt. Should access linger, organizations face the real danger of record tampering, device misconfiguration, or sabotage of scheduling systems—for example, a dismissed technician remotely logging in to reprogram infusion pumps or download patient records, leading to undetected compromise of medical data integrity or direct patient harm (*UVA Policy*, 2016).

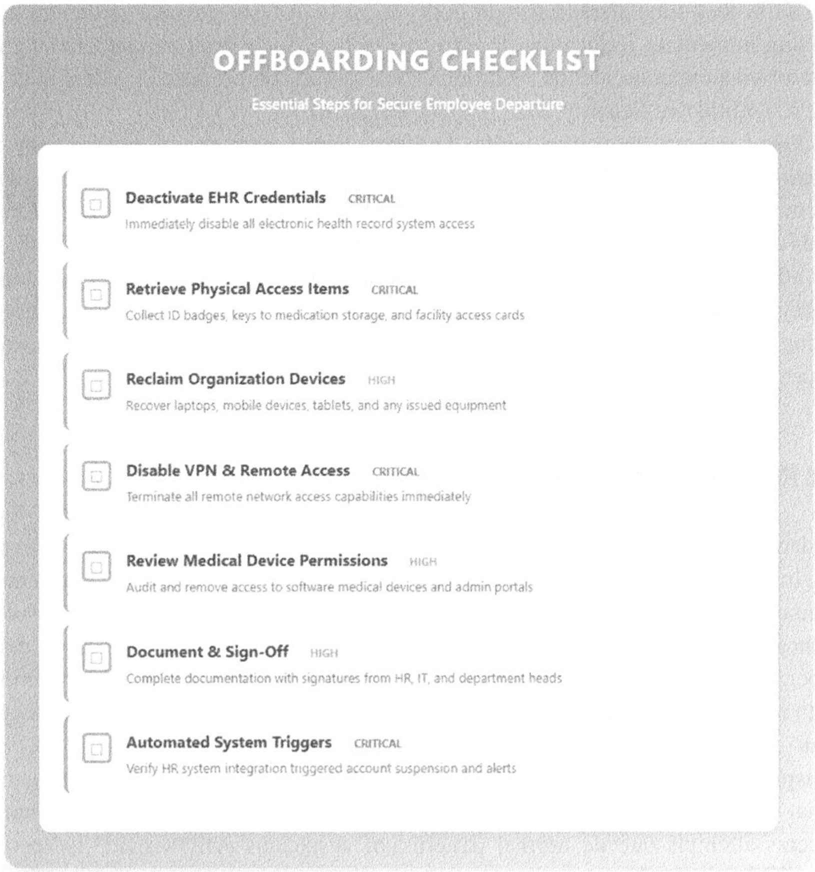

Periodic audits of user privileges form the next pillar of defensive practice. Over time, healthcare organizations accumulate dormant accounts, leftover administrator rights from legacy projects, and evolving staff assignments that blur intended role boundaries. Quarterly or biannual access reviews enable the identification of these risks by generating comprehensive lists of all user accounts, mapped to current job functions and privileges. Automated reporting systems spotlight dormant credentials, orphaned logins tied to separate users, or excessive privileges left over from group projects. Cross-team collaborations between HR, IT, and department heads are essential, so that access removal triggers, role changes, and review results are documented and acted upon promptly. For example, an audit might reveal that several research assistants who completed a drug trial project still retain access to experimental device firmware, highlighting a need for prompt corrective deprovisioning (*UVA Policy*, 2016).

Healthcare's reliance on short-term personnel, such as locum tenens physicians, traveling specialists, and contracted cleaning crews, makes scalable access policies crucial. Temporary users must have narrowly scoped access windows, strict time-bound credentials, and frequent monitoring. Policy should mandate that system access for each temp user is automatically configured to expire at the end of a contracted period, with technical controls enforcing these parameters. Monitoring software can identify out-of-band login attempts, flagging when, for instance, a contracted nurse's ID badge is used in a medical records terminal hours after their authorized shift. Real-time alerts can then notify security or IT of possible credential misuse or theft, enabling immediate verification or corrective action. Such strict parameters help mitigate risks of overprivileging or accidental retention of access that compromised insiders might exploit (*Leaving USC—Employee Gateway*, 2025; *UVA Policy*, 2016).

While diligent removal processes, audits, and time-bound permissions substantially mitigate exposure to insider threats, determined adversaries may still circumvent controls. Even in robustly managed environments, the unpredictable nature of human behavior and rapidly shifting operational needs demand that healthcare organizations supplement technical defenses with constant vigilance and adaptive incident planning. This foundational approach ensures that when gaps do arise, they are detected, investigated, and contained before evolving into significant breaches, underscoring the need for continuous monitoring and rapid incident response as the next critical defense layer.

Incident Response Involving Insiders

Checklist-driven offboarding and rigorous access management form critical foundations for preventing insider threats, but technical and procedural readiness to investigate and contain incidents is equally essential after safeguards fail. When an insider breach is suspected or confirmed in a healthcare environment, clear distinctions must be drawn between breaches that result from deliberate, malicious acts and those stemming from accidental or negligent actions. The approach to remediation, evidence collection, and future risk mitigation depends heavily on this assessment.

Upon suspicion of an insider threat, incident handlers swiftly implement investigation procedures that isolate affected accounts or devices. This often starts with flagging abnormal log entries, access attempts outside normal patterns, or suspicious activities tied to specific user IDs. For example, a nurse's account accessing high-profile patient records outside of scheduled shifts may prompt security tools to trigger an alert. At this early stage, detailed event logs are preserved according to institutional log retention policies, ensuring that evidence dating back days or weeks remains available for review in line with best practices highlighted in security guidance (*CMS Information Security & Privacy Group*, n.d.). Event correlation techniques come into play, cross-examining firewall logs, intrusion detection alerts, and application audit trails to construct a timeline of the incident and determine the extent of compromise. Ensuring synchronization of system clocks across all devices allows for precise reconstruction of sequences, which is critical both for the investigation and for any subsequent legal proceedings.

Containment proceeds in parallel with investigation to stop the breach from spreading. This may involve disabling compromised accounts, removing affected endpoints from the network, or temporarily restricting access to sensitive systems. In the case of a staff member suspected of tampering with EHRs, their access is suspended immediately while backup copies of all relevant data, audit logs, and system snapshots are secured. If a device, such as an infusion pump or diagnostic imaging workstation, is believed to have been accessed improperly from within the network, analysts take forensic images of the device's storage and relevant network traffic captures. Packet sniffers, configured with specific capture criteria, may be deployed to monitor ongoing suspicious activity, with care taken to comply with privacy requirements and minimize unnecessary data collection. Authorization is sought as required by organizational policy before activating such tools (*CMS Information Security & Privacy Group*, n.d.).

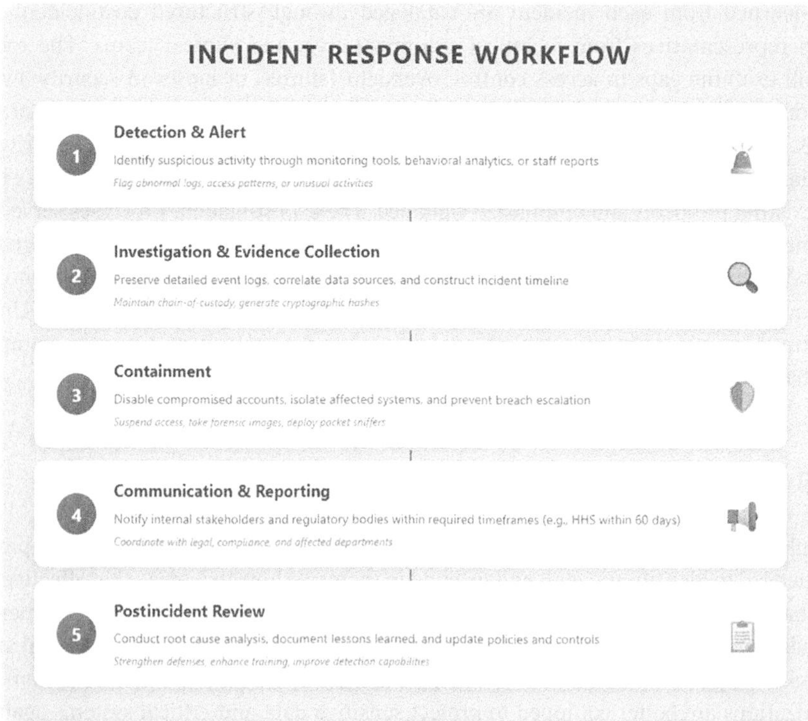

Evidence handling during the investigation follows strict chain-of-custody procedures. Every individual who collects, transports, or examines evidence documents their actions, timestamps, and the exact nature of what is handled. Audit trails capture every change or access to evidence files, and physical evidence—such as confiscated workstations or portable devices—remains sealed and signed for at every handoff. Electronic evidence is stored on secure, access-restricted servers, and cryptographic hash values are generated immediately to allow for later verification of integrity. For example, when imaging a workstation belonging to a suspected insider, a forensic specialist may log the device's serial number, photograph connection ports, and record all file hashes before transferring the image to a central evidence locker. These procedures meet both internal compliance rules and requirements set forth under legal frameworks such as the Federal Rules of Civil Discovery (University, n.d.).

Accurate, timely communication with internal management and external stakeholders is essential as the investigation unfolds. Healthcare organizations are obligated to report breaches to regulatory bodies—such as the Department of Health and Human Services—within specific timeframes, often no later than 60 days after discovery for incidents involving Personally Identifiable Information or Protected Health Information. Disclosures must include incident details, mitigation steps taken, and estimated impact scope. Coordination among security, legal, compliance, and affected clinical departments ensures that messaging is clear, disciplined, and consistent, reducing the risk of misinformation. Failures in prompt reporting or incomplete documentation have led some organizations to face regulatory penalties, lawsuits, and significant reputational loss. For example, missing key deadlines or underreporting the scope of an EHR access incident may result in substantial financial penalties under the Health Insurance Portability and Accountability Act (HIPAA) or similar laws.

Lessons learned from each incident are cataloged through structured postincident reviews, drawing in representatives from technical, administrative, and clinical teams. The root cause analysis will examine gaps in access control, oversight failures, or lapses in security awareness. Remediation plans might include updating access policies, improving log retention or analysis capabilities, deploying more advanced endpoint detection tools, and strengthening security around high-risk databases or devices. Staff training sessions, drawing directly on the incident's findings, reinforce existing protocols and emphasize vigilance. These postincident processes serve not only to repair immediate harm but also to iteratively build the organization's capacity to detect, prevent, and respond to future insider threats by systematically updating procedures and technologies to address the latest tactics and risks (*CMS Information Security & Privacy Group*, n.d.; University, n.d.). Continuous improvement, enabled by rigorous incident review and process optimization, forms the backbone of a resilient cybersecurity posture.

Bringing It All Together

Having explored the multifaceted nature of insider threats in healthcare and the comprehensive strategies needed to identify, prevent, and respond to them, we now recognize that effective defense requires a balanced integration of advanced technical controls, rigorous policy enforcement, and a culture of vigilance supported by leadership commitment. With insights into behavioral analytics, strict access management, robust incident response, and fostering open communication, healthcare organizations are better equipped to protect sensitive data and critical systems against both malicious and accidental insider risks. Moving forward, this foundation enables security professionals, device engineers, and compliance officers to implement adaptive, context-aware measures that evolve alongside emerging threats, ensuring resilient and secure healthcare environments that safeguard patient safety and privacy in an increasingly complex digital landscape.

References

Account Misuse | Anti-Forensics. (2025). Insider Threat Matrix. https://insiderthreatmatrix.org/articles/AR5/sections/AF024.

Account Obfuscation | Anti-Forensics. (2025). Insider Threat Matrix. https://insiderthreatmatrix.org/articles/AR5/sections/AF024/subsections/AF024.001.

CMS Information Security Privacy Group. (n.d.). Security.cms.gov. https://security.cms.gov/policy-guidance/risk-management-handbook-chapter-8-incident-response-ir.

Leaving USC—Employee Gateway. (2025, July 31). Employee Gateway. https://employees.usc.edu/work-life/leaving-usc/.

Mastering Cybersecurity Policy: Analysis and Implementation. (n.d.). OERTX. https://oertx.highered.texas.gov/courseware/lesson/4779/overview.

Sari, P. K., Handayani, P. W., Hidayanto, A. N., Yazid, S., Aji, R. F. (2022, December 14). *Information Security Behavior in Health Information Systems: A Review of Research Trends and Antecedent Factors.* Healthcare. https://doi.org/10.3390/healthcare10122531.

University, C. M. (n.d.). *Incident Response Plan—Information Security Office—Computing Services—Carnegie Mellon University*. Www.cmu.edu. https://www.cmu.edu/iso/governance/procedures/IRPlan.html.

UVA Policy. (2016). Virginia.edu. https://uvapolicy.virginia.edu/policy/SEC-037.

Video Modules Podcasts Managing Human Risk Requires More Than Just Awareness Training. (2019). ISACA. https://www.isaca.org/isaca-digital-videos/podcasts/managing-human-risk-requires-more-than-just-awareness-training.

13

Incident Detection, Response, and Recovery

Shortly after the start of a busy hospital shift, an alert flashed on the security operations center monitor: unusual data traffic originating from a connected infusion pump in the intensive care unit. Simultaneously, nursing staff reported intermittent malfunctions with networked heart monitor tracking critical patient vitals. Despite attempts to isolate the issue, delays in recognizing and responding to these anomalies allowed the problem to escalate, risking both patient safety and clinical workflow integrity. The incident exposed gaps in detection capabilities and communication between technical teams and frontline clinicians, underscoring the complex challenges healthcare organizations face when managing cybersecurity events involving medical devices. This situation reflects the urgency of establishing effective processes that not only identify threats swiftly but also enable coordinated action and recovery within tightly regulated clinical environments.

Medical Device Incident Detection Technologies and Practices

Anomaly detection algorithms form a cornerstone of advanced monitoring strategies for medical device security in clinical settings. These algorithms monitor the activity of medical devices such as infusion pumps and heart monitors by first learning what normal behavior looks like—establishing statistical baselines built on patterns of device communication, user access, and operational schedules. If an infusion pump suddenly starts transmitting data at times when it is normally idle or communicates with unfamiliar systems, or if a heart monitor's telemetry patterns shift in ways mismatched to patient trends, the anomaly detection system flags these deviations for scrutiny. Traditional rule-based methods rely on explicit lists of prohibited actions or known bad indicators, which means they perform reliably for well-defined, static threats. In contrast, machine learning and data-driven approaches continuously refine their understanding of what should be considered typical, adapting to evolving hospital workflows and emerging threat types. This adaptive capacity allows healthcare organizations to detect subtle, unknown forms of compromise and misuse sooner than static rules could allow, raising alerts even for attacks lacking a known signature (Doménech et al., 2024).

Moving beyond standalone monitoring, Security Information and Event Management (SIEM) systems play a vital role in healthcare-specific cybersecurity defense. SIEM platforms collect and analyze diverse security-relevant logs and events from medical devices, electronic health records systems, endpoints, and hospital network infrastructure. By bringing together diverse data points such as user authentication attempts, medical device event logs, records of configuration changes, and network traffic patterns, SIEM systems allow security teams to see connections and interdependencies that would remain invisible within siloed data streams. For example, if a hospital experiences multiple failed login attempts on ventilator control software at the same time as suspicious outbound traffic from networked infusion pumps, SIEM's correlation engines help analysts piece together the broader story—potentially identifying a lateral movement attack targeting critical patient care devices. Aggregating threat intelligence feeds into the SIEM provides additional context, enabling organizations to detect global threats as they emerge and understand their relevance within the local clinical environment (Brandao, 2025). The efficiency and speed with which SIEM systems surface incidents make them indispensable in situations where attackers' dwell times can drastically impact patient safety outcomes.

Integrating cybersecurity alerting with clinical alarm management systems addresses a critical challenge uniquely present in healthcare: the risk of overwhelming clinicians with excessive, irrelevant, or nonactionable alerts. When security tools issue too many alerts—especially false or low-priority ones—clinicians may become desensitized and overlook high-impact threats, a phenomenon closely related to alarm fatigue. Effective integration hinges on filtering and prioritizing alerts so that clinicians receive actionable warnings without unnecessary disruption. For instance, a ransomware signature detected on a radiology workstation would be escalated above routine vulnerability scans, ensuring patient care is not interrupted unless necessary. Consolidating security and clinical alerts helps ensure that only clear, high-risk events reach the critical decision-makers in a timely fashion. Organizations achieve this balance by mapping security alerts to patient safety risks and matching alert urgency to the clinical context, maintaining a seamless flow of essential information that never undermines direct patient care (Doménech et al., 2024).

The persistent challenge of false positives requires ongoing attention. Advanced monitoring systems often generate thousands of alerts daily in busy hospitals, of which only a fraction represent real threats worth investigation. To address this, security teams must constantly tune detection thresholds and refine rulesets. This process involves analyzing historical alert data, eliciting operational feedback from clinical staff, and carefully calibrating escalation protocols. For example, if a new model of insulin pump needs to regularly update its software, initial alert rules might flag this as unusual communication. Security staff can then adjust thresholds, recognizing the update pattern as legitimate, ensuring that routine operations are not obstructed by security controls. Each tuning cycle not only reduces unnecessary distractions but also strengthens clinical trust in the alerts that do arrive, so responders remain vigilant without being overstretched (Brandao, 2025).

Healthcare organizations face distinctive pressures to interpret and prioritize security threats rapidly while maintaining high standards of patient care and regulatory compliance. These practices—adaptive anomaly detection, centralized SIEM analytics, alert integration with clinical systems, and rigorous tuning of detection parameters—strengthen the ability to identify, categorize, and escalate cybersecurity events in real time. Each capability depends on a clear understanding of operational requirements and threat scenarios within clinical environments, as well as close collaboration between security engineers and frontline staff. Maintaining this vigilance ensures that organizations can recognize incidents at the earliest moment, respond with clarity, and recover while minimizing harm. Strong detection and alerting technologies create the firm foundation upon which the broader incident response process can securely and reliably build (Brandao, 2025; Doménech et al., 2024).

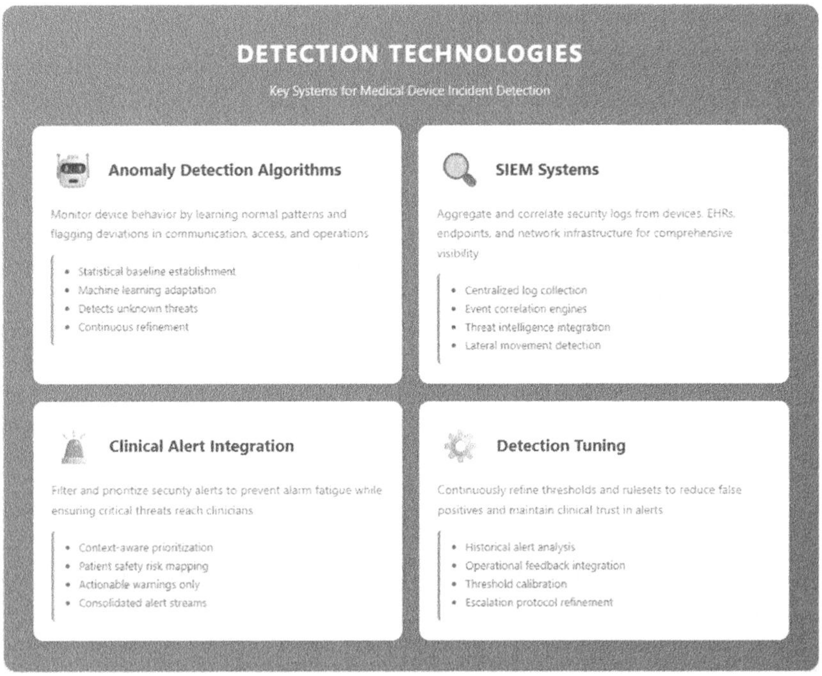

Immediate First Response Actions After a Breach

Immediate Device Isolation

The moments immediately following the detection of a cybersecurity event involving medical technology require precision and urgency. Effective containment begins with targeted isolation of affected devices to halt threat spread and safeguard patient care. Physical methods, such as disconnecting network cables or removing devices from power sources, protect both the compromised technology and the broader clinical environment from malicious activity. Logical separation techniques, including segmentation of network traffic using virtual local area networks (VLANs) or applying emergency firewall rules, prove essential when disconnecting the device is not feasible. For example, an infusion pump flagged with suspicious behavior must be severed from both wired and wireless connectivity, and its operational status verified. Segregating that device, either physically by removing it from the care area or logically through network controls, significantly curtails attacker access and prevents further propagation across interconnected electronic medical record systems or other critical assets (*How to Build a Data Breach Incident Response Plan | Censinet*, 2025; Williams & Woodward, 2015).

Applying immediate isolation protocols also requires suspending affected user credentials, blocking suspicious network traffic, and disabling automated transfers. For instance, upon detection of unusual data transmissions from a connected ventilator, security personnel should revoke remote access permissions and halt all data transfers linked to the device. These steps serve not only to contain a potential breach but also to protect real-time patient data and ensure other clinical systems remain functional.

Internal Communication and Coordination

Response effectiveness hinges on swift, well-orchestrated communication between technical and clinical teams. After isolating compromised devices, it is critical to notify the designated incident response team as well as affected clinical departments. Information security staff require clear preestablished communication channels and escalation paths, ensuring essential stakeholders—such as clinical engineering, IT, infection control, and patient safety officers—have situational awareness. Assigning responsibility for notifications drives clarity under stressful conditions.

Defined roles and rapid information sharing enable prompt assessments of clinical impact. Take, for example, a networked drug infusion system. When a compromise is suspected, IT staff and nursing leadership must immediately collaborate to decide if safe device replacement is possible or if alternative administration methods are needed. Clinicians with firsthand knowledge of device operation and patient dependency are indispensable for ensuring safe transitions and maintaining uninterrupted care (Williams & Woodward, 2015). Without such coordination, containment efforts risk inadvertently interrupting life-sustaining therapies or delaying urgent interventions.

Initial Evidence Collection and Chain of Custody

Accurate, timely collection of digital evidence lays the foundation for effective downstream investigation and regulatory compliance. The incident response team must preserve system logs, device configuration files, volatile memory states, and forensic images of affected devices before shutting down or altering systems. For example, when a multifunction patient monitor exhibits abnormal network activity, forensic imaging of device storage and collection of access logs become the first priority. Capturing this data at the earliest juncture prevents evidence loss and supports incident reconstruction.

Maintaining an unbroken chain of custody is critical. Detailed records of who collected, handled, and transferred evidence—as well as exact timestamps—reinforce evidentiary integrity (*How to Build a Data Breach Incident Response Plan | Censinet*, 2025). Storage of forensic copies must adhere to organizational and regulatory requirements, such as encrypted archives for logs and tamper-evident containers for removable storage media. Any system actions, such as enabling or disabling device functions during the incident, also require documentation. Failing to meticulously track these activities jeopardizes the validity of internal investigations and potential legal proceedings.

Preserving Patient Safety During Containment

While technical containment progresses, maintaining patient safety supersedes all other objectives. Incident response plans must enable clinical care teams to quickly access alternative devices or enact emergency workflows when technology is taken offline. Protocols for immediate device replacement, rerouting therapy through manual methods, or transitioning patients to unaffected equipment prevent adverse clinical outcomes. For example, if a ventilator in an intensive care unit requires isolation, rapid deployment of backup units and prompt notification of respiratory therapists ensure continuous respiratory support.

Collaboration between clinical and IT security teams proves vital. IT should provide clinicians with up-to-date system status, while clinical leadership must identify and prioritize critical devices for rapid assessment. This joint decision-making guarantees that response activities, such as device shutdowns or treatment rerouting, do not compromise patient well-being (Williams & Woodward, 2015). Incident scenarios may require dynamic staffing adjustments, expedited supply retrieval, or even temporary changes to standard workflow, underscoring the importance of practiced incident response exercises and scenario-based planning within healthcare organizations.

Setting Up for Effective Postcontainment Investigation

The discipline established during early response is essential for the investigative phase. With containment achieved and evidence secured, incident response teams shift attention to root cause analysis. This requires extensive review of system logs, network traffic, and user access records, using retained forensic artifacts to reconstruct the breach timeline and understand attack vectors. Coordination with regulatory authorities becomes critical, as incident documentation and evidence must satisfy compliance obligations and inform postincident recommendations (*How to Build a Data Breach Incident Response Plan | Censinet*, 2025).

Ongoing review of response protocols and communication strategies follows every incident. Lessons learned refine both technical controls and workflows, providing a continual feedback loop that adapts to new cybersecurity threats while supporting clinical resilience (Williams & Woodward, 2015).

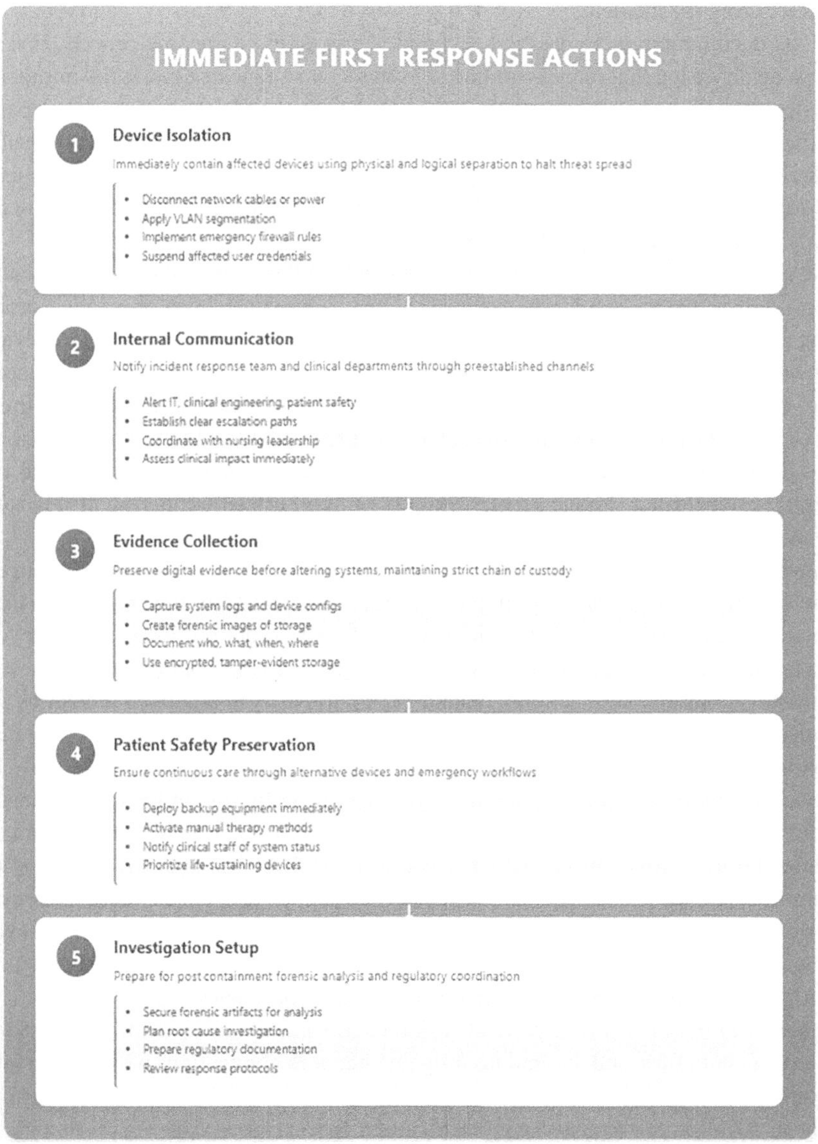

Forensic Investigation Methodologies for Compromised Devices

Healthcare cybersecurity professionals who respond to compromised medical devices know that once containment actions are complete, the focus must shift to securing digital evidence critical to a reliable forensic investigation. Organizations working in highly regulated environments like healthcare need to image device storage immediately. Exact digital images captured from internal drives, removable media, or volatile memory ensure the preservation of all original data, including hidden or deleted files that might contain traces of malicious activity. Teams accomplish this with forensic software capable of sector-by-sector duplication, producing checksums or hash values for both source and image. Imaging must occur with the aid of write-blockers, hardware or software tools that guarantee evidence remains unmodified throughout the acquisition. For example, a forensics analyst in a hospital will connect a medical infusion pump's onboard storage to a hardware write-blocker when duplicating its contents, preserving the state of files and logs as encountered during the incident.

Meticulous documentation forms the backbone of any credible imaging process. Investigators maintain an evidence log that records unique identifiers for storage devices, serial numbers, dates, times, the personnel present, imaging tool settings, hash values, and details of every step performed. In a healthcare setting, documentation might also include details particular to medical devices such as device model, firmware version, and its location within the facility. These practices not only reinforce investigative credibility but also provide traceability if questions arise about evidence authenticity or potential contamination (Team, 2023).

Effective chain of custody protocols determine whether collected digital evidence will withstand scrutiny in court or before regulatory authorities. Teams must establish an explicit protocol for controlling access to all physical and digital artifacts relevant to the compromised device. Each piece of evidence receives a label with a unique reference number. Investigators record every instance of evidence transfer in a chain of custody log that documents the names of individuals accessing or receiving the evidence, reasons for access, precise date and time, and conditions under which the handover took place (Team, 2023). Secure evidence storage, using locked cabinets, digital vaults, or dedicated storage servers, prevents unauthorized tampering or environmental degradation.

For example, a security analyst collecting a forensic image of a wireless heart monitor creates an entry in the chain of custody log stating that at 15:04 on June 14th, the analyst transferred the device's memory card to the digital forensics team lead. The log details that the memory card was stored in a tamper-evident bag in the facility's secure evidence room. Teams maintain these records until all investigations and legal actions conclude, often for years in regulated industries.

Root cause analysis stands as a core objective of any forensic investigation. Security teams use a combination of static and dynamic analysis to reconstruct the incident and identify the specific method of compromise. Investigators analyze system configuration files, access logs, kernel or firmware logs, and security event management outputs. They correlate timeline data to determine when abnormal activity began, which accounts or applications were involved, and whether local or remote exploitation occurred (katharina.kiener-manu, 2019). For example, a forensics team may discover, through network log analysis, that an attacker exploited an outdated transport layer security (TLS) library in a drug-dispensing system. By tracing logins and command executions, the team identifies lateral movement into patient information repositories.

A thorough examination involves correlating forensic artifacts such as modified registry entries, new or altered system files, and anomalous outgoing network connections with known indicators of compromise. Teams further leverage memory analysis to discover signs of malware residing exclusively in RAM or to reconstruct a crash caused by malicious code. Each finding receives documentation with clear linkage to the evidence chain, preserving analytical integrity (Team, 2023). The investigation's results inform both the immediate remediation—such as closing exploited vulnerabilities and removing persistence mechanisms—and longer-term risk mitigation strategies like access restriction, network segmentation, or mandatory security updates for medical device firmware.

Coordinating with law enforcement is often essential, particularly when attacks have legal, regulatory, or public safety implications. Healthcare cybersecurity professionals must understand which incidents require legal reporting, the jurisdictional boundaries involved, and which procedures must be observed to maintain admissibility of evidence. Internal teams should engage with law enforcement agencies through clearly designated points of contact while maintaining mutual awareness of ongoing investigative actions. For example, when a hospital uncovers a ransomware incident affecting medical imaging devices, its internal cyber incident response team prepares sanitized copies of digital evidence for law enforcement while retaining originals under internal control and separate chain of custody.

Protocols specific to healthcare environments govern how investigators respond to subpoenaed evidence, ensuring no disruption to patient care and the integrity of clinical systems. Coordinated interviews, joint analysis of technical findings, and collaboration on risk-mitigation measures allow internal and external stakeholders to advance their investigations simultaneously and avoid operational conflicts. Each action benefits from documentation and legal consultation, preserving regulatory compliance (katharina.kiener-manu, 2019; Team, 2023).

Through these best practices—imaging evidence with precision, maintaining unbroken chain of custody, conducting thorough root cause analysis, and coordinating with legal authorities—healthcare organizations can maximize investigative reliability. This measured approach empowers teams to validate systems, rebuild device integrity, and restore safe patient care environments with confidence.

Protocols for Recovery and Safe Restoration to Operation

Once forensic analysis and evidence preservation are complete in a healthcare breach scenario, restoring compromised medical devices and systems is crucial for patient safety and clinical continuity. This segment of the incident response cycle serves as the bridge from initial containment to the safe resumption of patient care and system-dependent clinical tasks. It demands rigorous adherence to technical, regulatory, and operational standards to ensure that no remnants of malicious activity persist and that healthcare delivery remains uncompromised.

Clean Media Restorations

Thorough and secure media restoration removes malware persistence mechanisms and eliminates the potential for subsequent reinfection. Secure wiping, using industry benchmarks such as NIST SP 800-88 for data sanitization, begins the process (Abdul, 2025). All storage media in compromised devices must undergo this process, whether using built-in manufacturer utilities or validated third-party tools. For example, after a ransomware incident on a networked infusion pump, technical teams use verified wiping tools to sanitize internal storage before reinstalling firmware. It is critical that reinstallation employs firmware or software images sourced from manufacturer-verified repositories or cryptographically signed—never from generic or outdated backups. Audit trails from these procedures provide evidence of compliance for devices such as patient monitors, MRI workstations, and laboratory analyzers, supporting regulatory and legal requirements through recorded step-by-step restoration logs (*Best Practices—Your Personal Device | University Information Security and Data Privacy*, 2025).

This approach also ensures traceability, particularly for regulated environments, by documenting the serial number of devices restored, checksums of the software or firmware installed, and signatures from responsible personnel. Steps extend to network configurations and application installations; none must be restored from potentially tarnished sources, restricting restoration to vetted assets only. This disciplined restoration not only blocks malware recurrence but also fortifies the organization's defenses in line with regulatory expectations.

Validation Procedures

After technical restoration, comprehensive validation procedures confirm that devices are safe for operational use. Before rejoining a network or resuming patient interaction, every device requires multilayered verification. This involves technical assessments—such as vulnerability scans, penetration testing, and checksum validation of system files—as well as functional testing that replicates common clinical workflows. For example, a CT scanner recently rebuilt after compromise undergoes simulated scans with test phantoms and security-focused stress tests to ensure imaging integrity and operational reliability.

Cybersecurity controls embedded in the device are examined to confirm proper configuration. Credential management, audit logging, and anti-malware mechanisms are checked for exploitation or misconfigurations. These validation activities align with standards from agencies like the FDA's premarket and postmarket cybersecurity guidance as well as IEC 80001 for IT network risk management in health settings. Full documentation of test methodologies, outcomes, retest requirements, and approval signatures builds the assurance necessary for clinical stakeholders and external auditors (Abdul, 2025).

This validation is not merely technical; operational validation involving nursing and clinical engineering staff ensures that devices function correctly in active clinical contexts. By involving frontline users in this portion of the process, organizations can better confirm that the devices perform seamlessly in complex, high-stakes environments.

Coordinating Downtime Minimization

Restoration must be executed with careful attention to clinical disruption. Effective minimization of downtime involves synchronizing technical repairs with the realities of patient care, ensuring that essential workflows continue with as little interruption as possible. Establishing detailed contingency plans, such as temporary deployment of backup ventilators in the ICU, allows critical care delivery to persist while primary systems are restored. Coordinators create clear notification protocols that communicate to clinical staff the exact timelines, affected systems, and restoration progress.

Prioritization is key; restoration workflows must first target high-impact, life-sustaining equipment. For example, a cardiac monitoring system would receive expedited restoration compared to a noncritical administrative workstation. Transparent communication with stakeholders—clinical teams, device users, and information security officers—facilitates a coordinated response, allowing resumption of services according to medical necessity and workflow priorities. These approaches uphold the dual imperatives of patient safety and organizational resilience (*Best Practices—Your Personal Device | University Information Security and Data Privacy*, 2025).

Postincident Recertification

Following restoration and successful validation, postincident recertification is essential before devices reenter service. Regulatory authorities and internal corporate compliance teams require artifacts such as logs documenting media wipes, validation test reports, and updated configuration records. Engaging with regulators early in the process, especially when high-profile or life-supporting systems are involved, expedites recertification and return to clinical use. Standard pathways for recertification, such as those enforced by FDA or regional agencies, involve reviewing technical remediation steps, verifying documentation, and in some cases conducting independent safety checks.

Notification to government agencies strengthens institutional credibility. Credible, well-documented remediation shortens investigative timelines, improves organizational transparency, and assures patients and staff of ongoing safety.

By tightly integrating these protocols—clean media restoration, validation, downtime management, and recertification—healthcare organizations establish foundations for both immediate recovery and future resilience. Documenting every step offers a rich source of institutional learning, feeding robust postincident review sessions focused on iterative improvement and proactive risk mitigation (Abdul, 2025; *Best Practices—Your Personal Device | University Information Security and Data Privacy*, 2025).

Postincident Reviews and Organizational Learning

In the high-stakes landscape of healthcare cybersecurity, organizations must use every incident as a foundational learning opportunity, embedding robust mechanisms for extracting actionable insights. The aftermath of a security event demands not only skilled technical restoration but also a systematic, blame-free examination of the factors underlying the incident. Blame-free postmortem reviews enable more honest disclosure of mistakes and system weaknesses. By consciously fostering psychological safety, organizations ensure staff at every level feel secure in reporting errors, unusual device behaviors, or procedural lapses without fear of censure. For instance, if a hospital detects unauthorized access to infusion pumps, responders who know they will not face punishment are far more likely to share step-by-step details that led to the event, highlight overlooked security controls, and suggest alternative actions. Psychological safety thus becomes a catalyst for comprehensive root cause discovery and supports the development of a durable organizational memory (Jervis, 2022).

Encouraging curiosity and objective analysis within review processes requires explicit managerial support and structured formats for discussion. It's essential to establish ground rules that shift the emphasis from "who did it" to "what conditions made the event possible." Adopting regular, standardized postincident reviews—such as monthly security case conferences or cross-functional debriefs—reinforces the organization's commitment to continuous, collective learning rather than punitive postmortems. Participants from IT, clinical engineering, compliance, and administration bring diverse contexts to each event, dissecting technical, human, and policy contributors to the breach. When staff see that leaders prioritize deep understanding over discipline, reporting rates and disclosure honesty measurably improve, which directly benefits the quality of actionable intelligence harvested from every incident (Jervis, 2022).

Comprehensive root cause documentation underpins effective remediation. Successful organizations track not just the immediate triggers of incidents, but all contributing environmental, technical, and procedural factors. Accurate, detailed incident reports chronicle event timelines, the sequence and rationale of technical actions, device configuration details, anomaly logs, and even communication patterns between stakeholders. Specialized vulnerability registries aggregate device-specific flaws and procedural lapses discovered during investigations, allowing IT teams and medical device manufacturers to prioritize remediation based on incident frequency and potential impact. For example, after analyzing an incident where a networked ventilator was compromised due to outdated firmware and insufficient network segmentation, the investigation record would describe both the technical exploit and the organizational policy that delayed critical updates. Such precise documentation not only fulfills internal learning objectives but also serves as an auditable trail for regulatory compliance and legal reviews, which is particularly vital in highly regulated healthcare settings (*Common ca Database by the Linux Foundation*, 2023).

Maintaining a living archive of incident records enables rapid root cause comparison when new anomalies arise. Analysts can trace patterns across incident histories, accelerating forensic investigations and informing preemptive actions. A digital repository of redacted incident summaries, including remediation measures and residual risks, aids ongoing security posture assessment and future training efforts. Well-structured process maps translate lessons learned from complex postmortems into visual workflows, exposing chokepoints and systemic weaknesses for targeted improvement.

Continuous improvement demands that postmortem discoveries trigger systematic revision of organizational controls, technical architectures, and policy frameworks. Thorough incident analysis reveals which processes require optimization. For instance, if a phishing attack enabled credential compromise due to weak authentication protocols and infrequent staff training, documentation should lead to deploying multifactor authentication and integrating adaptive user training modules. Technical control enhancements might include expanding device network segmentation and automating patch deployment cycles to mitigate vulnerabilities exposed by the event. Organizations should review and update response playbooks after each postmortem, aligning procedures with the latest threat intelligence and organizational experience. Deploying iterative updates based on actual incidents integrates resilience into daily practice and cultivates a security culture of vigilant adaptation (*Common ca Database by the Linux Foundation*, 2023; Jervis, 2022).

Knowledge sharing within and beyond the organization transforms isolated incident experience into sector-wide resilience. Internal dissemination can employ security bulletins, regular cross-departmental briefings, and digital platforms that centralize access to incident summaries, remediation guides, and updated policies. For example, after a significant device compromise, a hospital might conduct a postincident symposium that brings together biomedical technicians, software engineers, and clinical staff to walk through the technical and procedural factors in detail. Publishing anonymized incident learnings via secure channels with industry partners or contributing to regulatory incident reporting programs extends the feedback loop, enabling other healthcare providers and device manufacturers to preempt similar risks (*Common ca Database by the Linux Foundation*, 2023).

Each phase in the feedback journey—from blame-free investigation to comprehensive documentation, policy revision, and knowledge sharing—reinforces the healthcare organization's readiness for future threats. By embedding these practices into daily operations, incident response becomes not just reactive, but a powerful engine for continuous improvement across the entire digital health ecosystem (*Common ca Database by the Linux Foundation*, 2023; Jervis, 2022).

Lessons Learned

Having examined the essential phases of cybersecurity incident management for medical devices—from detection and immediate containment to forensic investigation, recovery, and organizational learning—it is clear that a comprehensive, coordinated approach is vital for protecting patient safety and complying with regulatory requirements. By integrating adaptive monitoring technologies, structured response protocols, meticulous evidence handling, and rigorous recovery practices, healthcare organizations can not only address incidents effectively but also build resilience against future threats. Moving forward, embedding continuous improvement through systematic postincident analysis and knowledge sharing will enable healthcare professionals, engineers, and compliance officers to anticipate evolving risks and strengthen security measures proactively. This proactive posture ensures that medical device cybersecurity evolves alongside technological advances and emerging threats, safeguarding both clinical operations and patient well-being in an increasingly connected healthcare environment.

References

Abdul, S. (2025, April 24). *What to do After You have been Hacked*. Cybersecurity Guide. https://cybersecurityguide.org/resources/cyber-incident-guide/.

Best Practices - Your Personal Device | University Information Security and Data Privacy. (2025). Harvard.edu. https://privsec.harvard.edu/best-practices-your-personal-device.

Brandao, P. R. (2025, June 23). *Exploring the Role of Artificial Intelligence in Detecting Advanced Persistent Threats*. Computers; MDPI AG. https://doi.org/10.3390/computers14070245.

Common CA Database by the Linux Foundation. (2023). Ccadb.org. https://www.ccadb.org/cas/incident-report.

Doménech, J., Martin-Faus, I. V., Mhiri, S., Pegueroles, J. (2024, November). *Ensuring Patient Safety in IoMT: A Systematic Literature Review of Behavior-based Intrusion Detection Systems*. Internet of Things. https://doi.org/10.1016/j.iot.2024.101420.

How to Build a Data Breach Incident Response Plan | Censinet. (2025). Censinet.com. https://www.censinet.com/perspectives/how-to-build-a-data-breach-incident-response-plan.

Jervis, R. (2022, January 13). *Why Postmortems Fail*. Proceedings of the National Academy of Sciences of the United States of America; National Academy of Sciences. https://doi.org/10.1073/pnas.2116638118.

katharina.kiener-manu. (2019, March). *Cybercrime Module 4 Key Issues: Standards and Best Practices for Digital Forensics*. Www.unodc.org. https://www.unodc.org/e4j/en/cybercrime/module-4/key-issues/standards-and-best-practices-for-digital-forensics.html.

Team, T. G. (2023, June 26). *A Guide to Digital Forensics and Incident Response (DFIR)*. Graylog. https://graylog.org/post/a-guide-to-dfir/.

Williams, P., Woodward, A. (2015, July). *Cybersecurity Vulnerabilities in Medical Devices: A Complex Environment and Multifaceted Problem*. Medical Devices: Evidence and Research. https://doi.org/10.2147/mder.s50048.

14

Third-Party Vendor and Supply-Chain Security

How can healthcare organizations ensure that the medical devices they rely on remain secure throughout their complex supply chains? What processes are essential for evaluating and managing third-party vendors to prevent vulnerabilities from entering clinical environments? How can collaboration between providers and suppliers enhance resilience against evolving cybersecurity threats? This chapter addresses these critical questions by examining the strategies, controls, and partnerships necessary to protect medical devices from risks associated with vendor relationships and supply-chain management. It considers how systematic assessment, secure update procedures, logistical safeguards, ongoing cooperation, and coordinated incident response contribute to maintaining device integrity and patient safety in an increasingly interconnected healthcare landscape.

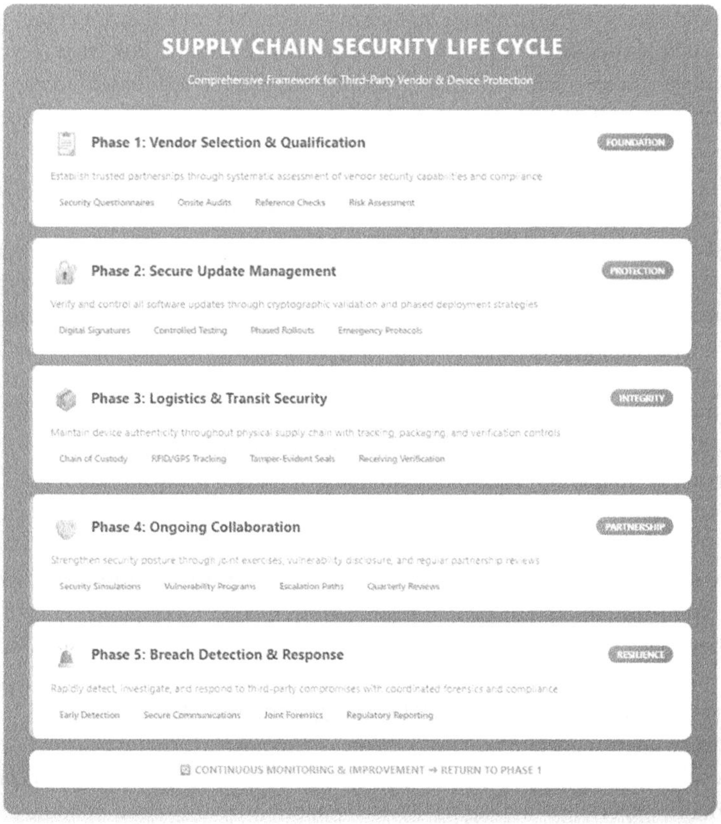

Systematic Vendor Assessment and Qualification

Selecting reliable vendors is crucial for protecting patient safety and data within healthcare organizations. Vendors involved in medical device supply chains play a pivotal role in determining the security posture of both the devices and the information ecosystems they connect to. Assessing a vendor's qualifications and readiness to adhere to rigorous security standards is foundational not only for regulatory compliance but also for responding to the evolving threat landscape that targets healthcare environments (Ostermann et al., 2025). Each method used for vendor assessment contributes unique insights that illuminate both strengths and vulnerabilities in a vendor's security approach.

Developing robust security questionnaires equips organizations to gauge whether prospective or current suppliers meet the specific technical and legal standards demanded by clinical and hospital settings. A systematic approach starts with identifying the precise obligations imposed by healthcare regulations such as the Health Insurance Portability and Accountability Act (HIPAA) or the FDA's cybersecurity guidelines and then translating these into focused questions. Questions should probe for evidence that fundamental security controls are in place. For instance, one might ask vendors to detail their protocols for encrypting patient data at rest and in transit or to provide documentation of device software update procedures meeting FDA requirements. Another question might require vendors to explain their response timelines and escalation procedures for detected security incidents relevant to hospital networks.

Adopting a standardized questionnaire format streamlines evaluation, allowing reviewers to compare responses objectively. Consistency is especially important, given the diverse range of suppliers contributing software, firmware, or hardware components; it ensures that assessment is systematic and comprehensively addresses potential risk domains. For example, standardized sections could cover topics such as authentication mechanisms, role-based access controls, and third-party code audits, reflecting issues flagged as gaps in existing regulatory guidance (Ostermann et al., 2025). By requiring clear documentation and evidence, organizations not only filter out vendors lacking in cybersecurity readiness but also encourage supply-chain transparency that strengthens compliance across the board.

On-site audits complement questionnaire-based assessments by enabling reviewers to observe and verify the implementation of declared security measures. These audits often expose the practical reality behind paper-based promises. A security team might conduct a walkthrough of a manufacturer's facility, checking physical access points, evaluating the use of biometric controls, and confirming the existence of segregated storage areas for sensitive device components. Inspections of production lines reveal whether tamper-evident procedures have been embedded and whether access to firmware their staff handle is adequately logged and monitored.

Direct observation frequently uncovers issues not apparent in documentation. For example, despite written assurances, an audit could reveal that legacy workstations with weak authentication protocols are still connected to production networks. Such findings highlight the necessity of moving beyond written declarations to direct, evidence-based risk evaluation. Regular audits build accountability into the relationship, giving organizations greater leverage to demand remediation of vulnerabilities that could compromise patient safety or data confidentiality. These measures echo research on real-world incidents showing that gaps in practical implementation, rather than policy alone, trigger many severe breaches (Ostermann et al., 2025).

Reference checks offer important perspectives on a vendor's performance history, reliability, and response to security challenges. Healthcare organizations benefit from engaging peer institutions that have firsthand experience managing devices supplied by the same vendor, particularly in critical situations. For example, a reference check might uncover an instance in which a vendor delayed the release of a security patch after a significant vulnerability surfaced, leaving deployed medical devices susceptible to attacks across multiple hospitals. Such revelations not only speak to the vendor's technical effectiveness but also its culture of transparency and willingness to collaborate during incidents.

This peer-sourced intelligence frequently reveals trends that may not surface in questionnaires or audits. A pattern of delayed incident notification, repeated policy violations, or a strong track record in rapid, well-managed responses influences the risk calculus for future engagements. Reference checks build a fuller picture, extending the assessment beyond the vendor's assertions to include the lived reality of their broader client base.

Sustaining high standards requires ongoing monitoring of vendor compliance after initial qualification. Tools for automated performance tracking enable the review of patch release cycles, access-control modifications, and adherence to updated security policies in near real time. Organizations can schedule recurring reviews to extract reports on security maintenance activities, then benchmark these against commitments made during onboarding. Automated alerts flag policy deviations—such as missed patch deadlines or irregular authentication events—so stakeholders can respond before minor issues escalate.

Continuous compliance review is particularly valuable as regulatory obligations, attacker tactics, and device technologies shift. By embedding feedback loops and escalation protocols, healthcare IT teams can work with vendors to resolve nonconformities swiftly, reinforcing the expectation that security vigilance does not end at contract signing (Baz et al., 2023; Ostermann et al., 2025). This persistent oversight proves critical for software updates and patching, where selecting vendors with proven records of delivering authenticated, timely updates shields organizations from emerging exploits and maintains operational resilience.

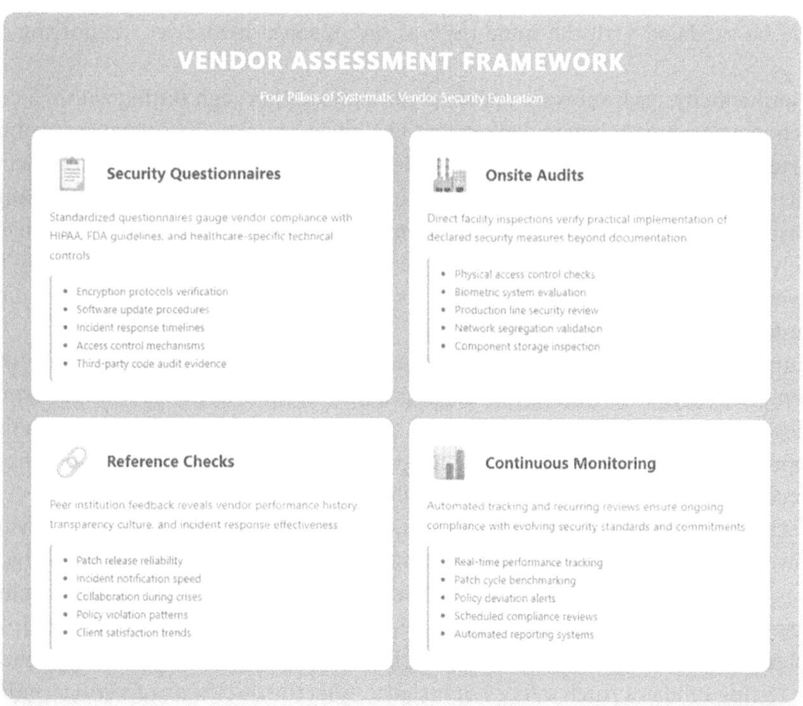

Secure Management of Updates and Patches from Vendors

A trusted supply chain depends on rigorous vendor evaluation and qualification. Selecting vendors based on their security track record forms the first line of defense, but due diligence extends to all interactions, especially software updates for medical devices. Updates from external vendors can introduce serious risks if not verified and managed correctly. Ensuring only authentic, vendor-approved updates reach connected medical devices is vital for safeguarding functionality and protecting patient safety, and this begins with the implementation of robust digital signature verification measures (House, 2025).

Digital signature verification acts as a gatekeeper for every software package received from vendors. Vendors sign update packages using their private cryptographic keys. When a medical device receives an update, the device or its management platform uses the vendor's public key to verify the signature and confirm the package's integrity and legitimacy. This process blocks any updates that have been altered or created by unauthorized sources. For example, security teams once intercepted a compromised patch for a monitoring device: signature verification prevented installation, revealing that the legitimate package had been replaced by a malicious actor in transit. Without verification, the compromised code could have enabled attackers to access sensitive patient data or disrupt device operation, highlighting the dire consequences of neglecting this crucial step (*Solved—How to Ensure Security Updates in a Fleet of Servers?*, 2021).

A mandatory verification workflow protects against such threats. Security teams should ensure that no updates progress beyond the initial receipt stage until digital signature checks are complete. Devices must reject unsigned updates and those with failed verifications. Strong integration between update management tools and vendor key management prevents workflow bypass. Medical institutions should periodically review and update trusted public keys held by their systems, minimizing the risk of key compromise. When new vendors are onboarded, their keys should be validated with the same rigor as the updates themselves, reinforcing the chain of trust.

Beyond authenticity, each software update must undergo thorough testing within a controlled, nonproduction environment before deployment in clinical contexts. This practice involves setting up isolated testing labs that replicate production device configurations and workflows. Security and biomedical engineering teams install and run the update, monitoring for operational changes, compatibility issues, or performance degradation. Controlled environments allow detection of adverse effects safely, without risking patient care. For example, an update that unexpectedly increased device power consumption was identified and corrected during a controlled trial, avoiding potentially hazardous failures in actual patient rooms.

A structured, risk-limited rollout strategy ensures minimal disruption across fleets of devices. Teams often use phased deployments: an initial group receives the update, with performance and security closely monitored before extending the deployment to broader device populations. This gradual approach permits rollback if issues emerge and prevents widespread impact. Change management records every stage, supporting auditability and continuous improvement.

Emergencies require swift, disciplined adaptation. Emerging threats like zero-day vulnerabilities can leave devices exposed, and prompt patching becomes necessary. To balance urgency with oversight, established fast-track approval workflows come into play. Security and risk management leaders review critical vulnerabilities, prioritize patch testing, and coordinate with clinical, IT, and biomedical staff. Limited rapid deployment to high-risk or exposed devices can commence as soon as baseline testing validates patch safety. For instance, after the disclosure of a critical vulnerability in infusion pumps, a healthcare system's emergency response team quickly tested and deployed the vendor patch to all affected pumps within hours, averting the risk of exploitation. Cross-functional coordination and clear role assignment are indispensable for integrating emergency patches at speed without neglecting quality controls (*Solved—How to Ensure Security Updates in a Fleet of Servers?*, 2021).

Communication also plays a decisive role. Organizations must establish trusted, authenticated notification channels with vendors for disseminating update and patch information. Options include secure vendor web portals, encrypted email communication, or dedicated application programming interfaces (APIs) for automated alerts. Authentication ensures only official notifications prompt action, neutralizing the threat of social engineering or spoofed alerts. Internally, information about required updates should be promptly routed to responsible technical teams with clear instructions and summaries of the associated risks. Timely, reliable communication ensures updates are neither delayed nor mishandled, supporting proactive risk management.

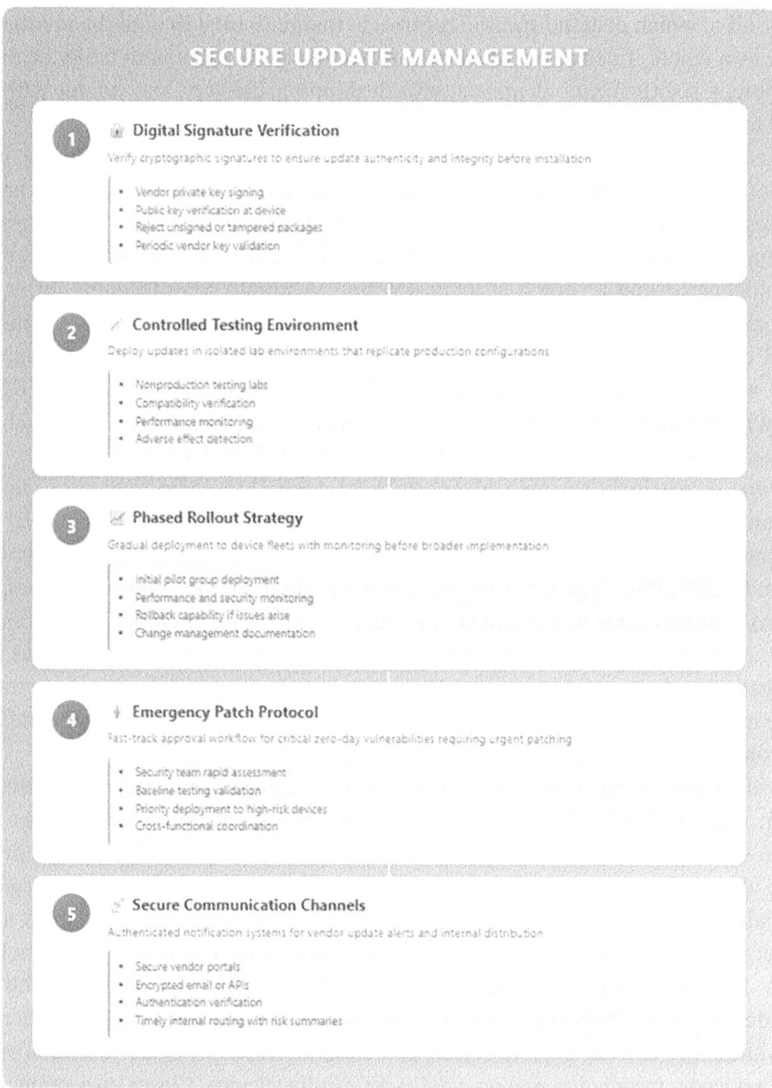

Every process step, from initial receipt through final deployment, must be built on diligence and a culture of security. These technical controls for digital updates align with the broader necessity for secure device logistics and transport. As updates flow through digital channels, their security reinforces and depends on the strength of physical supply-chain safeguards. Coordinating both digital and physical assurances creates a comprehensive defensive posture against risks at every point in a device's life cycle (House, 2025).

Securing Logistics and Device Transit

Medical device security extends beyond digital protections and robust software updates, requiring constant vigilance as devices physically move through multiple hands and environments before clinical use. At each stage, the authenticity and integrity of devices are threatened by a range of security risks, all of which demand specific controls to maintain trust in both the products and the supply chain as a whole. Interference during transit can expose patients to faulty or maliciously altered equipment, so organizations must approach shipping, delivery, and storage with the same rigor applied to digital security measures.

Physical custody of medical devices represents a critical point of vulnerability for healthcare organizations. To defend against unauthorized manipulation or substitution, supply-chain professionals must establish and maintain a detailed, end-to-end chain of custody. This process begins at the point of manufacture and tracks every handoff, storage facility, and transportation event until the device is delivered and accepted at its destination. Accurate recordkeeping forms the basis of this approach. Digital tracking systems and traditional custody logs provide a time-stamped history of all critical tracking events (CTEs) the device undergoes, such as packing, shipping, transferring, and receiving. These records typically include key data elements (KDEs)—the who, what, where, when, and why for each event—capturing vital context and enabling swift investigation if discrepancies emerge (*GS1 Global Traceability Standard | GS1*, n.d.).

Organizations increasingly leverage advanced technologies like radio-frequency identification (RFID) tags and GPS-based tracking to maintain real-time visibility of shipments. RFID can monitor both individual devices and bulk containers, alerting teams to any unexpected movement or break in custody. GPS allows logistics teams to pinpoint a shipment's location at all times, detecting detours or unscheduled stops that might signify theft or tampering attempts. If a shipment strays from its planned route or experiences unscheduled delays, real-time monitoring allows for immediate investigation and remediation. Integrating digital sensors into packaging or containers can even provide environmental data—such as temperature and humidity—to detect conditions that could compromise sensitive medical equipment.

Tamper-evident packaging serves as a vital control to deter, detect, and document potential manipulation during transit. These packages incorporate visual indicators, such as breakable security tape, colored seals, or holographic labels that irreversibly reveal any attempt to access or open the device. Some manufacturers employ uniquely engineered mechanical locks or electronic seals requiring single-use codes, ensuring that any compromise triggers an explicit alert. Upon arrival at a healthcare facility, trained personnel must carefully inspect packaging for irregularities. Effective controls require that all individuals involved in receiving and handling devices can identify subtle signs of tampering—like glue residue, resealed flaps, or altered security labels—and know the correct steps to report and escalate these findings. In one hospital case, vigilant staff noticed a misaligned seal on a delivered cardiac device. Closer inspection, combined with a review of the electronic shipping log, revealed that the box had been secretly opened and resealed at an unauthorized transfer point. Because all events had been logged, investigators traced the breach to one handler, leading to supplier remediation and improved packaging protocols.

Trust in third-party logistics providers is paramount, as these companies physically possess the devices during transit and storage. Security teams must thoroughly vet logistics companies before onboarding. This vetting might involve comprehensive background checks, reviews of past incident histories, and assessments of the provider's information security and physical safeguards. Contracts should clearly specify security requirements: documenting expectations around route planning, custody protocols, driver training, vehicle security, and contingency plans for emergencies. Ongoing assessment continues after initial selection. Healthcare organizations can monitor provider adherence through periodic audits, digital monitoring of shipments, and review of incident response records. When deviations from expected protocols occur—such as unexplained route changes or lax handling procedures—incident response procedures must activate promptly. The provider must immediately document the event, notify all stakeholders, and support a comprehensive investigation. This structured approach ensures that even if a violation occurs, any damage can be quickly contained and a repeat incident prevented.

Rigorous procedures govern device receiving and acceptance at healthcare sites. Staff verify that the delivery matches all accompanying documentation, with serial numbers and product identifiers cross-checked against shipping manifests and prior records. Any discrepancies—such as mismatched batch numbers or unexpected damage—prompt additional investigation and reporting. The receiving process includes formal documentation of each inspection step, direct notification of the responsible security or supply-chain team, and, if needed, quarantine of the device pending further review. This practice preserves a transparent audit trail and upholds end-to-end traceability, as described in supply-chain security standards (*GS1 Global Traceability Standard | GS1*, n.d.).

Every phase, from packaging and transit through final acceptance and storage, operates as part of an interconnected risk management process that sustains authenticity and device integrity. Collaborative efforts with logistics providers, vendors, and internal teams underpin the success of these controls, forming the foundation for a continuous, adaptive approach to supplier relationship management and ongoing medical device security.

Collaborative Supplier Relationship Management for Ongoing Security

A strong vendor relationship serves as the backbone for enduring medical device security as devices move from deployment into operational environments. The ongoing necessity to uphold trust and transparency with vendors after safe receipt of devices underpins a strategy of continuous collaboration and vigilance. While device movement controls ensure initial integrity, robust collaborative practices with vendors remain essential for managing the persistent security risks emerging from the evolving threat landscape and the complexity of healthcare networks (Abrahams et al., 2024; *Validate User*, n.d.).

Joint Security Exercises

Hospitals and medical device manufacturers increasingly rely on joint security exercises, which are structured simulations of real-world cyber incidents that can compromise critical assets. Tabletop simulations invite leaders from both organizations to discuss and rehearse how they would respond to incidents like a ransomware outbreak targeting clinical devices during peak hours. Live simulations escalate the realism, involving actual deployment of simulated malware on a training network that mirrors the production environment. These exercises compel teams to test processes and communication channels, clarify responsibilities, and identify procedural bottlenecks.

For example, a hospital IT department coordinates with its device supplier to simulate the rapid spread of malware through a fleet of networked infusion pumps. During the simulation, IT staff and vendor engineers use the designated out-of-band communications channel to discuss initial findings and collaboratively isolate the affected pumps from the hospital network. Stakeholders document required technical actions, such as patch deployments, firmware rollbacks, and device reimaging procedures. Postexercise reviews uncover gaps in role definition and technical support documentation, resulting in both organizations updating incident playbooks and conducting follow-up training to enhance operational readiness (Abrahams et al., 2024).

Shared Vulnerability Disclosure Programs

Collaborative vendor relationships require robust mechanisms for identifying and addressing device vulnerabilities. Hospitals and vendors establish mutual disclosure programs with designated portals or secure communication channels where newly discovered weaknesses can be safely reported. When a biomedical engineer in a hospital detects an anomalous behavior in a device's remote diagnostics protocol that exposes patient monitoring data, the engineer leverages the disclosure portal to file a detailed report, which triggers an immediate triage workflow by the vendor's incident response team.

Within hours, vendor analysts verify the issue, prioritize risk, and develop a patch, communicating clear timelines and mitigation measures to the hospital's IT staff. Throughout the process, both parties share progress updates, ensuring that device operators and clinicians receive timely advisories and guidance for affected models. Program governance covers tracking of open vulnerabilities, traceability of fixes, and documentation of lessons learned, preventing recurrence and reinforcing a culture of openness (Abrahams et al., 2024; *Validate User*, n.d.).

Escalation Procedures

Defined escalation pathways are indispensable to prevent unclear responsibilities or operational delays during security incidents. Manuals and partner agreements specify primary and backup contacts for routine and after-hours incidents, with contact hierarchies that cross-reference both the hospital's IT security operations center and the vendor's technical response division. Each pathway includes response expectations, authentication steps, and channels that bypass ordinary ticketing queues for urgent incidents.

Consider an attempted after-hours breach involving suspicious traffic detected on a diagnostic imaging server connected to peripheral devices. The on-call IT manager at the hospital identifies the event and activates the agreed escalation plan, contacting the vendor's incident lead directly. The vendor provides immediate remote diagnostics, correlates the event with recent threat intelligence, and, jointly with the hospital IT team, executes containment actions. This clear procedural architecture eliminates ambiguity, ensures rapid information flow, and supports audit trails for compliance and regulatory requirements (Abrahams et al., 2024).

Regular Partnership Reviews

Sustaining strong vendor relationships involves more than incident response; it requires an ongoing cycle of assessment and improvement. Both parties conduct regular partnership reviews—often quarterly—to evaluate security performance against key contractual benchmarks, monitor compliance with updated healthcare regulations, and address the impact of new threats. These meetings bring together risk managers, technical leads, and compliance officers to review joint incident logs, audit results, and outstanding remediation efforts.

For instance, in one review session, both sides discover that the turnaround time for patching a widely used cardiology monitor consistently exceeds the window specified in the contract. Open, data-driven discussion prompts renegotiation of service level agreements and investment in automated firmware deployment mechanisms. These reviews strengthen mutual accountability, establish metrics for ongoing improvement, and provide a proactive forum for aligning responses to emerging regulatory demands (Abrahams et al., 2024; *Validate User*, n.d.).

The effectiveness of these collaborative approaches rests on the reliability of communication, the detail of technical protocols, and the willingness to adapt in response to experience. Each mechanism, from exercises to reviews, prepares organizations to respond rapidly and cohesively to cybersecurity breaches detected anywhere in the third-party life cycle. The ability to detect issues promptly and coordinate remediation across all involved parties becomes the next area of focus as organizations strive for comprehensive medical device security.

Responding to Third-Party Breaches Affecting Medical Devices

Effective vendor engagement and real-time data sharing play a critical role in preparing for external incidents impacting medical devices. Regular, structured collaboration ensures that both healthcare organizations and third-party partners maintain heightened awareness of emerging risks, exchange threat intelligence, and participate in simulation exercises that keep response plans relevant. Joint reviews of network logs, vulnerability assessments, and device inventories enable each partner to recognize evolving attack vectors. Shared readiness drills, wherein both vendors and care providers rehearse simulated ransomware or data breach scenarios, can reveal policy gaps or response bottlenecks early, enabling continuous improvement. This proactive alignment means when a real compromise occurs, teams will have preestablished trust, familiar communication routines, and mutual understanding of escalation thresholds, reducing delays and confusion during a crisis (Elendu et al., 2024).

Early Detection in the Vendor Ecosystem

Healthcare supply chains often span geographically and technologically diverse environments. Upstream monitoring of third-party data flows, calendarized device scans, and ingestion of specialized threat intelligence feeds tailored for healthcare enable hospitals to spot irregularities as they emerge. Security teams often deploy anomaly detection tools within vendor-accessible network segments, flagging unexpected system pivots, irregular patching cycles, or abnormal data exfiltration attempts. For instance, if a network-connected imaging system suddenly begins communicating with a known suspicious domain after a vendor maintenance session, these early warning systems will trigger automated alerts. Integrating medical device vulnerability data directly into monitoring platforms can further sharpen awareness; assets with unpatched firmware are assigned higher alerting thresholds, ensuring that potential supply-chain compromises affecting critical devices, such as infusion pumps or remote monitoring modules, receive immediate human attention (Elendu et al., 2024; Toll, 2025).

Threat Intelligence Workflow

A systematic detection approach often involves layering threat intelligence feeds from both healthcare-specific sources and broader industry databases. These feeds are cross-referenced with network traffic logs to spot indicators of compromise tied to recent third-party breaches. For example, security teams can set predefined actions, such as quarantining certain network segments when new malicious signatures tied to vendor software updates appear, providing rapid containment before patient safety or data integrity is threatened (Toll, 2025).

Secure Communications and Timely Notification

Once a potential incident is detected, secure and rapid communication channels are indispensable for coordinated action. Encrypted messaging platforms and authenticated API endpoints form the backbone of inter-vendor notification systems, protecting sensitive patient and proprietary device information. Notification templates, standardized using formats such as Structured Threat Information Expression (STIX), give all parties clear, concise summaries of what occurred, the devices or systems implicated, and recommended next steps.

A breach workflow typically begins with automated alert generation, escalating from security information and event management (SIEM) platforms to designated incident response coordinators in both the healthcare provider and the vendor team. For example, when a vendor system compromise impacting imaging devices is discovered, the SIEM triggers encrypted alerts to key clinical engineers, procurement officers, and the vendor's own security operations. Response teams then use secure collaboration suites, allowing them to triage the incident in real time while preserving evidence and confidentiality. These systems restrict access to breach details, ensuring only authorized parties receive the notification, helping minimize risk if the notification itself is intercepted (Toll, 2025).

Coordinated Forensics and Evidence Handling

Preserving the integrity of evidence across several organizations is paramount when investigating vendor-related device incidents. Teams agree on standardized digital forensics protocols before an event occurs—these cover everything from log collection formats to disk imaging procedures for compromised medical devices. Cross-organization task forces work from shared playbooks, capturing volatile logs, device states, and configuration snapshots in a chain-of-custody compliant manner. During an investigation into malware found on a fleet of infusion pumps, for example, hospital forensic analysts extract system images while vendor specialists supply device firmware histories and patch records. All evidence is hashed and time stamped, with custody transfers documented at each step.

Synchronizing the forensic timeline is achieved by correlating device activity logs, patch deployment records, and network traffic from each organization. Tools that automate log correlation and timeline generation aid in consolidating sequences of compromise across disparate IT environments. Secure cloud-based analytical environments allow both parties to analyze artifacts jointly while ensuring data remains encrypted and access-controlled, supporting both transparency and compliance (Elendu et al., 2024).

Navigating Notification Regulations and Legal Complexities

Healthcare cybersecurity laws demand timely and accurate incident reporting, especially in multi-vendor and cross-border supply-chain incidents. Teams must compare internal notification policies with international regulations like the General Data Protection Regulation (GDPR), which requires

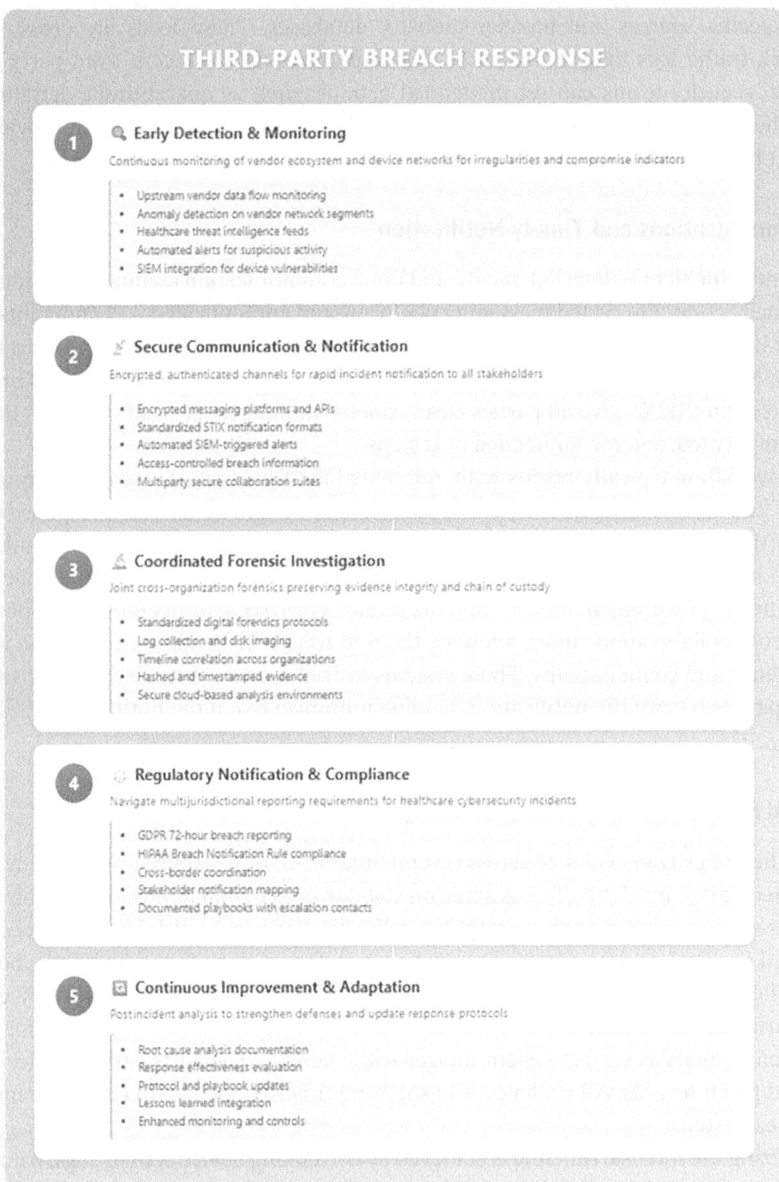

reporting significant breaches within 72 hours, and the United States' HIPAA Breach Notification Rule, which specifies the affected parties and authorities to notify (Elendu et al., 2024). Mapping notification obligations against these standards, compliance officers review device and data relationships to clarify which stakeholders require alerts and in what sequence.

When a device breach spans various countries, compliance and legal experts coordinate parallel notifications to national cybersecurity centers, device regulators, and institutional partners. They must also resolve conflicts that may arise, such as differences between a hospital's internal disclosure timeline and a vendor's responsibility under contract or differing legal jurisdictions. Documented playbooks with predefined escalation contacts and notification templates facilitate alignment, ensuring all parties act within established legal and contractual frameworks while adapting to newly published supply-chain security standards.

A comprehensive, systematized approach to detection, notification, forensics, and compliance provides the most resilient defense against malicious supply-chain incidents while ensuring regulatory adherence and continuity of patient care (Elendu et al., 2024; Toll, 2025).

Bringing It All Together

Now that we understand the critical processes for securing medical devices through vigilant third-party vendor management and supply-chain controls, healthcare organizations are better positioned to anticipate and mitigate evolving cybersecurity threats. By integrating systematic vendor assessments, rigorous update verification, secure logistics, collaborative partnerships, and coordinated incident response, stakeholders can establish a resilient defense that safeguards patient safety and data integrity throughout the device life cycle. Moving forward, continuous adaptation of these practices in response to emerging risks, advances in technology, and regulatory developments will be essential for maintaining trust and operational reliability in increasingly interconnected healthcare environments.

References

Abrahams, T. O., Farayola, O. A., Kaggwa, S., Uwaoma, P. U., Hassan, A. O., Dawodu, S. O. (2024 10, January). *Reviewing Third-party Risk Management: Best Practices in Accounting and Cybersecurity for Superannuation Organizations*. Finance Accounting Research Journal. https://doi.org/10.51594/farj.v6i1.706.

Baz, A., Ahmed, R., Khan, S. A., Kumar, S. (2023, January 1). *Security Risk Assessment Framework for the Healthcare Industry 5.0*. Sustainability. https://doi.org/10.3390/su152316519.

Elendu, C., Omeludike, E. K., Oloyede, P. O., Obidigbo, B. T., Omeludike, J. C. (2024, September 27). *Legal Implications for Clinicians in Cybersecurity Incidents: A Review*. Medicine; Wolters Kluwer. https://doi.org/10.1097/md.0000000000039887.

GS1 Global Traceability Standard | GS1. (n.d.). Www.gs1.org. https://www.gs1.org/standards/gs1-global-traceability-standard/current-standard.

House, T. W. (2025, January 16). *Executive Order on Strengthening and Promoting Innovation in The Nation's Cybersecurity | The White House*. The White House. https://bidenwhitehouse.archives.gov/briefing-room/presidential-actions/2025/01/16/executive-order-on-strengthening-and-promoting-innovation-in-the-nations-cybersecurity/.

Ostermann, M., Mathias, R., Jahed, F., Parker, M. B., Hudson, F. D., Harding, W. C., Gilbert, S., Freyer, O. (2025, July 15). *Cybersecurity Requirements for Medical Devices in the EU and US—A Comparison and Gap Analysis of the MDCG 2019–16 and FDA Premarket Cybersecurity Guidance.* Computational and Structural Biotechnology Journal; Elsevier. https://doi.org/10.1016/j.csbj.2025.07.024.

Solved—How To Ensure Security Updates In A Fleet Of Servers? (2021, April 20). The FreeBSD Forums. https://forums.freebsd.org/threads/how-to-ensure-security-updates-in-a-fleet-of-servers.79977/.

Toll, W. (2025, June 23). 15+ Network Segmentation Control Examples to Strengthen Healthcare Security. Elisity.com; Elisity. https://www.elisity.com/blog/15-network-segmentation-control-examples-to-strengthen-healthcare-security.

Validate User. (n.d.). Academic.oup.com. https://academic.oup.com/cybersecurity/article/7/1/tyab024/6448501.

15

Penetration Testing and Red Team Exercises

More than 60% of healthcare organizations reported experiencing a cyberattack targeting medical devices in the past year, with many incidents exploiting vulnerabilities that traditional security audits failed to detect. In one notable case, attackers gained access to a hospital network through a seemingly minor misconfiguration of a wireless infusion pump, leading to prolonged unauthorized control before detection. Such breaches highlight that theoretical assessments and compliance checklists alone cannot capture the complex, interconnected risks inherent in clinical environments. As healthcare systems become increasingly reliant on digital technologies embedded directly within patient care workflows, the ability to simulate real-world adversarial tactics—while balancing operational safety—has emerged as a vital component of cybersecurity strategy. This chapter examines how offensive testing methods challenge assumptions about security, reveal hidden weaknesses, and drive targeted improvements across devices, networks, and organizational processes in healthcare settings.

The Case for Offensive Testing in Healthcare

Offensive security testing in healthcare environments requires methodologies beyond policy reviews and theoretical risk assessments. Red team exercises and penetration tests probe the real-world resilience of medical devices and hospital networks by imitating tactics, techniques, and procedures used by actual adversaries. Conventional audits, even when rigorous, cannot simulate the dynamic and contextual threats facing connected medical infrastructure. For example, a paper-based control assessment for an infusion pump might confirm cryptographic requirements on data flows at rest or in transit, yet fail to consider a scenario in which a misconfigured wireless authentication permits a remote attacker to gain persistent access over the device. Simulated adversarial testing surfaces this kind of chain of weaknesses—misconfiguration, network protocol flaws, and deficient device isolation—demonstrating how seemingly minor issues can be combined in ways that threaten patient safety or lead to catastrophic breaches (Mindgard, 2025; *Red Team Exercises in Cybersecurity: Benefits & Examples*, 2024).

Red team exercises consistently expose vulnerabilities in healthcare environments that evade traditional controls. While certification and documentation reviews excel at identifying missing controls or obvious compliance gaps, they are less effective when vulnerabilities stem from oversights in device implementation or operational configuration. Typical examples include medical endpoints shipping with hardcoded credentials, undocumented debug ports left enabled in production, or legacy firmware with exploitable cryptographic routines. Consider a situation where a radiology workstation connected to a hospital network was penetrated during a red team exercise due to an open management interface; this flaw was missed by static code reviews because it was introduced through operational misconfiguration after deployment. In another example, an adversarial simulation targeting a wireless insulin pump discovered exploitable authentication mechanisms because the test chained together overlooked issues in network segmentation and access point misconfiguration. These findings underscore the need for adversarial simulations, as they reveal the operational realities and interconnected vulnerabilities that accumulate across heterogeneous, complex device ecosystems (Mindgard, 2025).

Such offensive testing does more than uncover technical weaknesses. Red team and penetration exercises place entire healthcare organizations under stress, evaluating how staff, procedures, and systems interact in the face of simulated active threats. Incident response plans crafted on paper often fail to account for latent breakdowns in communication or responsibility. During a stealth ransomware simulation, a typical scenario involves patient monitoring devices exhibiting anomalous behavior; when red team operators escalate the attack, the real-world drill exposes delays as clinical engineering and IT security teams struggle with ambiguous escalation protocols or limited situational awareness. These exercises illuminate critical gaps: a missed alert in a clinical device security dashboard, uncertainty about which stakeholders have decision authority, or confusion about when to involve external partners. Practical recovery and containment strategies improve only when these scenarios are simulated and lessons internalized before a real-world event occurs (*Red Team Exercises in Cybersecurity: Benefits & Examples*, 2024).

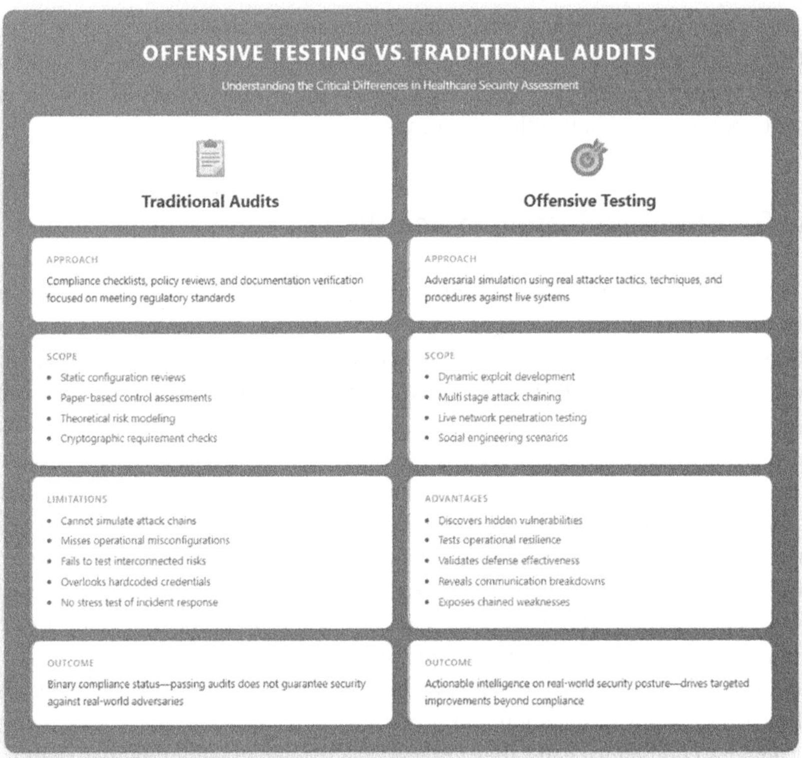

Regulatory frameworks have made offensive testing a cornerstone requirement in many healthcare settings. The Health Insurance Portability and Accountability Act (HIPAA), the European Medical Device Regulation (MDR), and the US Food and Drug Administration (FDA) premarket guidance all encourage or mandate adversarial assessments for critical systems. Organizations demonstrate compliance in risk management programs when they regularly conduct and document penetration or red team tests. For instance, evidence of a coordinated red team simulation that uncovers systemic weaknesses in access control and results in improved patching and documentation can be directly submitted as part of an annual audit or regulatory review. This not only helps organizations pass audits but also contributes to a legally defensible security posture. By proactively demonstrating that controls are effective in practice, healthcare entities reduce liability and shore up their ability to demonstrate due diligence to regulators, auditors, and patients alike (Mindgard, 2025; *Red Team Exercises in Cybersecurity: Benefits & Examples*, 2024).

Hands-on adversarial testing transforms defensive postures. It empowers healthcare organizations to move beyond prescriptive checklists and instead develop adaptive, threat-informed controls that withstand real-world attack scenarios. This approach does not conclude with the assessment of technical controls alone. It integrates security into clinical workflows, elevates team readiness, and supports enduring compliance in a complex regulatory environment. The act of structuring and implementing penetration tests with patient safety at the forefront drives the need for collaborative scoping, disciplined tool selection, and robust communication between IT, clinical, and executive leaders. The efficacy of these exercises depends on careful coordination to minimize system disruption while maximizing the fidelity and applicability of findings, ensuring that security outcomes support—not undermine—continuity of clinical operations.

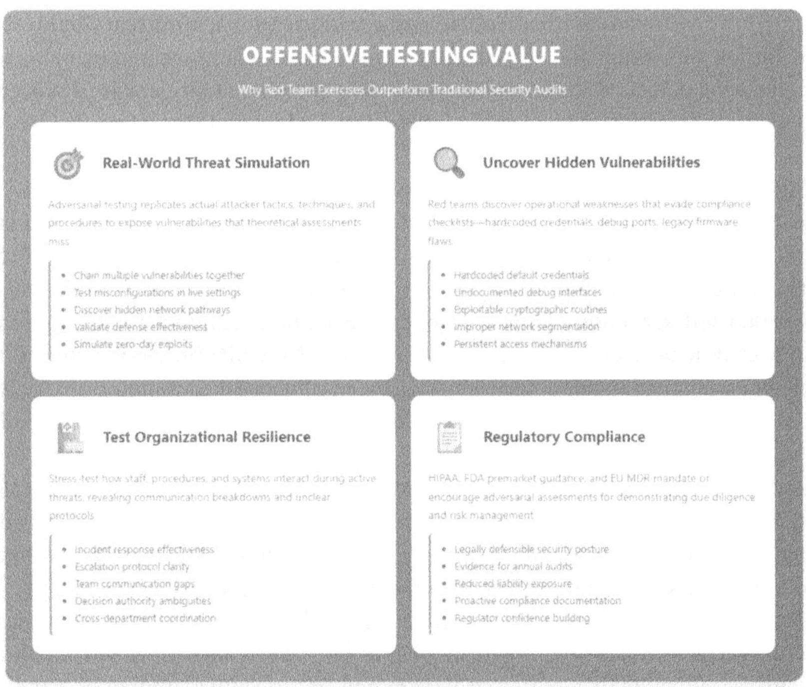

Structuring a Medical Device Penetration Test

Defining the scope of a penetration test is the foundation for safe and effective security validation in healthcare environments. Unlike generalized infrastructure testing, medical device penetration tests must carefully balance technical thoroughness against clinical safety requirements. Healthcare systems operate nonstop and directly support patient lives; inappropriate test actions risk serious harm. Establishing precise scope boundaries—such as restricting tests to backup firmware on lab devices rather than production units in active use—protects patient care while allowing meaningful assessment. For example, isolating a test insulin pump from live inventory ensures that assessment of firmware vulnerabilities does not disrupt therapy, even if device routines fail during analysis. Clear documentation of what is included and what must be excluded—like production ventilators or in-use monitoring beds—shields the organization from accidental disruption and provides legal clarity for all participants. Scope definition also anticipates operational realities revealed by digital health readiness tools, where scores indicate communication openness and teamwork are strengths, but continuous learning around test implications may require further investment (Ibrahim et al., 2024).

Once scope is finely defined, tool and technique selection must match the unique traits of healthcare device ecosystems. Medical devices often run proprietary firmware on embedded hardware with constrained interfaces and wireless communication protocols. Automated scanners designed for enterprise networks often fail to reach deeper device logic, making specialized firmware analysis tools and manual review essential. Firmware analyzers—capable of extracting and decompiling device logic—expose issues such as hardcoded credentials or insecure update mechanisms. Wireless sniffers, configured to observe traffic on clinical monitoring networks, identify weak protocol implementations in real-world situations. For instance, conducting protocol fuzzing on a wireless infusion pump in a controlled environment can reveal faulty input handling that would escape both signature scanning and compliance-based audits. These approaches, inspired by risk assessment models emphasizing clinical decision-making and technology reliability, ensure discovered flaws have real operational context and can support prioritized mitigations for systems most critical to care delivery (Ibrahim et al., 2024).

Successful penetration testing in healthcare hinges on early, ongoing, and collaborative engagement with clinical stakeholders. Testing timelines must align with the ebb and flow of patient care, recognizing that even short disruptions can create risk. Clinical leadership involvement at the project planning stage secures buy-in and shields security teams from organizational push-back. Penetration test scheduling outside of high-acuity hours, combined with up-front notification to impacted departments, compresses operational risk while preserving clinical capacity. For example, a preplanned wireless security assessment of networked monitoring devices might coincide with regular equipment maintenance windows, avoiding interference with patient care routines. Preauthorized downtime waivers, along with clear communication channels for triggering emergency pause of testing in the event of unexpected clinical events, are practical safeguards. Having real-time point-of-contact arrangements, as advised by digital health adoption assessments focusing on clinician engagement and patient acceptance, reassures clinical staff and reduces anxiety around unfamiliar or disruptive events (Dermody et al., 2025; Ibrahim et al., 2024).

The impact of a device penetration test is reflected in how its results are translated into actionable improvements across technical and organizational levels. Deliverables must span both detailed, technically precise vulnerability reports and clear, jargon-free executive summaries. For developers and engineers, vulnerability reports should map findings to each affected component, include risk ratings, and provide practical recommendations for remediation that account for device-specific constraints. Reports must avoid template language and convey exploit narratives that resonate with the daily workflows of health technology teams. At the same time, executive leaders require concise, outcome-focused summaries that highlight patient safety implications, estimated remediation effort, and potential regulatory repercussions. Conducting structured debriefing sessions—where technical, operational, and clinical teams review test outcomes together—ensures shared understanding, builds cross-functional trust, and establishes accountability for follow-up actions. This collaborative approach supports the continuous learning and leadership engagement shown as essential facets in broader safety culture surveys (Ibrahim et al., 2024).

Concrete examples affirm these practical steps. Limiting a test to offline insulin pumps can expose preinstalled vulnerabilities without endangering active patients. Scheduling wireless protocol assessments during device maintenance prevents operational delay while uncovering real design weaknesses. Sharing multilayered reports, followed by joint debriefs, supports both rapid technical fixes and long-term learning that shape future device procurement and life cycle management. These elements collectively contribute to a culture of proactive risk management and patient safety.

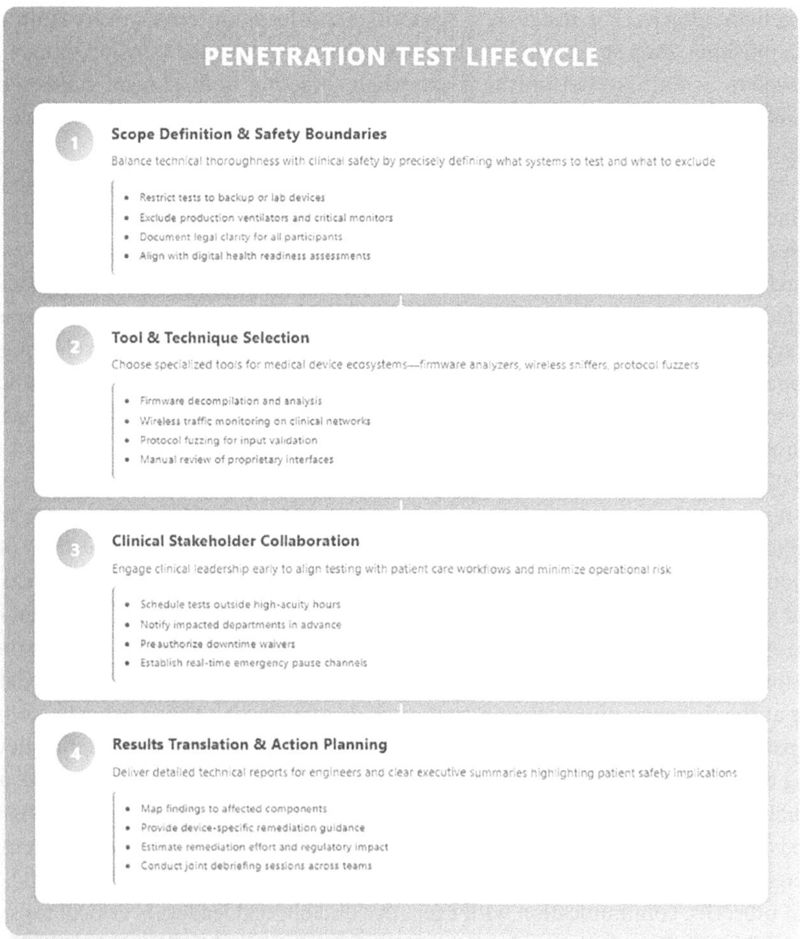

As these structured processes evolve, stakeholders must remain cognizant of the additional dimensions posed by large-scale red team exercises across hospital networks. These escalate the scope of simulated attacker scenarios, requiring nuanced scenario design, enhanced communication with executive leadership, and even more rigorous safeguards to sustain both patient safety and operational integrity. This shift introduces new levels of complexity, where device-specific lessons transform into organization-wide defenses, and simulation realism aligns with the ultimate goal of resilient healthcare delivery (Dermody et al., 2025; Ibrahim et al., 2024).

Red Teaming Hospital Networks

The design of red team exercises for hospital environments starts with building scenarios that mimic genuine threats relevant to healthcare operations. These attack scenarios need to align with current threat intelligence, practical risk assessments, and the specific technological landscape of clinical settings. Red team leaders draw from data such as recent ransomware outbreaks against hospitals and security alerts highlighting vulnerabilities in medical devices and hospital networks. Using this intelligence, they create exercises that represent real-world adversarial techniques instead of generic attacks. For instance, a scenario could focus on ransomware infiltrating the network via phishing, then spreading to vulnerable, internet-connected infusion pumps. Another example involves a multilayered phishing campaign targeting both clinical staff—who might receive emails made to look like urgent care communications—and IT personnel who control privileged systems access (Aldosari, 2025; He et al., 2021).

Effective red teaming leverages multifaceted attack vectors, pushing beyond simple technical exploits to include social engineering, business process manipulation, and third-party supply-chain compromises. Social engineering exercises—such as simulated SMS phishing to clinicians about on-call schedules or deceptive vendor notifications to IT staff—demonstrate how attackers prey on human factors, which remain leading causes of breach incidents in healthcare (He et al., 2021). Designing scenarios with a balance of technical and psychological challenges thoroughly tests organizational resilience. To create relevant and believable exercises, teams should study actual incidents, emerging attack strategies, and documented weaknesses within hospital workflows, including how high staff workload and stress increase susceptibility to deception and mistakes (He et al., 2021).

Realism must remain carefully balanced with operational safety. Hospitals cannot risk interrupting patient care or compromising sensitive data. As a result, red team organizers often turn to simulation tools, digital twin environments, or isolated test beds that replicate real medical devices and workflows without impacting production systems (Aldosari, 2025). When exercises involve live environments, stringent safeguards are needed. Network segmentation ensures attack traffic remains isolated. Defining strict stop conditions empowers operators to immediately halt an exercise at the first sign of negative impact, such as unexpected device behavior or increased latency on patient-critical systems. Simple strategies, such as deploying nonproduction devices and establishing manual overrides, enable teams to intervene quickly, further reducing risks to patient care. Manual overrides on connected medical devices, for example, allow technicians to revert test-induced changes on command, should systems appear unstable. These techniques ensure realism in testing does not override the primary imperative of patient safety (He et al., 2021).

Clear and proactive communication with executive and clinical leaders is critical for successful and safe offensive exercises. Red team leaders must prepare concise briefing materials that explain the intent, objectives, phases, and possible implications of the test. These documents outline all anticipated risks, specify impact windows, clarify the chain of command, and map the precise involvement of IT, clinical, and support teams. By articulating test details before execution, leaders set organizational expectations and gain crucial support for timely resource allocation and incident response (Aldosari, 2025). For example, a briefing document might state, "Scenario 1 will simulate network malware propagation over a three-hour period, affecting only redundant lab systems, and includes planned failovers to backup networks tested during a 90-minute window." Such clarity fosters trust, aligns all stakeholders, and ensures a rapid, coordinated response if adverse events arise.

Disruptions may still occur, and efficient response strategies are essential. Teams create comprehensive recovery plans that enable swift restoration of systems and workflows following incident simulation. Rollback scripts for firmware changes on medical devices, preconfigured network switch settings, and backup virtual machine snapshots are all examples of preparatory steps that speed recovery. Staging redundant network infrastructure provides additional assurance that service continuity remains intact, even if adversarial tests temporarily interrupt regular operations (Aldosari, 2025; He et al., 2021).

Documentation during red team exercises is not only about technical forensics but also about comprehensive incident and near-miss recording. Every abnormal response, unintended disruption, or workaround triggered by the exercise should be captured in reports. This practice embeds organizational learning in hospital culture and drives future improvements. For example, if an exercise uncovers that a critical vendor interface lacks effective failover—but a workaround developed in real time maintains care—documenting this discovery leads to permanent process enhancements and better preparedness for actual incidents (He et al., 2021). Careful records from past exercises form the foundation for effective structured analysis, robust issue tracking, and strategic remediation planning.

In sum, thorough scenario creation, vigilant safety methodologies, authentic stakeholder engagement, and detailed evaluation practices combine to make red team exercises a powerful catalyst for strengthening hospital cyber resilience and protecting patient safety (Aldosari, 2025; He et al., 2021).

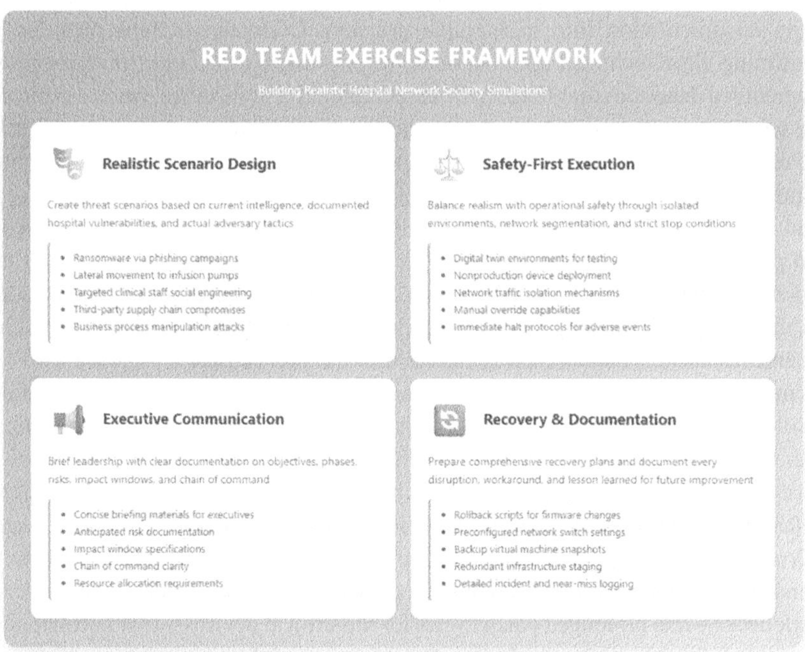

Postexercise Analysis and Remediation Planning

Effective cybersecurity improvement within healthcare begins with the meticulous documentation of penetration testing and red team exercise findings. Every vulnerability identified must be described in precise detail, outlining the nature of the weakness, reproducible steps, and tangible supporting evidence. Presenting risk ratings contextualized for patient safety impact establishes immediate priorities for remediation. For example, a firmware flaw in an infusion pump, if left unaddressed, may allow unauthorized access that could alter dosage delivery; this risk description should pair technical exploitability with patient health consequences. Supporting materials, such as packet captures or exploit demonstration videos, provide essential corroboration for technical and oversight stakeholders, strengthening the case for corrective action and serving as a reference point during future audits (*Preparing for the Post-Quantum Era: Quantum-Ready Architecture for Security and Risk Management (QUASAR)—a Strategic Framework for Cybersecurity*, 2025).

Categorizing vulnerabilities by risk type, severity, operational context, and potential patient impact guides the prioritization of mitigation. Segregation of risk domains—medical devices, electronic health records, network infrastructure—enables teams to direct immediate focus toward those issues most likely to affect patient welfare or disrupt clinical services. Assigning unique identifiers and tracking findings through each verification cycle ensures no gaps emerge in communication or accountability. Maintaining clear documentation trails not only bolsters compliance with mandates on traceability and change management (*GAO-08-1029G, Federal Information System Controls Audit Manual (FISCAM): Exposure Draft*, 2008), it underpins effective audit practices and underlines organizational commitment to improvement. In practice, discovering a remote code execution vulnerability in bedside monitors warrants comprehensive logged evidence, cross-referenced with inventory records and incident tracking systems, so that each subsequent test and remediation step remains fully visible.

Development of actionable remediation roadmaps follows as the practical mechanism for transforming documentation into measurable progress. Corrective actions must be precisely defined, outlining clear technical specifications, responsible parties with unambiguous ownership, and scheduled deadlines that reflect risk-based urgency. For instance, when a critical wireless encryption weakness is uncovered in remote monitoring systems, the roadmap should assign firmware patching to the device manufacturer by a specified date, while instructing the hospital IT team to reinforce access controls and monitor postpatch behavior. Sequencing tasks according to severity and resource capacity in this way sustains momentum, ensuring that high-risk items are remediated promptly while also enabling parallel progress on secondary issues. Integration with broader organizational security initiatives—such as embedding remediation timelines for device classes within a hospital's ongoing IT risk reduction program—fosters holistic improvement and avoids piecemeal solutions that address symptoms but not root causes.

Progress measurement hinges on the establishment and ongoing tracking of meaningful metrics. Defining tailored key performance indicators—such as mean time to remediate, proportion of recurring to new issues, and decline in high-severity findings across successive exercises—creates an objective basis for evaluating program effectiveness (*Preparing for the Post-Quantum Era: Quantum-Ready Architecture for Security and Risk Management (QUASAR)—a Strategic Framework for Cybersecurity*, 2025). Trend analysis applied to these metrics enables security teams to identify persistent weaknesses, such as recurring credential reuse in medical imaging systems, or reveals bottlenecks—like prolonged patch cycles for life-critical devices—that necessitate targeted resource augmentation or process redesign. Data-driven insights derived from these measures support ongoing optimization in line with evolving standards and regulatory requirements, and inform decisions such as when to renegotiate maintenance and update agreements with vendors who consistently underperform.

Engaging stakeholders in the improvement process requires tailoring communications to audience expertise and strategic priorities. Reports for executive leadership should distill technical findings into plain-language summaries focused on patient safety, institutional reputation, and compliance obligations, while technical appendices provide implementation teams with actionable detail. Transparency about both successes—such as verification of effective network segmentation—and enduring risks builds trust in the security program and fosters buy-in for upcoming initiatives. Briefings for clinical staff must emphasize workflow impacts and practical security advice, addressing how remediation steps affect day-to-day practice. Formal

communication mechanisms, such as regular progress reports, feedback channels, and clear escalation paths, ensure all stakeholders remain informed, aligned, and equipped to support risk reduction efforts (*GAO-08-1029G, Federal Information System Controls Audit Manual (FISCAM): Exposure Draft*, 2008).

Sustained cyber resilience in healthcare depends upon processes that reinforce ethical and legal responsibility through auditability, continuous improvement, and active engagement of the entire organization. Persistent evaluation through structured documentation, dynamic remediation plans, quantitative performance tracking, and rigorous stakeholder dialog forms the backbone of secure, patient-centric healthcare programs prepared for the evolving threat landscape.

Ethical and Legal Boundaries in Adversarial Testing

Ensuring that penetration tests and red team exercises are conducted responsibly in healthcare environments is critical to safeguarding the well-being of patients, the integrity of clinical operations, and organizational assets. In adversarial assessments, technical proficiency must be matched by adherence to ethical and legal frameworks, especially considering the sensitive nature of healthcare data and the direct implications on patient safety. Successfully embedding these proactive security measures requires carefully constructed policies and protocols that prioritize transparency, participation, and risk management at every point.

Informed Consent Procedures

An effective offensive security program in healthcare is predicated on robust informed consent procedures. It begins with the structured involvement of all stakeholders who might be affected by testing activities, including hospital administrators, IT and incident response teams, operational leadership, and, for some scenarios, clinical staff. In certain circumstances where patient data or device operation could be impacted, explicit communication with relevant patient advocacy representatives is warranted. The informed consent process includes full disclosure about the testing's objectives, its scope (which departments, devices, or systems will be tested), the anticipated risks, and all mechanisms in place to minimize possible disruptions or exposure of sensitive information. Stakeholders should receive detailed explanations of proposed risk mitigation, such as maintaining system backups or activating failover protocols, and assurances on how their responsibilities and roles intersect with the testing schedule.

Critical to this process is ensuring that consent is voluntary, fully informed, and clearly documented, mirroring the principles found in clinical research ethics and regulatory practice (Shah et al., 2024). Consent documentation must be comprehensive and carefully archived for compliance verification; in addition, all parties should have access to signed protocol summaries and contact points for ongoing questions or withdrawal, ensuring continuous understanding and control (*Legal, Ethical, and Safety Issues | Clinical Center*, 2019).

Legal Notifications

Before commencement of offensive security testing, legal notification procedures are often necessary to align the organization's activities with relevant state, federal, and industry regulations. Certain types of testing—such as those targeting systems involved in patient care, regulated medical devices, or those operating within the purview of governmental mandates—may trigger requirements for notification or coordination with oversight bodies. These can include health departments, the FDA, or, in some cases, law enforcement, especially if testing could be misinterpreted as malicious activity or if it could cause regulatory reporting thresholds to be crossed.

Organizations must consult legal counsel to determine precisely what must be communicated, to whom, and under which jurisdiction, taking into account nuanced situations like testing during active healthcare delivery or on systems hosting protected health information (Shah et al., 2024). Meticulous documentation of all notifications, correspondence, and regulatory determinations helps minimize exposure to liability and demonstrates due diligence in the event of posttest audit or investigation.

Safe Testing Environments

The protection of patient safety and system reliability is central to any adversarial exercise in healthcare. Rigorous selection and detailed preparation of the testing environment are foundational risk mitigation strategies. When practical, security teams utilize sandboxed laboratory networks, purpose-built to emulate critical clinical infrastructure without risk to live operations. Devices under test—such as infusion pumps or imaging workstations—are often mirrored or emulated with synthetic data and simulated workloads, ensuring test realism while shielding actual patient care processes.

If testing must take place in a live environment, additional controls become mandatory. These include deep network segmentation to insulate testing targets, real-time monitoring by both IT and clinical staff, and the definition of hard stop conditions should unexpected disruptions occur. Comprehensive communication plans are established so that clinical workers receive timely notifications about testing windows, impacted services, and escalation contacts, maintaining situational awareness and prioritizing continuity of care (*Legal, Ethical, and Safety Issues | Clinical Center*, 2019).

Data Handling During Testing

Meticulous attention to the management of sensitive data is essential throughout offensive security activities. Healthcare settings contain multiple classes of sensitive data, each subject to specific regulatory and privacy mandates. For example, patient records are strictly protected under HIPAA, while device telemetry and operational logs may fall under broader privacy and security policies. All testing artifacts—whether logs, screenshots, memory dumps, or intercepted traffic—should be stored in encrypted containers and accessed only by authorized personnel, with access controls precisely mapped to least privilege requirements.

Data transferred between systems during testing must be encrypted using strong standards, and secure deletion procedures must be applied postexercise to avoid the inadvertent retention of regulated data. In addition, the use of anonymization or de-identification techniques further mitigates reidentification risks when retention or analysis is required (Shah et al., 2024). Documented data life cycle protocols cover every step, from data collection through secure erasure, and underscore the severe organizational, financial, and reputational impacts of data mishandling. These practices uphold both ethical research requirements and the broader duty to protect patient confidentiality (*Legal, Ethical, and Safety Issues | Clinical Center*, 2019).

By building each of these practices into offensive security operations, healthcare organizations maintain an ethical and legal shield that enables robust, actionable risk discovery without compromising trust or safety.

Final Thoughts

Now that we understand the vital role of offensive security testing in healthcare, organizations can confidently adopt structured penetration tests and red team exercises to proactively identify and address vulnerabilities within medical devices and hospital systems. By integrating these adversarial approaches with careful planning, ethical considerations, and detailed remediation processes, healthcare professionals can enhance patient safety while maintaining compliance and operational integrity. Moving forward, continued collaboration among security experts, clinical teams, engineers, and regulators will be essential to evolving resilient defenses that adapt to emerging threats, ensuring that healthcare environments remain secure and capable of delivering safe, uninterrupted care in an increasingly complex digital landscape.

References

Aldosari, B. (2025, May 6). *Cybersecurity in Healthcare: New Threat to Patient Safety*. Cureus; Springer Science and Business Media LLC. https://doi.org/10.7759/cureus.83614.

Dermody, G., Rania Shibl, Wang, M., Ward, A., Watson, J., North, K., Blake, J., Mealy, E., Koay, A. M., Sirinthip Roomkhan, Yeoh, P. L., Fritz, R. (2025, April 2). *Multi Perspective Considerations for Health Smart Home: Early Phase Exploratory Study*. Journal of Advanced Nursing; Wiley. https://doi.org/10.1111/jan.16945.

GAO-08-1029G, Federal Information System Controls Audit Manual (FISCAM): Exposure Draft. (2008, July 31). Gao.gov. https://www.gao.gov/assets/a77127.html.

He, Y., Aliyu, A., Evans, M., Luo, C. (2021). *Healthcare Cyber-security Challenges and solutions Under the Climate of COVID19: A Scoping Review*. Journal of Medical Internet Research; ncbi. https://doi.org/10.2196/21747.

Ibrahim, A. M., Alenezi, I. N., Mahfouz, A. K. H., Mohamed, I. A., Shahin, M. A., Abdelhalim, E. H. N., Mohammed, L. Z. G., Abd-Elhady, T. R. M., Salama, R. S., Kamel, A. M., Gouda, R. A. K., Eldiasty, N. E. M. M. (2024, December 19). *Examining Patient Safety Protocols Amidst the Rise of Digital Health and Telemedicine: Nurses' Perspectives*. BMC Nursing; Springer Science and Business Media LLC. https://doi.org/10.1186/s12912-024-02591-8.

Legal, Ethical, and Safety Issues | Clinical Center. (2019). Nih.gov. https://www.cc.nih.gov/patient-info/legal.

Mindgard. (2025, January 28). *Red Team Attack Simulation: How Does It Work?* Mindgard.ai; Mindgard. https://mindgard.ai/blog/red-team-attack-simulation.

Preparing for the Post-Quantum Era: Quantum-Ready Architecture for Security and Risk Management (QUASAR)—A Strategic Framework for Cybersecurity. (2025). Arxiv.org. https://arxiv.org/html/2505.17034v1.

Red Team Exercises in Cybersecurity: Benefits Examples. (2024, September 9). SentinelOne. https://www.sentinelone.com/cybersecurity-101/services/red-team-exercise-in-cybersecurity/.

Shah, P., Thornton, I., Turrin, D., Hipskind, J. E. (2024, November 24). *Informed Consent*. National Library of Medicine; StatPearls Publishing. https://www.ncbi.nlm.nih.gov/books/NBK430827/.

16

Medical Device Cybersecurity in the Era of AI and Machine Learning

The integration of artificial intelligence (AI) and machine learning (ML) into medical devices is revolutionizing healthcare by enhancing diagnostic accuracy, personalizing treatment, and enabling real-time clinical decision support. These technologies allow medical devices to analyze complex data patterns and adapt over time, far surpassing the capabilities of traditional static systems. However, as medical devices become more intelligent and interconnected, they also face new and complex cybersecurity challenges. The complexity of AI algorithms, continuous model updates, and opaque decision-making processes increase vulnerabilities that can be exploited by malicious actors, threatening patient safety and data privacy.

This chapter examines the transformation of medical devices through AI and ML, focusing on their integration methods and the unique security risks that arise. It discusses emerging AI-driven attack vectors such as data poisoning, model theft, adversarial inputs, and bias, highlighting the need for specialized defense strategies. The chapter explores how securing AI training pipelines, ensuring data provenance, and maintaining human oversight contribute to robust protection. Additionally, it reviews regulatory and ethical considerations shaping AI-driven device development, including transparency requirements and cross-border legal challenges. Finally, it addresses future-proofing practices that combine adaptive defenses, threat intelligence, and automated security operations to sustain resilience in an evolving threat landscape.

Integrating AI and ML into Medical Devices: Capabilities and Complexities

Artificial intelligence, or AI, refers to computational systems designed to perform tasks that would typically require human intelligence. In medical devices, AI includes everything from automated image classifiers to real-time clinical decision support systems. Machine learning, or ML, is a subset of AI in which a system learns patterns from data instead of following explicitly programmed rules. The distinction between AI and ML is crucial. While AI may incorporate rule-based logic, ML focuses on algorithms that adapt over time with exposure to new inputs, making them particularly well-suited for continuously improving medical technologies (Hanna et al., 2025).

Medical devices integrated with AI and ML enable a broad range of established and emerging applications that extend capabilities far beyond earlier programmable or static systems. Diagnostic imaging is a key area where AI-driven algorithms analyze radiology scans, detecting anomalies such as tumors with a level of consistency and accuracy that often exceeds experienced human practitioners. In diabetes care, ML models optimize insulin dosing by analyzing patterns in a patient's glucose levels, insulin use, meals, and activity. These models respond in near real time to subtle physiological changes, reducing the risk of hypoglycemic events and improving outcomes. Other predictive algorithms process data streams from bedside monitors to identify precursors to patient deterioration, triggering rapid response interventions and saving lives in critical care scenarios. Each example highlights improvements in speed, accuracy, and adaptability, but also demands rigorous technical controls to ensure reliability, safety, and security (Rabbani et al., 2025).

Maintaining the performance of AI and ML models deployed in medical environments poses new technical challenges. Algorithms exposed to changing patient populations, evolving clinical protocols, or shifts in data collection can become less effective—a risk known as model drift. Model drift results when a model's underlying assumptions no longer match real-world inputs, causing inaccuracies that can jeopardize patient safety. To mitigate this, device manufacturers employ remote update mechanisms that allow them to recalibrate or retrain models over-the-air. For instance, if a radiology image analysis model underperforms in a new demographic group, engineers can transmit updated model parameters across the hospital's secure network directly to the affected imaging devices. A typical remote calibration workflow includes capturing new labeled data, training an improved model in a secure environment, validating the model's safety, and then transmitting encrypted updates to field devices. Some systems implement continuous learning cycles, using ongoing data feedback to refine models dynamically while ensuring robust oversight to prevent accidental bias or malfunction (Hanna et al., 2025).

The integration of sophisticated AI and ML architectures into medical devices significantly increases codebase complexity. Algorithms for deep learning may include millions of interconnected parameters, with additional software layers coordinating data ingestion, integrity checks, user interfaces, and connectivity protocols. This ballooning complexity broadens potential attack surfaces, as every additional module or interface can introduce a new vulnerability. For example, a hospital networked monitoring suite might combine patient tracking, analytics, and alerting—all powered by AI models that communicate in real time. As subsystems become tightly coupled, unintended interactions between an AI-driven predictive model and other device modules can result in unsafe behavior or create new paths for exploitation by attackers. The breadth of features and intricate dependencies also complicate testing and quality assurance, making it easier for both bugs and security flaws to escape detection before deployment (Rabbani et al., 2025).

Medical device AI models face another challenge: the opacity of decision-making, known as the "black box" problem. Clinicians may receive a device alert suggesting urgent action but remain unable to verify why the AI deemed the situation critical. In one scenario, an intensive care monitor's algorithm might flag deterioration, prompting a disruptive workflow change, yet delayed diagnosis occurs if clinical staff lack insight into what drove the model's conclusion. Trust is eroded when device users cannot understand or interrogate AI decisions, while regulatory specialists and legal investigators struggle to establish accountability if a device error leads to patient harm. To address these difficulties, developers are adopting interpretable models where feasible, as well as algorithm auditing tools that log decision rationales or provide human-interpretable explanations. For example, certain image classifiers now highlight regions in radiographic images that contributed most to a cancer diagnosis, giving clinicians much-needed context alongside the AI's prediction (Hanna et al., 2025).

As AI and ML reshape the core functions of medical devices, the intersection of updating mechanisms, growing code complexity, and reduced explainability creates new vulnerabilities. Algorithm updates flowing over the network present opportunities for malicious actors to intercept or compromise models. Extensive code and subsystem integration mean a single overlooked flaw can propagate throughout a networked ecosystem. Meanwhile, opaque decision-making frustrates both detection and response to failures or cyberattacks. These advances deliver vast improvements in clinical performance and efficiency while at the same time demanding vigilant, specialized defenses—a reality that defines the next frontier in medical device security (Hanna et al., 2025; Rabbani et al., 2025).

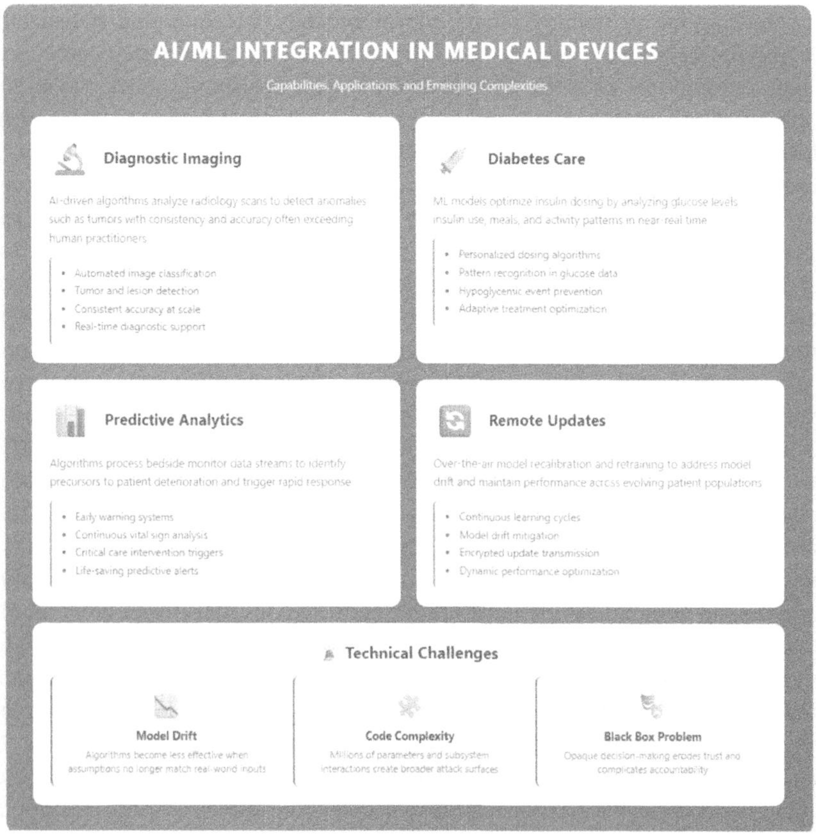

Identifying Unique AI-Driven Attack Vectors in Medical Devices

Modern medical devices increasingly rely on integrated AI and ML models, deepening the complexity of their internal architectures and expanding the potential attack surface. As devices such as smart infusion pumps, AI-enabled imaging platforms, and autonomous diagnostic agents grow more interconnected, the vulnerabilities specific to their AI components demand rigorous examination. These vulnerabilities stem both from the technical intricacies of model design and from the dynamic, autonomous qualities of learning systems embedded in clinical workflows.

A critical threat to medical device security comes from data poisoning attacks. In these scenarios, threat actors inject false or manipulated data into the training pipeline, intentionally skewing the AI model's learning process. This corrupts foundational algorithms, causing unpredictable behavior or systematically degrading clinical performance. For example, consider an infusion pump designed to titrate medication dosages using real-time patient feedback and historic datasets. If adversaries introduce tainted cases—such as false high-dosage requirements—into the pump's training data, the model may begin to recommend dangerously inappropriate treatments. The insidious aspect of data poisoning lies in its subtlety; a small fraction of manipulated records can evade common validation checks, inserting unsafe or biased logic deep within the model's learned weights. Most organizations lack mechanisms to proactively detect poisoning, leaving risks latent until manifested in clinical harm (Ellis, 2025; *What Is AI Security? Key Threats and How to Stop Them*, 2025). Addressing this vulnerability requires organizations to implement provenance tracking for all training inputs, robust anomaly detection, and statistically driven audits to screen for anomalous examples. Ensuring data integrity at every step—from acquisition through preprocessing—serves as the primary defense against poisoning attacks (Ellis, 2025).

Model inversion and theft present another class of vulnerabilities, particularly relevant to medical devices deployed in distributed care environments or exposed to remote queries. Through repeated and carefully crafted queries, adversaries can infer proprietary training data or even reconstruct sensitive details about included patients. In the context of a diagnostic algorithm embedded within a portable screening device, an attacker may extract patient attributes—such as disease markers or demographic features—by observing subtle variations in model outputs. This not only exposes confidential health information but also places the intellectual property underpinning complex medical logic at risk. Such attacks undermine both patient privacy and the competitive value of proprietary models. Technical defenses include encrypting model parameters and outputs, rate-limiting or authenticating external queries, and implementing strict monitoring for anomalous access patterns (*What Is AI Security? Key Threats and How to Stop Them*, 2025). These approaches aim to prevent extraction, but persistent adversaries may still slowly glean information unless the device rigorously compartmentalizes sensitive data and enforces layered access controls.

Adversarial input generation represents a significant and practical risk within AI-powered medical technologies. Here, malicious actors design inputs engineered to exploit model weaknesses—often by introducing minimal, nearly undetectable perturbations that trigger gross misclassifications or unsafe behaviors. In clinical imaging, for instance, subtle modifications to a radiological scan can cause a diagnostic AI to overlook malignant lesions or to falsely identify healthy tissue as cancerous. Such adversarial examples capitalize on the high-dimensional, nonlinear mappings learned during training, where small input changes can lead to large output variations. This susceptibility is not accidental; it arises from the inherent way ML models generalize and the lack of robust, human-understandable decision boundaries. Defending against adversarial inputs in medical devices requires multiple layers of countermeasures, including input preprocessing, adversarial training routines that expose models to perturbed cases during development, and ensemble modeling approaches that reduce single-point model failures (Ellis, 2025; *What Is AI Security? Key Threats and How to Stop Them*, 2025).

Bias and fairness risks materialize when medical AI systems inherit or amplify disparities present in their underlying datasets. These can manifest as systematically worse outcomes or lower sensitivity for certain demographic groups. In scenarios such as AI-powered triage and prioritization systems, models may deprioritize minority populations if those groups were underrepresented in the training corpus. The downstream effects can include unjust delays in care, inferior diagnostic accuracy, or inappropriate therapy recommendations. Beyond operational risks, such bias exposes organizations to regulatory, legal, and ethical liability. AI in healthcare remains subject to strict expectations around equity and privacy, necessitating structured audits for demographic fairness and ongoing adjustments to reflect evolving population diversity (*What Is AI Security? Key Threats and How to Stop Them*, 2025).

The normalization of these risks within daily device operations reflects the evolving threat landscape in medical technology. As AI moves toward continuous learning and autonomous operation, proactive, structured strategies for AI security become vital elements in ensuring clinical safety and compliance, setting the stage for the development and deployment of robust technical countermeasures.

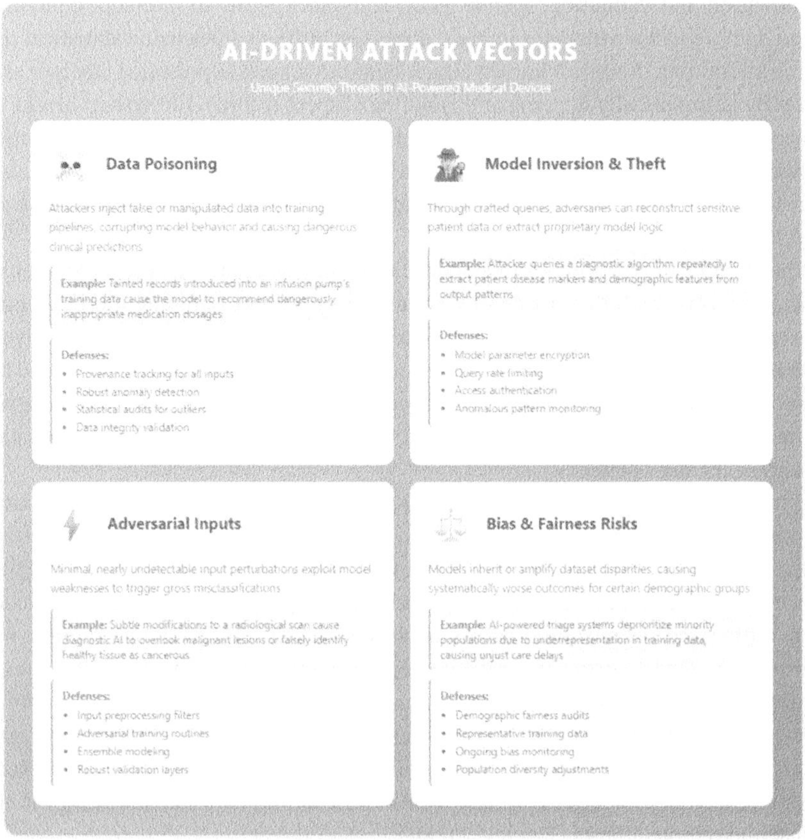

Securing AI Training and Operational Pipelines for Medical Devices

AI-enabled medical devices present new security challenges and require tailored strategies to defend against threats like adversarial manipulation, data poisoning, and embedded bias. The first layer of defense starts with model security, which addresses how adversaries might exploit neural networks to influence device behavior. For example, attackers can craft subtle input changes that trick diagnostic imaging AI into misclassifying X-rays. Healthcare organizations must rigorously test models with adversarial datasets and enforce routine robustness evaluations during the development life cycle. Equally, persistent bias in training data can cause unsafe or inequitable device performance. Bias auditing—using representative population datasets and diversity metrics—counters this risk so medical devices consistently serve all demographics (Rawles, 2025; *The Risk Assessor's New Role in the Age of AI: Are You Ready?*, 2025).

Data provenance forms the foundation of trust in AI-powered pipelines. Every step from data collection to model training and deployment must have a verifiable record. For example, clinical guidelines, payer-specific requirements, and expert human reviews informing device recommendations are all captured within a provenance record. In infusion pump safety systems, source verification and audit trail creation mark the origin and transformation history of each data point, enabling traceability if a device functions unexpectedly (Rawles, 2025). Source verification, such as assigning cryptographic signatures to validated lab results, ensures that only trustworthy datasets enter the model's training corpus. If a vendor supplies physiological signals, automated cross-referencing and digital attestation confirm legitimacy before the signals impact learning cycles.

Data sanitization and validation operate as critical checkpoints to intercept contaminated or manipulated data. Practical strategies include removing outlier values using statistical thresholds and scanning all training inputs for known attack signatures, such as poisoned labels or anomalous image noise. By automatically flagging and isolating suspect records, the system blocks adversarial data injection attacks that may otherwise degrade model acuity. In blood glucose monitoring devices powered by AI, automated dataset validation detects and excludes corrupted clinical trial records, ensuring only high-confidence data guides algorithmic predictions (*The Risk Assessor's New Role in the Age of AI: Are You Ready?*, 2025).

Continuous monitoring platforms, embedded at every pipeline stage, further shield sensitive systems. These platforms track the provenance of incoming training samples, alerting engineers to anomalous spikes or behavioral drift. In a remote cardiac monitoring device, anomaly detection mechanisms analyze streaming patient telemetry for patterns that diverge from validated baselines, helping responders intervene long before model performance declines. Integrated dashboards combine real-time alerts, anomaly statistics, and provenance details to provide a unified threat surface for device engineers and operators.

Securing model delivery and storage is essential to prevent theft, tampering, or unauthorized modifications. All model binaries must use strong cryptographic encryption when transmitted over hospital networks or uploaded to edge devices. AES-256 encryption secures model files, while secure transport protocols such as TLS 1.3 prevent interception during network uploads. Once stored, only authorized users should access deployed models, enforced via granular role-based access controls on device firmware. Multifactor authentication mitigates risk when clinical engineers approve firmware updates. Secure enclave technology and hardware security modules store cryptographic keys and critical models, ensuring only authorized application code loads and decrypts them, even if an attacker compromises other device layers. In connected insulin pumps, secure enclaves protect dosing algorithms and training datasets, reducing the impact of both physical and remote attacks (Rawles, 2025).

Human-in-the-loop oversight remains an indispensable safeguard, even in highly automated device ecosystems. Clinical domain experts review algorithmic outputs in real time or at scheduled intervals, providing expert assessment of prediction accuracy and identifying potential sabotage. For example, in AI-driven anomaly detection on infusion pump flow rates, technicians regularly audit flagged events, reviewing system logic against their clinical knowledge before approving device responses. These review cycles establish baseline performance metrics, track system drift, and support rapid manual intervention if automated alerts miss subtle failures. Performance monitoring dashboards consolidate expert input, capturing not only system responses but also the rationale behind human overrides, which developers can use to recalibrate future models (*The Risk Assessor's New Role in the Age of AI: Are You Ready?*, 2025).

Ensuring auditability and transparency further supports security by equipping engineers and regulators with detailed logs of device decision-making. Model audit trails must record input data hashes, applied model versions, clinical context, feature activations, and output recommendations. Explainable AI tools help clarify why a model generated a specific output, presenting model logic and confidence scores in clear language. These technologies enhance audit capacity, helping compliance teams, engineers, and clinicians trace the sequence of events that led to medical device actions. Transparent records also facilitate regulatory submissions, with detailed logs supporting postmarket device surveillance, incident investigations, and regulatory reviews required by the Health Insurance Portability and Accountability Act (HIPAA) and U.S. Food and Drug Administration (FDA) guidelines (Rawles, 2025; *The Risk Assessor's New Role in the Age of AI: Are You Ready?*, 2025).

Security, auditability, and oversight must operate within a robust governance framework addressing legal, regulatory, and ethical obligations. HIPAA mandates end-to-end audit trails for access and use of protected health information, while FDA Software as a Medical Device rules require full provenance documentation and ongoing performance validation. Adhering to these frameworks not only optimizes patient safety but also protects organizations from compliance failures and reputational damage, uniting technical rigor with ethical responsibility to govern the future of AI-driven medical devices (Rawles, 2025; *The Risk Assessor's New Role in the Age of AI: Are You Ready?*, 2025).

Navigating Regulatory and Ethical Issues for AI Medical Devices

Regulatory agencies such as the FDA and the European Medicines Agency (EMA) have recognized that AI and ML present new regulatory challenges for medical devices. Their evolving guidelines emphasize that static approval processes often fail to address the dynamic, adaptive behavior characteristic of AI-driven models, particularly for systems that continue to learn after initial deployment. Both agencies require rigorous premarket review for safety and effectiveness, requiring device manufacturers to provide evidence from clinical data, risk assessments, and robust validation of model performance. However, unlike traditional devices, AI-powered systems also undergo continuous postmarket monitoring. Regulators require ongoing surveillance to detect performance drift, unanticipated risks, or emerging biases once AI tools operate in real-world clinical environments. For example, if a diagnostic algorithm retrains on local patient data, its risk profile may shift, prompting the need for additional validation or regulatory reassessment. This risk-based framework enables tighter oversight on higher-risk functions—such as autonomous triage or disease prediction—while supporting incremental model improvements through controlled updates.

Transparency in AI decision-making, or explainability, is increasingly mandated by regulators as a principle for medical AI. The FDA and EMA instruct manufacturers to document how algorithmic decisions are reached, requiring that reasoning pathways be interpretable by clinical staff. Explainability is crucial in healthcare settings, where clinicians must understand and trust recommendations generated by AI tools, especially if these tools influence high-stakes diagnoses or treatments. For instance, an AI-powered radiology tool that highlights visual markers and provides rationales for flagging images facilitates clinician review and correction. Regulatory requirements stipulate features such as algorithmic auditing, version controls, and regular reporting, all aimed at ensuring real-world system behavior aligns with clinical expectations. These mechanisms also allow for swift identification and rectification of errors arising from unforeseen model adaptation or data distribution changes.

Informed Consent and Medical Ethics

AI-enabled devices introduce new ethical duties for healthcare organizations and providers, most notably around the process of informed consent. When AI systems participate in diagnosis or treatment recommendations, patients must be made aware of the technology's role and characteristics. Effective consent processes demand disclosure of an AI's abilities, limitations, and potential risks or uncertainties, especially when automated decisions may impact outcomes for conditions such as cancer staging or precision medication dosing. Ensuring patient understanding supports ethical principles of autonomy and promotes shared decision-making between patients and clinicians. Regulatory frameworks now embed requirements that AI-driven tools provide clear explanations for their outputs and allow patients to ask questions or seek human review, mitigating the risk of harm from unexamined automated decisions (Pham, 2025).

Ethical routines extend beyond consent to include nondiscrimination, equity, and respect for patient rights, which are foundational principles reflected in the Universal Declaration of Human Rights. These frameworks act as boundaries against AI-induced healthcare disparities, emphasizing that technology must not introduce or worsen inequities in care access or quality (Pham, 2025).

Cross-Border Data Movement and Legal Compliance Challenges

AI medical devices frequently operate in globally distributed environments, where models or patient-generated data flow across national borders. This creates intricate legal and compliance challenges due to varying regional standards for data privacy and protection. The European Union's General Data Protection Regulation (GDPR) enforces strict consents, transparency mandates, and limitations on cross-border health data transfer, with punitive measures for violations. In the United States, HIPAA imposes mandatory controls on health data, while Asian regulators, such as Japan's Pharmaceuticals and Medical Devices Agency (PMDA), combine innovation support with requirements on patient safety and controlled technology testing (Pham, 2025).

Deployment of a clinical AI system in both the EU and United States, for instance, means organizations must meet GDPR obligations for European patients while simultaneously adhering to HIPAA's rules for American users. This complexity increases for multinational device vendors whose AI models rely on cloud infrastructure or are retrained using international patient data. Legal exposure may arise from data leakage, insufficient consent when moving data into new regions, or failure to delete datasets processed without adequate permission (Pham, 2025). Intellectual property disputes may also emerge over who owns model improvements generated from diverse patient cohorts or collaborations between institutions.

Best practices in compliance include rigorous anonymization, encryption, and audit trails for data management, as well as regular oversight by multidisciplinary legal, technical, and clinical teams. Organizations must conduct legal impact assessments for new deployments and harmonize operational policies to align with the strictest applicable standards.

Collectively, these regulatory, ethical, and legal considerations define the constraints and opportunities for continued AI innovation in healthcare environments. Navigating this evolving landscape is vital for building sustained patient safety, institutional trust, and enabling future growth in increasingly intelligent medical devices (Pham, 2025). This detailed attention to frameworks and compliance shapes the foundation for anticipating and countering advanced security threats as technology's sophistication and interconnectivity develop.

Future-Proofing Security Strategies for Intelligent Medical Devices

Understanding the future of medical device cybersecurity hinges on embedding defense strategies that evolve alongside intelligent systems. Risk environments in healthcare continue to intensify as attackers increasingly target AI-powered medical devices. Security postures that rely solely on static rules are becoming obsolete under such conditions. Embedding adaptive defense mechanisms that autonomously learn from adversaries is now critical. AI-driven security controls in medical devices can process anomalous network patterns and user behaviors, flagging unexpected changes that bypass traditional signature-based rules. For instance, an AI-enabled intrusion detection system learns the expected communication flow of a networked insulin pump; it issues rapid alerts when an attacker attempts to reprogram dosage commands. In digital imaging infrastructure, behavioral analytics pinpoint suspicious use anomalies—such as a diagnostic tool being accessed from outside normal hours—triggering predictive containment workflows. Automated defensive updates, underpinned by current threat intelligence, allow these controls to evolve continuously. This agility ensures rules for segmentation, access permissions, or privilege escalation closely track the methods of real-world adversaries, reducing dwell time and opportunity for harm (Mohamed, 2025).

Medical device resilience pivots on continuous threat intelligence integration. Automated intelligence feeds ingest details about emerging malware, attack techniques, and threat actor signatures from global sources. Platforms synthesize this intelligence, compare it against real-time telemetry from devices, and refine internal security controls. When an external threat feed signals a new ransomware campaign exploiting smart device channels, a medical security platform immediately applies tightening quarantine rules to susceptible endpoints. Automated response processes isolate compromised infusion pumps before attackers achieve lateral movement, containing impact to patient care systems. The propagation of actionable threat intelligence among interconnected healthcare networks further increases systemic resilience. For example, when a European hospital identifies weaponized AI code targeting cardiovascular monitors, alerting and countermeasure recipes are instantly disseminated to partner facilities, blocking the threat before it disrupts patient safety elsewhere (Koppireddy, 2025; Mohamed, 2025).

Sustaining effective AI-driven protection in medical devices also depends on rigorous long-term data governance practices. AI models trained on high-fidelity telemetry excel at recognizing novel attack indicators, yet require meticulous stewardship of their underpinning datasets. Security policies must ensure that data provenance, quality, and access auditing are scrupulously managed. Protections for personal health information remain mandatory for regulatory compliance and business confidentiality throughout the data life cycle. When a healthcare provider refines behavior-based threat models for surgical robots, ongoing curation of anonymized usage data supports accurate model retraining, counteracts model drift, and exposes emerging tactics. Meanwhile, technological safeguards enforce access restrictions and encryption, ensuring only authorized security personnel retrieve historic or real-time model data for incident forensics or model validation. As advances in attack sophistication force continuous reevaluation of model fitness, robust data stewardship supports both patient privacy and system resilience (Mohamed, 2025).

AI-driven security operations optimize the day-to-day management of medical cybersecurity in increasingly automated clinical environments. ML techniques fortify vulnerable AI or ML components within medical systems by diagnosing model misuse, detecting attempts at adversarial manipulation, or flagging configuration weaknesses. Automated anomaly detection leverages baseline patterns to distinguish between legitimate device updates and unauthorized activity, such as a compromised diagnostic tool uploading data to an unknown server. Consider a radiology platform employing reinforcement learning to adapt its alert rules: it learns genuine workflow exceptions versus stealthy attack patterns based on feedback from incident responders. Intelligent analytics consolidate data from thousands of connected endpoints, transforming raw alerts into actionable insights. Security teams benefit from a reduced triage workload and improved prioritization, focusing their interventions on the most significant threats with context provided by advanced analytic frameworks. In healthcare delivery organizations that depend on fluctuating device networks, operationalized AI can spotlight cascading vulnerabilities before attackers exploit them, orchestrate cross-device patches, and automate system hardening routines without human delay (Koppireddy, 2025).

By bringing together adaptive defense, threat intelligence, resilient data stewardship, and AI-driven operations, the future of medical device cybersecurity becomes dynamic and preemptive. The convergence of automated learning, predictive analytics, and rigorous governance allows organizations to move from reactive posture to strategic anticipation, closing gaps that threat actors have traditionally exploited. Cybersecurity professionals, engineers, and medical IT leaders must consider these methodologies foundational for securing the next generation of intelligent healthcare systems (Koppireddy, 2025; Mohamed, 2025).

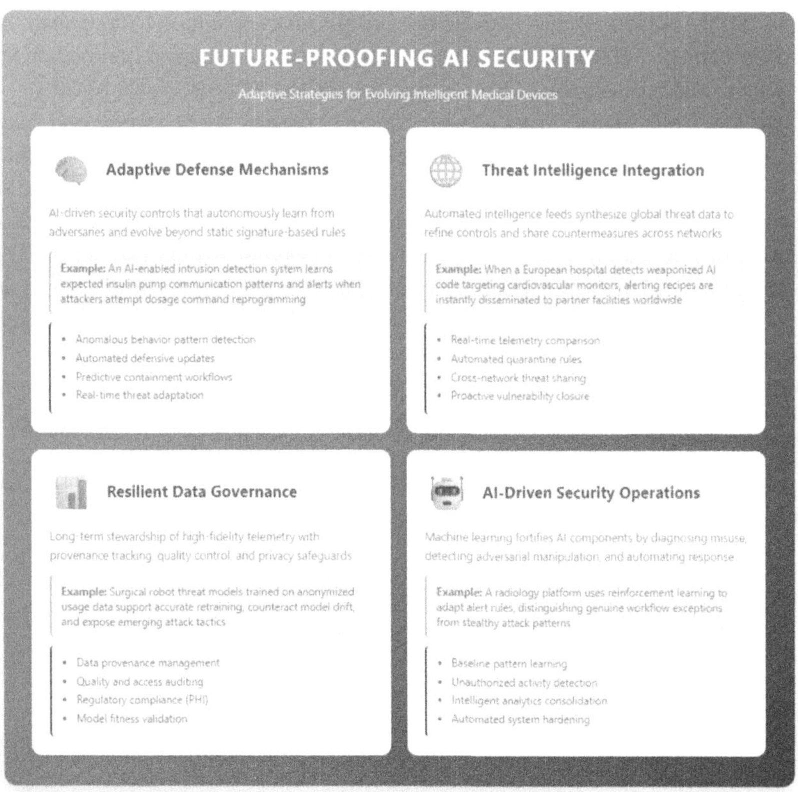

Final Thoughts

Now that we understand the profound impact of AI and ML on medical devices—including their capabilities, unique vulnerabilities, and complex regulatory environment—we can focus on developing adaptive, robust security frameworks that evolve alongside these technologies. Ensuring patient safety and data integrity requires integrating continuous monitoring, transparent audit mechanisms, proactive threat intelligence sharing, and strong governance to address dynamic risks such as model drift, adversarial attacks, and bias. By embracing these forward-looking strategies, healthcare cybersecurity professionals, device engineers, regulators, and IT managers can collaboratively safeguard intelligent medical systems against emerging threats, fostering trust and resilience in the future of connected healthcare.

References

Ellis, S. (2025, June 4). *Top 7 AI Security Threats and How to Protect Your Business.* Office1.com; Office1. https://www.office1.com/blog/top-ai-security-threats.

Hanna, M. G., Pantanowitz, L., Dash, R., Harrison, J. H., Deebajah, M., Pantanowitz, J., Rashidi, H. H. (2025, January 4). *Future of Artificial Intelligence (AI)—Machine Learning (ML) trends in pathology and medicine.* Modern Pathology; Elsevier. https://doi.org/10.1016/j.modpat.2025.100705.

Koppireddy, V. K. R. K. (2025, February 25). AI-driven Cybersecurity Integration: A Comprehensive Framework for Enterprise Security Automation and Threat Management. International Journal of Advanced Research in Engineering and Technology. https://doi.org/10.34218/ijaret_16_01_015.

Mohamed, N. (2025, April 30). *Artificial Intelligence and Machine Learning in Cybersecurity: A Deep Dive into State-of-the-Art Techniques and Future Paradigms.* Knowledge and Information Systems; Springer Science and Business Media LLC. https://doi.org/10.1007/s10115-025-02429-y.

Pham, T. (2025, May). *Ethical and Legal Considerations in Healthcare AI: Innovation and Policy for Safe and Fair Use.* Royal Society Open Science; The Royal Society. https://doi.org/10.1098/rsos.241873.

Rabbani, S. A., El-Tanani, M., Sharma, S., Rabbani, S. S., Yahia El-Tanani, Kumar, R., Saini, M. (2025, July 7). *Generative Artificial Intelligence in Healthcare: Applications, Implementation Challenges, and Future Directions.* BioMedInformatics. https://doi.org/10.3390/biomedinformatics5030037.

Rawles, T. (2025, June 14). *AI and LLM Data Provenance and Audit Trails for Healthcare Technology:* Building Trust Through Transparency. Onhealthcare.tech; Thoughts on Healthcare Markets and Technology. https://www.onhealthcare.tech/p/ai-and-llm-data-provenance-and-audit.

The Risk Assessor's New Role in the Age of AI: Are You Ready? (2025, July 14). Censinet.com; Censinet. https://www.censinet.com/perspectives/the-risk-assessors-new-role-in-the-age-of-ai-are-you-ready.

What Is AI Security? Key Threats and How to Stop Them. (2025, June 19). Mend.io. https://www.mend.io/blog/ai-security-guide-protecting-models-data-and-systems-from-emerging-threats/.

17

Wireless, Mobile, and Wearable Device Security

The rapid expansion of wireless, mobile, and wearable medical devices has transformed healthcare delivery by enabling continuous monitoring and greater patient engagement. However, this technological advancement comes with complex security challenges that differ significantly from traditional networked systems. The interconnected nature of these devices, their reliance on wireless communication protocols, and their integration within diverse environments create unique vulnerabilities. These risks are compounded by variations in user behavior, device management practices, and the blend of clinical and personal technology ecosystems.

This chapter examines the multifaceted security concerns associated with medical wearables and mobile health devices. It addresses the vulnerabilities inherent in their ecosystems, including wireless protocol weaknesses and app-related threats. Strategies for protecting sensitive health data through encryption, authentication, and incident response are explored in detail. Additionally, the chapter discusses best practices for managing device loss or theft and provides guidance on securely integrating personal devices into clinical and research settings. Together, these topics form a comprehensive overview aimed at strengthening the security posture of wireless and wearable medical technology in modern healthcare environments.

Understanding Threats in Wearable Device Ecosystems

Wearable and mobile health technologies reshape the boundaries of patient care, yet these gains come with an expanded array of cybersecurity threats. The complex networks underpinning health environments draw together wireless protocols, cloud ecosystems, user devices, and sprawling app stores, each introducing specific points of vulnerability. The risks associated with using medical wearables like heart monitors, glucose sensors, or fitness trackers hinge on the distinct environments in which these devices operate as well as the behaviors of their users.

In tightly controlled hospital settings, wearable devices typically transmit sensitive data over segmented, hospital-managed networks. Security operations teams implement strict monitoring, frequent system updates, and network-level protection to minimize the likelihood of unauthorized access. Examples include cardiac monitors that send patient vitals through encrypted wireless channels confined to the hospital's infrastructure. Device manufacturers and hospital IT departments cooperate on regular patch management and incident reporting, supporting device integrity. Staff are also trained to recognize unusual device behavior and follow set protocols for device isolation if a security breach is suspected. In these secure environments, network access is often limited to authorized endpoints, and firewall rules restrict communication with external parties, which dramatically narrows potential attack vectors (Ewoh & Vartiainen, 2024).

Transitioning to home or community settings, medical wearables face a radically changed risk landscape. Devices may pair with household Wi-Fi routers lacking proper configuration, exposing them to external probing or eavesdropping. Users often rely on default or weak passwords, rarely update firmware, and may not recognize the signs of compromise. While in the hospital, a pulse oximeter's data remains within a secure perimeter, at home the same device might upload readings via unencrypted channels or through an inadequately secured home network. Attackers could conduct man-in-the-middle attacks, intercepting streamed health data such as glucose levels from a Bluetooth-enabled sensor to a user's smartphone (Isreal et al., 2024). Elderly patients and those unfamiliar with security recommendations are particularly vulnerable, as irregular updating and unintentional disclosure of credentials may go unnoticed.

Wearable usage patterns also differ meaningfully. In clinical environments, continuous data monitoring is a byproduct of acute medical care, with round-the-clock usage and many access attempts by different caregivers. In contrast, home settings often see wearables used more sporadically, with data syncing whenever a device happens to connect to a user's phone or home network. The diversity of user devices adds another layer: a wearable might sync one day with a family member's older unpatched tablet and the next with a secure, modern smartphone. The weaker host device becomes an open gateway for malware or unauthorized access to sensitive records, further expanding the attack surface and complicating efforts to enforce consistent device security (Isreal et al., 2024).

App ecosystems supporting wearables and health devices introduce additional vectors for compromise. Companion applications handle data aggregation, analytics, and cross-platform sharing, but apps distributed via app stores or third-party sources may bypass device-level security assurances. Vetting standards differ between app stores; malicious or poorly designed applications sometimes slip through, or users intentionally sideload unverified health tracker apps. Technical examples include Android-based smartwatches infected by sideloaded apps that masquerade as legitimate pedometer tools but in reality exfiltrate heart rate data or login credentials to remote attackers. Such applications may access Bluetooth stacks, intercept data in real time, or install additional payloads upon pairing with susceptible hosts. Integration of third-party analytics or advertising network software development kit (SDKs) into wearable health apps adds further uncertainty, often facilitating unintended data sharing or introducing vulnerabilities due to inadequate regulation and oversight (Isreal et al., 2024).

Continuous, real-time streaming of health data by wearables means that sensitive information—personally identifiable data, biometric signals, and behavioral markers—remains perpetually exposed to interception. When a smartwatch relays a patient's blood pressure across Bluetooth with unencrypted or outdated protocols, eavesdroppers theoretically gain access to a rich data stream spanning health status and daily routines. End-to-end encryption becomes a critical requirement for each link in the chain, yet strong encryption sometimes conflicts with clinical demands for low-latency, always-available data access. Attackers may exploit these technical gaps, especially as data streams traverse less-protected points like home networks or sync with external services (Isreal et al., 2024).

Pairing and synchronizing wearables with personal devices introduce additional vulnerabilities. A blood glucose meter that communicates securely with a hospital's central records system can become compromised when paired at home with a phone running outdated software. The Bluetooth connection may be susceptible to exploits, allowing for the injection of malware or unauthorized access to device memory, jeopardizing not only the wearer's health data but also potentially the broader digital health infrastructure should the infected phone sync to cloud platforms. The variability in security posture across different devices—some current with security updates, others obsolete—demands greater attention to platform-agnostic security standards and robust threat detection methods (Ewoh & Vartiainen, 2024; Isreal et al., 2024).

Examining these numerous and shifting risks highlights the significant influence of operational context, user awareness, interconnected systems, and evolving device ecosystems on overall security. Understanding these interdependencies forms the groundwork for a deeper look at vulnerabilities found within wireless communication protocols, where protocol-level misconfigurations and emerging exploits remain ongoing challenges for the stability and confidentiality of medical device ecosystems.

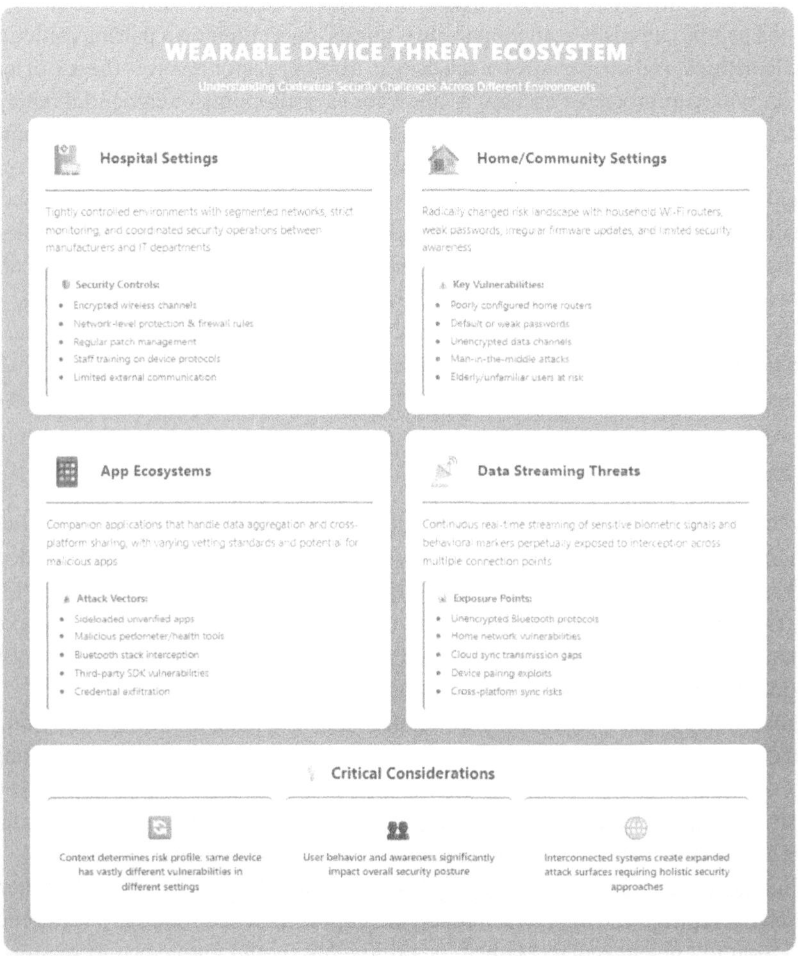

Risks in Bluetooth, NFC, and Short-Range Protocols

Short-range wireless protocols underpin the essential connectivity between wearable, mobile, and implantable medical devices. These links, while enabling efficiency and patient-centered care, also introduce distinctive vulnerabilities that adversaries can exploit if protocol defenses are incomplete or improperly configured. Each commonly used protocol—Bluetooth, Near Field Communication (NFC), and related wireless standards—presents attack surfaces that demand focused analysis and tailored configuration to protect sensitive clinical workflows and patient data.

Bluetooth Pairing Attacks

Bluetooth, and especially its low-energy variant BLE (Bluetooth Low Energy), is widely adopted for connecting glucose monitors, cardiac sensors, and infusion pumps to companion apps or hospital information systems. Device pairing represents a crucial step, where temporary trust relationships are established. Traditional pairing methods have employed static PIN codes, numeric comparisons, or, in some cases, require no authentication. Attackers exploit pairing weaknesses by executing "man-in-the-middle" or impersonation attacks, intercepting the pairing exchange when cryptographic protection is absent or user verification is weak (Cäsar et al., 2022).

For example, if clinical staff deploy wireless medication dispensers using default PINs, adversaries situated nearby can easily scan for pairing requests, inject their own pairing credentials using predictive algorithms, and seize control of the device. Another vector involves the use of legacy pairing protocols, where unprotected exchanges allow for eavesdropping or credential replay. Devices placed in patient-facing roles, intended to be "frictionless," may allow pairing with minimal user intervention, further reducing the barrier for attackers to execute interception or device mimicking.

To prevent such breaches, strong configuration practices are critical. Bluetooth Secure Connections pairing, available in BLE version 4.2 and above, leverages Elliptic Curve Diffie-Hellman cryptography and enforces authenticated pairing with user confirmation or numeric comparison. Device provisioning must avoid static PINs, instead enforcing unique, one-time-use passkeys. Additionally, administrators should disable legacy and insecure modes, restrict device discoverability, and log all pairing events for forensic review (Cäsar et al., 2022; Koutras et al., 2020). These measures insulate medical devices from compromise during and after the initial pairing process.

NFC Scan and Relay Threats

NFC protocols support rapid and contactless authentication, prescription tracking, and patient identification in clinical settings. Despite their short range, NFC-enabled medical devices face unique risks, notably relay attacks that artificially extend the communication perimeter. Attackers employ two colluding devices: one positioned near the target device or terminal, the other near the legitimate user. A wireless "tunnel" relays authentication traffic, tricking medical systems into accepting unauthorized transactions (Koutras et al., 2020).

For instance, a malicious relay could allow an unauthorized person outside a hospital to trigger medication dispensing by capturing data broadcast by a nurse's card in another area. Coupled with insufficient encryption, attackers may scan for unencrypted data, clone device credentials, or modify in-transit prescriptions. These attacks are exacerbated in environments where confidentiality and rapid workflows preclude exhaustive user checks at every authentication step.

The technical remedy for NFC-based vulnerabilities includes strict mutual authentication, session-based encryption, and deployment of cryptographic nonces—random values generated for each transaction to prevent replay. Medical environments should enforce short session timeouts and require user confirmation for high-impact transactions. Moreover, disabling passive NFC operation in favor of active authentication and using shielded access readers further reduce the threat window (Koutras et al., 2020).

Wireless Jamming and Interference

Jamming attacks on medical device protocols introduce operational risks that extend beyond data theft. By broadcasting noise within the same radio band, adversaries impede or disrupt communication between patient monitors and hospital networks, threatening both workflow integrity and patient safety. Tools used for these attacks include programmed radio transmitters that overwhelm channels or deploy targeted interference against specific devices.

For example, targeted jamming during the remote monitoring of a cardiac patient could delay arrhythmia alerts, impeding timely intervention. Even transient interference may disrupt wireless insulin pumps or IV administration, causing cascading delays. The broad adoption of wireless communication in hospital infrastructure escalates these risks, rendering reactive manual fallback insufficient.

Mitigation strategies integrate both physical and protocol-level measures. Spectrum monitoring tools can alert staff to abnormal radio activity, facilitating proactive mitigation or channel switching. Device firmware should include channel-hopping capabilities, enabling them to avoid occupied frequencies. Security testing against interference, rigorous electromagnetic compatibility (EMC) evaluation, and maintaining a robust fallback wired communication plan ensure resilient operation under sustained attack (Cäsar et al., 2022; Koutras et al., 2020).

Firmware Over-the-Air (FOTA) Risks

Modern medical devices frequently support over-the-air firmware updates, designed to roll out critical patches and feature updates. However, when FOTA mechanisms lack cryptographic integrity checks or proper authentication, adversaries can intercept update prompts, substitute rogue firmware, or execute persistent infection. Medical infusion pumps with unsecured FOTA channels provide a concrete example: an attacker could modify medication delivery algorithms by mimicking trusted update servers.

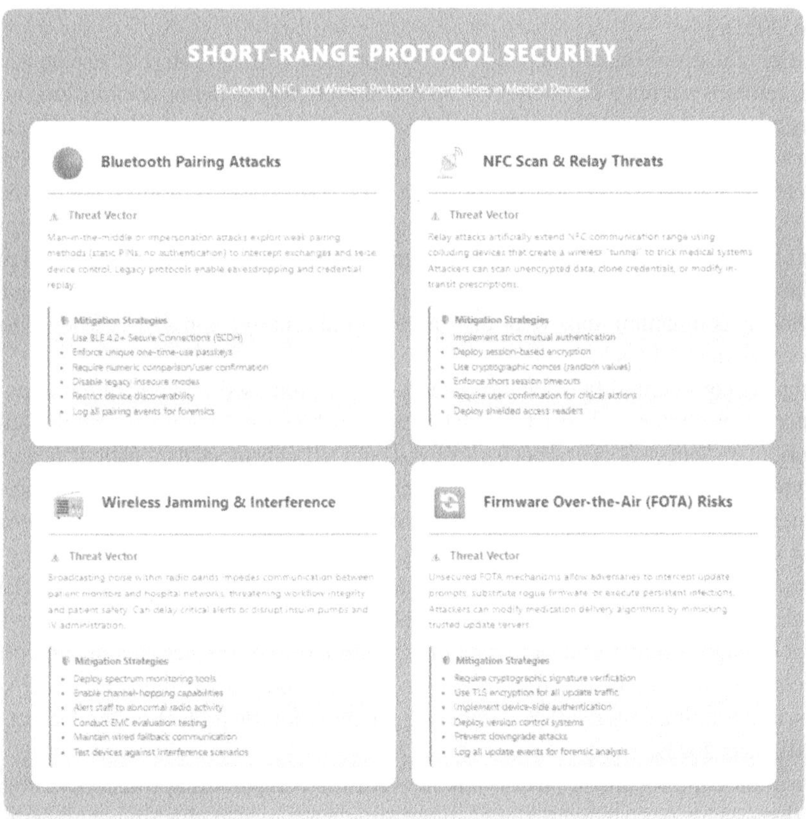

Key defenses for secure FOTA center on cryptographic signature verification, whereby only firmware signed by trusted developers will execute on the device. Transport Layer Security (TLS) must encrypt all update traffic, and device-side authentication ensures update servers present valid cryptographic credentials. Version control systems prevent downgrade attacks and log all update events, providing traceability for forensic analysis (Cäsar et al., 2022; Koutras et al., 2020).

As short-range protocols and mobile device interfaces become further integrated, scrutinizing not only the underlying radio security but also all interacting applications is essential. Robust technical defenses at protocol and device levels form the baseline for effective ecosystem protection, emphasizing the necessity of including companion applications and user-facing systems in holistic security governance.

Protecting Mobile Health Applications and Patient Data

App store vetting acts as a frontline defense for companion apps designed to interact with medical devices. Mobile app ecosystems frequently face threats not just from flaws in wireless protocols but also from malicious code and unsafe configurations at the application level. Effective app vetting means that each application is carefully checked before it becomes available for download, ensuring that potentially harmful software does not penetrate user devices. Rigorous app review typically involves automated static and dynamic analysis, paired with human assessment, to catch known malware signatures, suspicious behavior, and privacy issues. Continuous monitoring further strengthens this layer by catching problems that emerge after initial approval—such as malicious updates or new vulnerabilities—and pulling compromised apps before they cause harm (Median.co, 2025).

The security of users ultimately depends not only on technical controls, but also on user education, which remains a crucial component. End users, especially patients or clinicians, need guidance to only download authorized apps from official stores rather than using direct APKs, unofficial markets, or modded versions. For example, proactive in-app banners and onboarding pop-ups can alert users to the dangers of sideloading or using unauthorized apps, reinforcing correct behaviors at the moment of decision. Trusted bazaar programs or certification logos on approved apps help further by distinguishing genuine software from malicious lookalikes. Educational messages, delivered through clinic materials or device setup guides, have proved essential following incidents where malicious companion apps were distributed on alternative app stores, leading to theft of health records or device hijack.

Protecting the confidentiality of health data stored or processed by companion apps requires rigorous encryption strategies. Local data, such as medical histories, monitoring results, and configuration profiles, presents a prime target for adversaries if left unprotected on patient devices. Inadequate encryption exposes users to the risk of data exfiltration from lost, stolen, or malware-infected devices. Encrypting data at rest is essential. Healthcare apps should adopt AES-256 or a similarly robust symmetric encryption algorithm, with encryption and decryption running through platform-provided secure storage APIs such as Android Keystore or Apple's Secure Enclave. Relying on weak key management leads to predictable or retrievable encryption keys; for instance, storing keys in plain text alongside app data directly undermines the benefits of encryption. Best practice dictates generating keys within hardware security modules, setting appropriate key usage restrictions, and rotating keys regularly, alongside pin or biometric protection for decryption operations (Median.co, 2025).

Authentication controls play a pivotal role in elevating companion app security. Single-factor authentication—such as a password alone—offers little resistance to phishing or credential reuse.

Adding multifactor authentication (MFA), combining something the user knows (password), something the user has (authenticator app or secure token), and something the user is (biometric), provides layered defense. Biometric verification (using fingerprints or facial recognition) demands integration with operating system biometric services. This ensures that the biometric data never leaves the device or the secure enclave, protecting it from interception and replay attacks (Bob, 2025; Median.co, 2025).

The technical implementation of MFA in companion apps often involves time-based one-time passwords—delivered by third-party features like Google Authenticator—and push-based approvals. For critical workflows, such as device re-pairing or access to sensitive historical health data, an additional one-time password can be triggered. Usability poses a recognized challenge. Patients, especially those with disabilities or limited digital literacy, may find complex login hurdles discouraging. Solutions include adaptive authentication, which selectively applies high-security checks only when risk signals—like a new device or location—are detected, while allowing routine access through more streamlined touch or biometrics (Bob, 2025).

Auditing app permissions is another foundational security activity and has grown in importance as mobile operating systems expose more device capabilities to applications. Over-granting permissions or failing to revoke obsolete privileges can open up the attack surface.

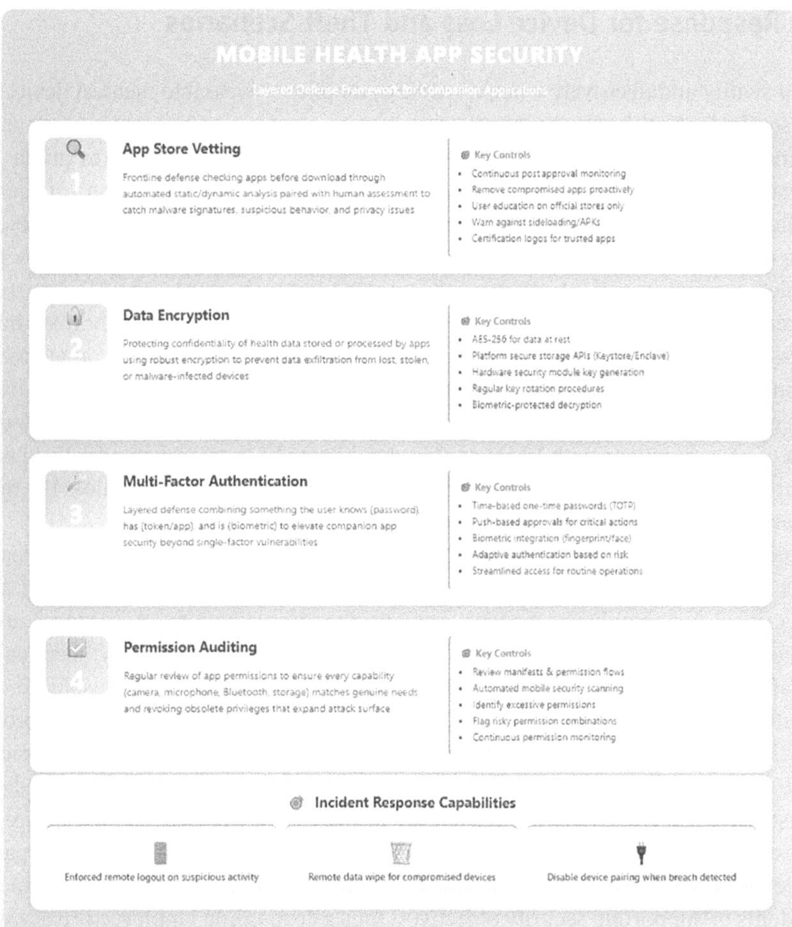

Every permission—access to camera, microphone, Bluetooth, or file storage—should match genuine needs. Regular audits require reviewing manifests and permission request flows, and using automated tools or mobile security monitoring solutions to identify excessive permissions and highlight risky combinations, like location access paired with internet data transmission. Industry examples reveal that breaches often exploited companion apps that were granted background microphone or file access, which attackers later abused for data leakage. Automated scanning and continuous permission review now form a core part of the defense for several regulated healthcare app developers, minimizing exposure as app features evolve (Median.co, 2025).

The possibility of device compromise, whether through theft, loss, or sophisticated malware, means that rapid risk-mitigation measures are vital. When compromise is detected, features such as enforced logout, remote data wipe, or disabling device pairing reduce the window for attackers. Crafting clear incident response playbooks ensures teams respond swiftly, limiting the damage from local breaches and preserving trust in the overall system. As the threat landscape expands around medical device companion apps, a robust and layered security posture—grounded in rigorous vetting, encryption, authentication, auditing, and incident readiness—is not just prudent, but necessary (Bob, 2025; Median.co, 2025).

Incident Response for Device Loss and Theft Scenarios

Responding swiftly and decisively when a wireless, mobile, or wearable medical device is lost or stolen is essential for healthcare organizations striving to protect sensitive patient data and uphold the integrity of their clinical operations. Robust incident response measures guide technical teams in managing such events, and remote wipe capabilities are a cornerstone of these efforts. When configured and deployed correctly, remote wipe allows for immediate, irreversible deletion of sensitive data from devices that are no longer physically secure. For example, if a physician loses a tablet containing patient charts while moving between wards, the IT department can activate a remote wipe, ensuring clinical information does not fall into unauthorized hands. The effectiveness of this measure relies on prerequisites such as persistent device connectivity—often requiring mobile devices to maintain regular contact with management services over Wi-Fi or cellular networks—and automation via enterprise mobility management (EMM) platforms. Integration of wipe functions with EMM enables healthcare IT to trigger standardized workflows and guarantee consistent response times, regardless of which staff member notices the incident or which department experiences the loss (*Mobile Device Security*, 2019).

Complementing remote wipe are advanced location tracking and event logging mechanisms. These tools work in tandem to pinpoint misplaced or stolen devices and construct a detailed audit trail critical for compliance reviews and incident investigations. A scenario in which a mobile device goes missing from a nursing station demonstrates their utility; location tracking can help staff pinpoint whether the device left the premises, was accidentally misplaced, or moved to an unauthorized area. Meanwhile, logging access attempts and device activity produces an immutable forensic record, aiding in the detection of tampering or illegal data extraction. For instance, if audit logs show several failed login attempts after the device disappeared, they can help reconstruct events and identify if sensitive records were exposed (*Mobile Device Security*, 2019). It's essential, however, to balance these capabilities with privacy obligations. Organizations should develop proportional tracking protocols, limiting location surveillance to responding to actual incidents or supporting security objectives, thus safeguarding employee privacy while still satisfying regulatory demands.

Auto-locking mechanisms provide another vital layer of protection. When configured effectively, these controls automatically lock devices after specified periods of inactivity or following repeated failed authentication attempts. This can dramatically reduce the opportunity for opportunistic abuse or data mining if devices are unattended for even brief moments in busy hospital settings. A device left on a nurse's cart is quickly locked, cutting off access before unauthorized individuals can exploit it, and repeated tampering—such as multiple failed PIN entries—can trigger administrative alerts or more comprehensive lockdowns (*Mobile Device Security*, 2019). Some devices also include tamper sensors that detect when casings are opened or attempts are made to reset system credentials, further increasing the security barrier. The challenge lies in striking a careful balance: policies must promote security without creating frustrating obstacles for clinicians who need rapid, reliable access to tools during urgent care situations. Customizable timeout settings and adaptive authentication thresholds help organizations tailor their approach, preserving usability while tightening security.

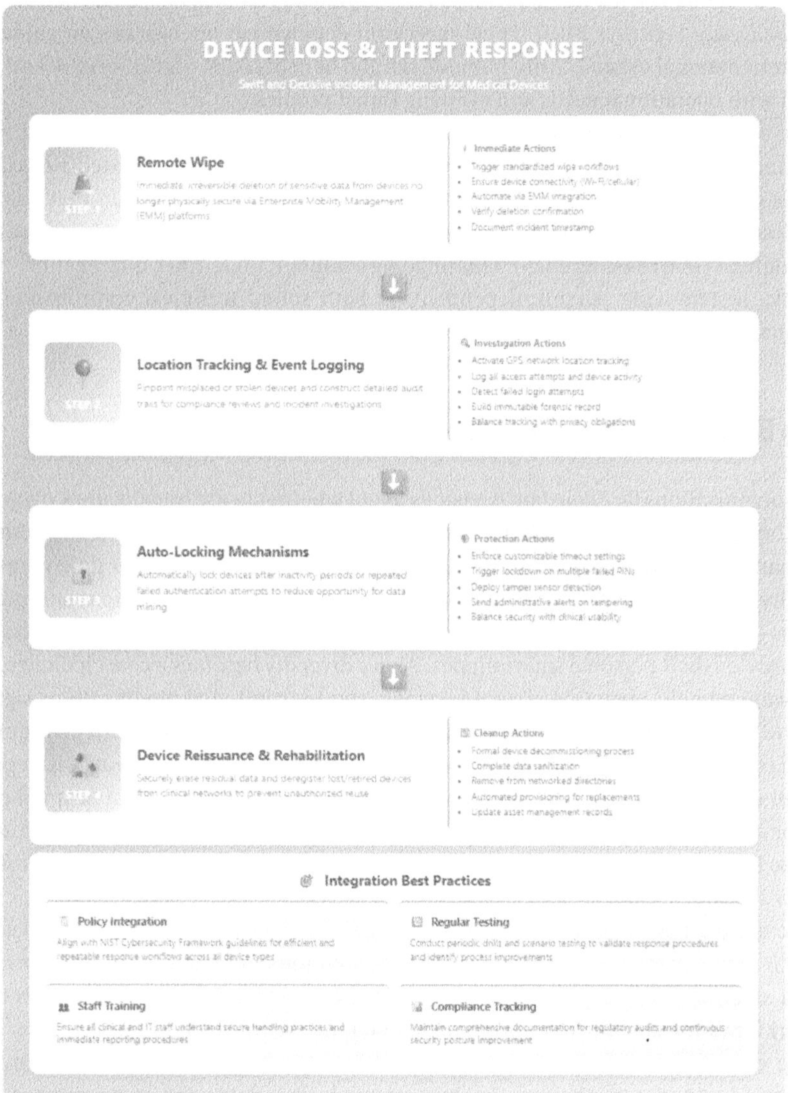

Attention to device reissuance and rehabilitation further fortifies the healthcare environment against lingering threats. Securely erasing any residual data and thoroughly deregistering lost or retired devices from clinical networks prevent unauthorized reuse or reentry into sensitive workflows. When devices are replaced or circulated to new users, identity and access management systems, combined with EMM platforms, support automated provisioning and credential reassignment, minimizing opportunities for manual misconfiguration or orphaned access rights (*Mobile Device Security*, 2019). Consistent organizational protocols ensure that every decommissioned device is formally recorded, wiped, and removed from all networked directories. During transition periods, staff training on secure handling practices, as well as up-to-date policy documentation, strengthens defenses against lapses that could inadvertently expose patient information.

Each of these technical controls—remote wipe, tracking and logging, auto-locking, and systematic device reissuance—works in concert to mitigate the impact of lost or stolen medical devices. Incident response workflows in healthcare depend not only on the presence of these features, but also on their integration with standards-based management platforms for efficiency and repeatability (*Mobile Device Security*, 2019). Their successful deployment often draws on guidelines and mappings from national cybersecurity frameworks and benefits from regular review and testing to stay aligned with operational needs and evolving threat profiles.

As device-centric controls become more sophisticated, attention must also extend to policies for integrating personal and wearable technologies used in clinical or research workflows. The complexities of securing both organization-issued and personal devices mean that risk management practices must evolve, encompassing broader asset classes and adapting processes to unique operating contexts (marks4sure, 2025). This ongoing evolution underlines how healthcare security is a dynamic, system-wide pursuit, dependent on both sound technical controls and adaptive organizational policies.

Securing BYOD in Clinical and Research Settings

Healthcare organizations face constant decisions about whether to use organization-owned devices or to allow bring your own device (BYOD) models, each carrying specific advantages and unique security, financial, and compliance trade-offs. Hospital-issued tablets can be preconfigured with strict security controls—such as remote wipe capabilities, enforced encryption, and preinstalled clinical apps—ensuring uniform compliance across the fleet. In contrast, when clinicians access patient records via their personal smartphones, device diversity becomes a core challenge. Variability in hardware models, operating system versions, and user-initiated modifications significantly increases the operational complexity of enforcing policy and ensuring security. Administrators must weigh cost savings from reduced device procurement and maintenance against a potentially increased attack surface and the need for specialized support. For example, a hospital provisioning iPads for all clinicians ensures technical homogeneity and straightforward policy enforcement but may face higher up-front costs and reduced flexibility for staff accustomed to Android devices (Wani et al., 2022).

Robust device management through Mobile Device Management (MDM) systems has emerged as a linchpin for securing diverse endpoints. MDM solutions, originally designed for traditional laptops and smartphones, now support an expanding ecosystem, including wearables that enter clinical workflows for patient monitoring or note-taking. With MDM, administrators can automate device enrollment, enforce minimum OS versions, push application and security updates, and trigger rapid response mechanisms such as remote lock or data wipe in the event of suspected compromise or loss. For instance, if a nurse's smartwatch used for patient alerts goes missing, MDM platforms enable the security team to revoke hospital app access and disable Bluetooth connectivity proactively. However, technical limitations become apparent when securing wearables; their restricted processing power and fragmented operating system support constrain the sophistication of security controls, sometimes precluding device-level encryption or real-time monitoring and requiring tailored risk mitigation strategies (Demanuele et al., 2022; Wani et al., 2022).

Integration of personal devices meaningfully supports agility in healthcare delivery, yet adds nuanced layers to workflow and security oversight. A physician documenting rounds on a personal tablet may access productivity-boosting apps unavailable on hospital-locked hardware, enhancing efficiency and provider satisfaction. Similarly, remote patient monitoring via BYOD wearable sensors allows real-time health data capture, encouraging patient engagement and enabling timely clinical interventions (Demanuele et al., 2022). Yet, these advantages hinge on strict controls to prevent data spillage, unauthorized access, or delayed incident response, particularly during emergencies. Role-based access restrictions and containerization solutions isolate clinical data from personal information on BYOD devices. Prompt biometric authentication and session expiry policies reduce exposure from unattended devices in fast-paced clinical settings. The configuration of frictionless, always-on VPNs helps restrict the transmission of protected health information (PHI) to secured channels only, but these controls must not delay emergency access or create barriers to urgent care.

Effective implementation of BYOD policies requires a combination of technical controls, administrative oversight, and change management tailored to healthcare's operational realities. Policies governing device eligibility clarify the range of permitted operating systems, minimum security posture, and acceptable use scenarios. Automated network access control tools assess device compliance in real time, denying access or placing noncompliant personal devices in restricted network zones. Regular compliance scans can detect outdated or jailbroken devices and trigger alerts or access suspension. Hospitals frequently supplement technical controls with targeted audit procedures, such as cross-checking mobile device logs against access records or monitoring for anomalous data transmission indicative of an application vulnerability or insider threat (Wani et al., 2022).

User education and structured change management underpin ongoing security success in BYOD environments. Many hospitals provide annual and situational training on secure device use, social engineering risks, and proper procedures for reporting incidents or device loss. These programs include simulation exercises that challenge staff to recognize phishing attempts and respond to simulated breaches, embedding a culture of vigilance throughout the workforce. Nevertheless, surveys indicate that while staff education is present in most organizations, it is rarely comprehensive or regular, which leaves gaps in user preparedness and overall threat readiness (Wani et al., 2022). Comprehensive onboarding for new applications or technology rollouts ensures clinicians can deploy new tools securely and without service interruption.

Continuous oversight, timely policy reviews, and rapid remediation capacity form the backbone of sustainable BYOD security. Hospitals conduct periodic policy assessments to adapt to evolving device trends, system vulnerabilities, and regulatory updates. Real-time alerting systems ensure that policy violations, such as attempts to install unapproved applications or disable security features, trigger immediate containment actions and incident investigations. Incident postmortems provide actionable feedback for refining training and technical controls. Throughout these practices, hospitals work to balance flexibility, productivity, and security, recognizing that ongoing stakeholder engagement and transparent policy communication are crucial for long-term compliance and the secure integration of BYOD and wearables within healthcare environments (Demanuele et al., 2022; Wani et al., 2022).

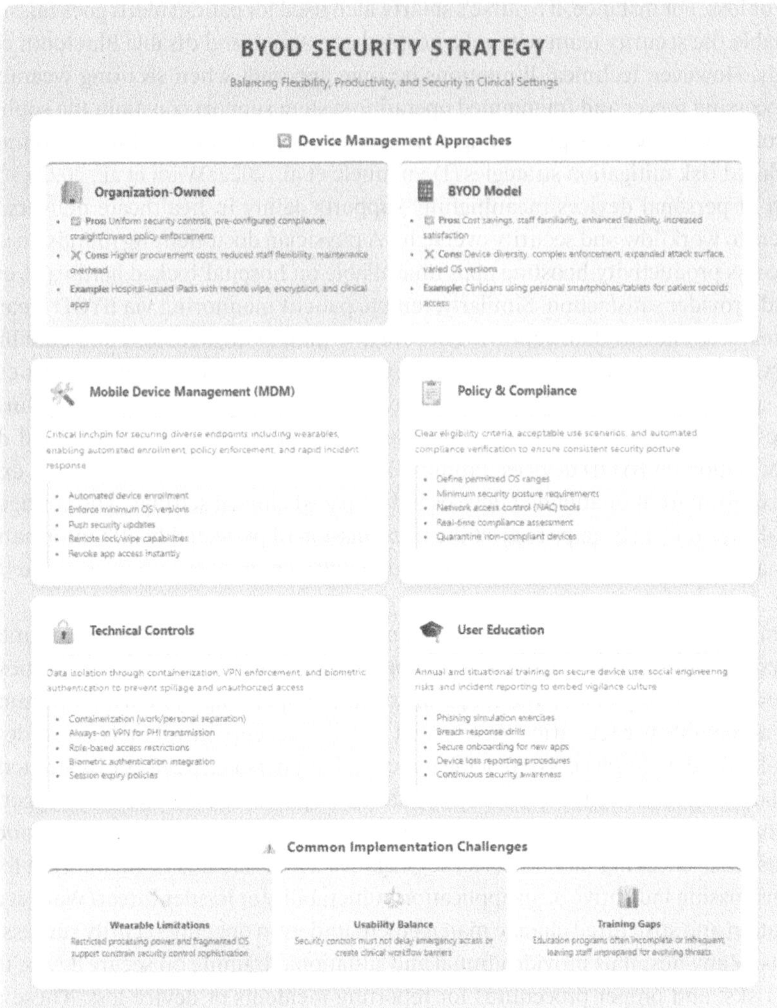

Final Thoughts

Having examined the complex security challenges inherent to wireless, mobile, and wearable medical devices—from ecosystem vulnerabilities and protocol weaknesses to data protection and incident response—it is clear that safeguarding these technologies demands a comprehensive, multilayered approach. As healthcare continues to embrace device integration and BYOD models, professionals must prioritize robust encryption, rigorous authentication, vigilant monitoring, and adaptive policy frameworks to mitigate evolving risks. Moving forward, collaboration between device manufacturers, healthcare providers, and cybersecurity experts will be essential to develop standardized, scalable defenses that keep pace with technological advances while preserving patient safety and data confidentiality. By applying the strategies outlined in this chapter, stakeholders can build resilient systems capable of anticipating emerging threats and ensuring secure, effective healthcare delivery in increasingly connected environments.

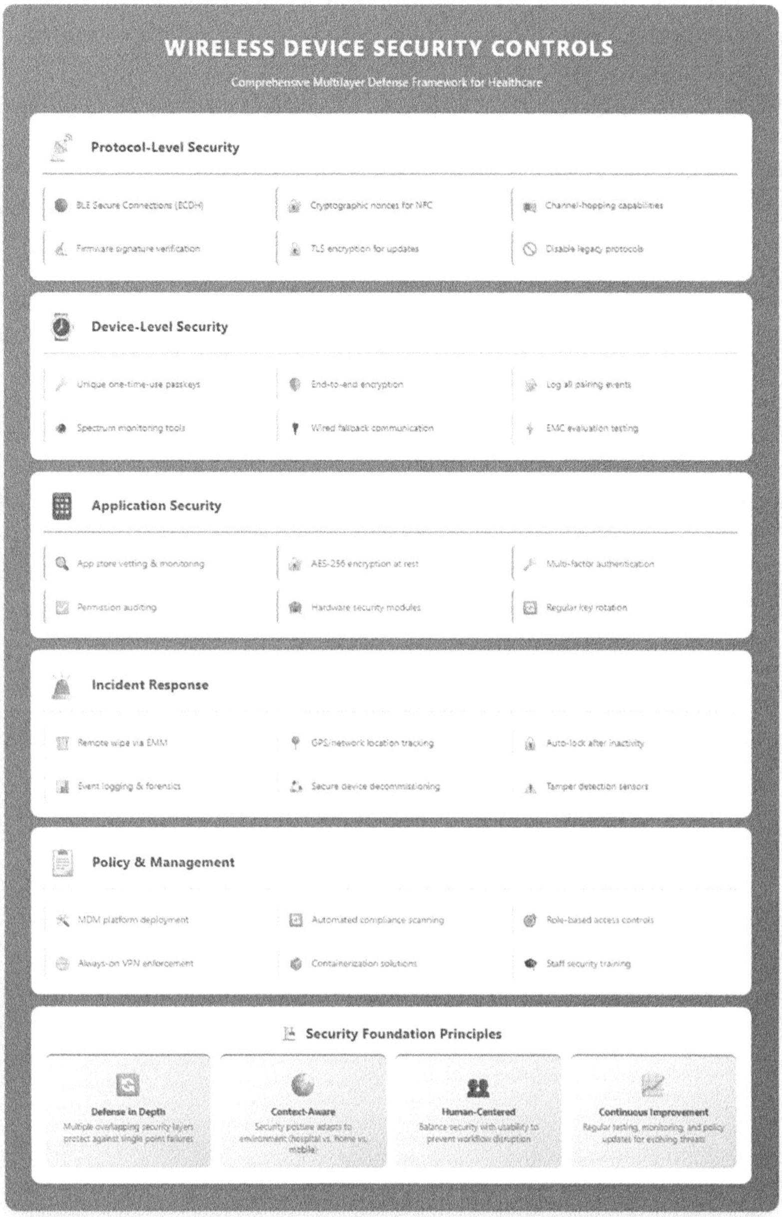

References

Bob. (2025, April 8). *Choosing the Right Authentication Methods for Mobile App Security*. Dogtown Media. https://www.dogtownmedia.com/choosing-the-right-authentication-methods-for-mobile-app-security/.

Cäsar, M., Pawelke, T., Steffan, J., Terhorst, G. (2022, March). *A Survey on Bluetooth Low Energy Security and Privacy*. Computer Networks. https://doi.org/10.1016/j.comnet.2021.108712.

Demanuele, C., Lokker, C., Jhaveri, K., Georgiev, P., Sezgin, E., Geoghegan, C., Zou, K. H., Izmailova, E., McCarthy, M. (2022). *Considerations for Conducting Bring Your Own "Device" (BYOD) Clinical Studies*. Digital Biomarkers. https://doi.org/10.1159/000525080.

Ewoh, P., Vartiainen, T. (2024, May 31). *Vulnerabilities, Cyberattacks and Socio-technical Solutions in Healthcare Systems: Systematic Review (Preprint)*. JMIR. Journal of Medical Internet Research/Journal of Medical Internet Research; JMIR Publications. https://doi.org/10.2196/46904.

Isreal, O., Rajuroy, A., Penzenstadler, B., Matthew, B. (2024, June 29). *Medical Data Breaches: Risks from Connected Wearables*. ResearchGate; unknown. https://www.researchgate.net/publication/393362145_Medical_Data_Breaches_Risks_from_Connected_Wearables.

Koutras, D., Stergiopoulos, G., Dasaklis, T., Kotzanikolaou, P., Glynos, D., Douligeris, C. (2020, January 1). *Security in IoMT Communications: A Survey*. Sensors. https://doi.org/10.3390/s20174828.

marks4sure. (2025). CompTIA SY0-701 New Questions—August updated SY0-701 Exam Release. Marks4sure.com. https://www.marks4sure.com/sy0-701-comptia-securityp-exam-questions.html.

Median.co. (2025). *Median.co*. Median.co. https://median.co/blog/best-practices-to-protecting-data-in-your-mobile-apps.

Mobile Device Security. (2019). Nist.gov. https://www.nccoe.nist.gov/publication/1800-4/VolB/.

Wani, T. A., Mendoza, A., Gray, K., Smolenaers, F. (2022, April). *Status of Bring-Your-Own-Device (BYOD) Security Practices in Australian—A National Survey*. Health Policy and Technology. https://doi.org/10.1016/j.hlpt.2022.100627.

18

Securing Medical Device Updates and Maintenance

How can healthcare organizations ensure that updates to medical devices are delivered securely without interrupting patient care? What strategies help manage the risks associated with device maintenance throughout their life cycle? How do medical teams balance the need for timely software patches with strict regulatory requirements and operational constraints in clinical environments? These questions highlight the complex challenges faced when securing medical device updates and maintenance.

This chapter offers detailed guidance on managing these challenges by exploring methods for secure update delivery, effective planning within clinical settings, proper decommissioning practices, continuous vulnerability management, and maintaining compliance with regulatory standards. The focus is on practical approaches that uphold device safety, reliability, and integrity while minimizing disruption to healthcare workflows. Readers will find comprehensive strategies designed to support cybersecurity professionals, medical device engineers, regulatory officers, and healthcare managers in protecting critical medical technologies throughout their operational life.

Best and Worst Practices for Update Delivery Mechanisms

Update Delivery Mechanisms: Direct vs. Over-the-Air

Medical device update strategies depend on the controlled, secure delivery of software and firmware to minimize risk and preserve safe clinical operation. There are two key mechanisms for delivering updates: direct physical connection and over-the-air (OTA) transmission. Physical update methods use hardware interfaces, such as USB, serial cables, or manufacturer-specific docking stations. These approaches reduce exposure to external, remote threats by isolating the device from hospital networks or the internet during the update. Their main benefit is that by requiring on-site physical access, opportunities for remote exploitation are constrained, supporting compliance efforts in highly sensitive areas, such as operating rooms and intensive care units (Infrastructure, 1997). However, these methods demand significant manual intervention. Staff must schedule time with field engineers or biomedical technicians to cycle through multiple devices, sometimes requiring device downtime, which can disrupt workflows and delay patient care. For example, a network-isolated infusion pump may undergo updates during planned, sterile maintenance windows using a technician's encrypted USB drive.

OTA update mechanisms, in contrast, offer remote management options that increase operational agility. Devices receive update packages over hospital wired or wireless networks, or even cellular links, which streamlines deployment and supports rapid scale across large fleets of connected assets. Such flexibility comes with increased attack surface; updates may be intercepted or spoofed by adversaries if secure communication protocols are not in use. Malicious actors can exploit vulnerabilities in endpoints or communication channels to inject unauthorized firmware or exfiltrate sensitive information. Assessment of environmental sensitivity is critical—the appropriateness of OTA updates depends on network segmentation, endpoint protections, and the presence of layered technical safeguards (Javaid et al., 2023). Diagnostic monitors in general ward settings, connected to a VLAN with strict firewall policies, may be suitable for OTA updates during nonclinical hours. In contrast, life-sustaining implantable devices in patients require much stricter controls, including multifactor authentication prior to instituting remote changes, or complete avoidance of OTA methods if residual risk is unacceptable.

Secure Channel Evaluation Criteria

Selection of an update channel relies on structured evaluation of multiple factors. The sensitivity of the clinical environment influences risk tolerance; devices in critical care benefit from isolated update routes that minimize network-based exposures. The presence of strong network controls, such as encrypted VPNs, network access control lists, and intrusion detection systems, affects the viability of OTA updates (Infrastructure, 1997). Evaluate whether each device supports modern encryption schemes (e.g., TLS 1.3) and ensure connection endpoints are regularly authenticated and logged. For physical updates, control the chain-of-custody for update media, logging every step of device access. Device criticality, combined with the need to avoid clinical disruption, often means a tailored hybrid approach: for some bedside diagnostic monitors, a fallback to physical update is mandated if remote authentication fails; for others, OTA may be viable if risk can be managed.

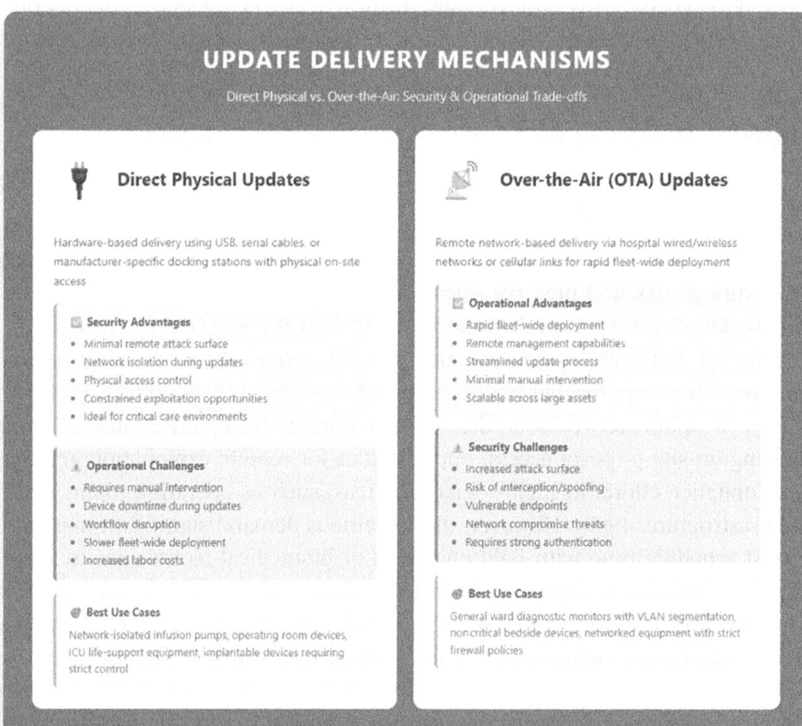

Authentication Protocols and Package Integrity Verification

Update authentication is central to a secure deployment strategy. The update process must verify both the authenticity and integrity of any incoming package. This is typically implemented using cryptographic signatures—each update is signed by the manufacturer's private key, and the receiving device checks the signature against a trusted public key embedded in secure hardware or firmware (Javaid et al., 2023). Any tampering or package alteration results in signature mismatch, automatically halting the process and alerting administrators. For implantable cardiac devices, public key infrastructure (PKI) systems enable secure delivery of firmware updates, ensuring only the vendor can authorize software changes. Device authentication processes should include the verification of digital certificates, checking issuer validity, expiration, and revocation status. Regularly update trust anchors on devices to counteract emerging threats such as certificate compromise or advances in cryptanalysis (Infrastructure, 1997). The authentication workflow involves obtaining the update package, checking its digital signature against a trusted certificate authority, validating package content hashes, then logging the update event for auditing.

Scheduling and Minimizing Care Disruption

Applying updates requires careful timing and scheduling to protect clinical workflows. Update windows should be coordinated with periods of minimal clinical activity, such as shift handovers, weekly maintenance cycles, or scheduled device downtime (Javaid et al., 2023). Planners should involve clinical stakeholders, giving them input into scheduling decisions. For instance, firmware refreshes for bedside monitors may be executed during early morning shift changes, avoiding direct impact on patient monitoring. Devices must allow for staged rollouts with rollback capability—should an update fail, the equipment should maintain core operational functions in a fail-safe state, preserving patient safety and clinical uptime. Emergency exceptions, triggered manually or automatically, should allow aborting updates and reverting to prior stable versions.

Automated vs. Manual Update Controls

Automation streamlines deployment, enhances coverage, and reduces human error, but it can also propagate operational risks if not appropriately managed. Automated systems can push timely updates across hundreds of devices simultaneously, but a malfunctioning or malicious package may affect all assets at once. Manual approaches, by contrast, are slow and resource-intensive, but allow critical review and local oversight. Device management systems should provide flexible policies, letting authorized staff approve, schedule, defer, or reject updates as clinical needs dictate (Infrastructure, 1997). In intensive care or high-acuity environments, end-user approval is vital; automated updates should be restricted, and clinical teams must confirm device readiness before proceeding. Training programs reinforce the importance of timely updates, help end users recognize and properly respond to update prompts, and improve procedural compliance across disciplines.

Technical and logistical strategies work together for minimal clinical disruption and robust device protection. By aligning security practices with real-world workflows, healthcare organizations can ensure updates safeguard both device integrity and patient safety (Infrastructure, 1997; Javaid et al., 2023).

Planning and Executing Safe Updates in Clinical Use

Secure delivery channels and trusted authentication, while foundational, are only the beginning of risk-managed device updates in real-world healthcare environments. Once an update is securely transferred and cleared for execution, precisely engineered logistical and technical solutions are required to prevent service interruption and ensure both patient safety and operational continuity. The approach begins with granular control over deployment through staged rollouts, further protected by comprehensive backup procedures, coordinated team communication, and robust automated rollback mechanisms.

Staged Rollouts for Risk Containment

Staged rollout strategies partition the deployment process, incrementally pushing updates to selected device subsets rather than en masse hospital-wide. Often guided by criteria such as

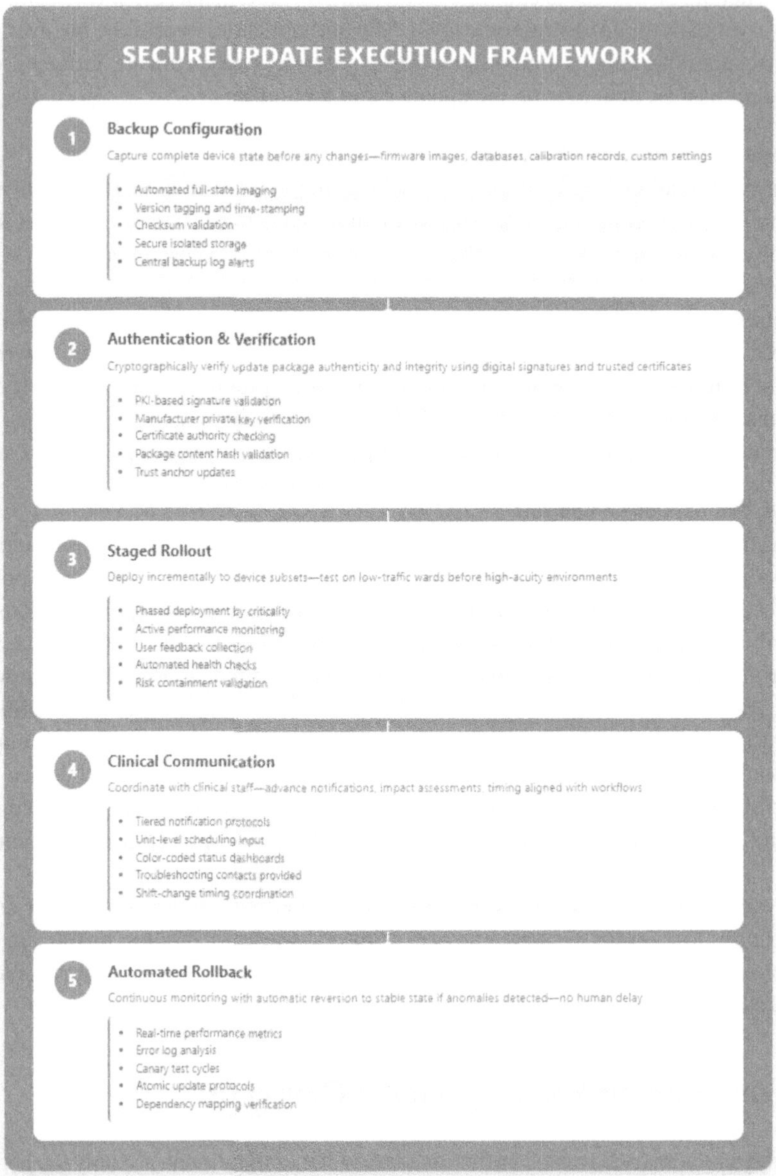

device criticality, patient load, or department scheduling, this phased approach allows for active monitoring during early stages. A limited rollout exposes a smaller population to any undetected issues, provides an environment for observation, and isolates potential failures before they can propagate. For example, an infusion pump update might be first installed on low-traffic wards or backup units monitored closely for a defined period. Success at this stage is validated through performance logs, automated health checks, and user feedback before the update moves to devices in high-acuity settings.

If anomalies emerge—remote device communication errors, unexpected clinical application behavior, or delayed system responses—rollback procedures are immediately enacted. Rollback depends on system-side snapshots created just before the update; these capture all application binaries, configuration files, and operational settings. Devices disconnect from the update channel, restore the stable environment using these snapshots, and then reenter the monitored operating pool. This combination prevents the initial failure from impacting broader clinical workflows and maintains service predictability, which is essential for dependent care applications (Ştefan et al., 2024).

Backup Configurations as a Foundation for Safe Change

Prior to any device update, capturing the existing configuration and operational state is a paramount safety step. Backup protocols involve duplicating firmware images, database files, calibration records, and custom user settings onto secure, isolated media. Automation of this process reduces the likelihood of manual omissions and accelerates recovery if needed. For example, a networked ventilator scheduled for an operating system patch has its full-state imaged beforehand, using an automated agent embedded in the hospital's device provisioning pipeline. Should the postupdate validation detect a parameter drift or system boot issue, restoration simply reloads the preupdate configuration, instantly returning respiratory support to its verified baseline.

Automated backup integration into update platforms also enables version tagging and time-stamping, which streamline regulatory audits and incident response. In some deployments, backup logs are transmitted to central control panels, alerting technicians to anomalies in the backup process itself—such as incomplete files or checksum mismatches. This provides a critical second line of defense, allowing clinical care to continue seamlessly while errors are analyzed and fixes are deployed (Ştefan et al., 2024).

Coordination and Communication with Clinical Staff

The human factor in clinical settings is as crucial as the technical. Minimizing disruption relies on transparent and advanced notification to all personnel dependent on device availability. Best practices entail a tiered communication protocol: stakeholders receive scheduled update notifications well ahead of time, with messaging calibrated to unit-level needs—such as during shift changes, after rounds, or in elective procedure lulls. Each notice specifies the duration, anticipated impacts, functional changes, and troubleshooting contacts. Real-world examples include color-coded dashboards in central nurse stations displaying device statuses or integration with staff scheduling software that defers updates outside peak patient volumes.

Staff input is actively solicited during the operational impact assessment. For instance, ICU managers may request updates be deferred during periods of heightened critical activity, while surgical teams may align device update windows with regular equipment sterilization cycles. Clear lines of communication reduce operational uncertainty and facilitate contingency planning, as personnel are primed to anticipate and report postupdate symptoms. This readiness, fostered by early alerts and participatory planning, diminishes adverse outcomes and supports smooth transitions between software versions (Ştefan et al., 2024).

Automated Rollback and Fault Detection Technologies

To further mitigate patient risk, medical devices increasingly incorporate automated rollback technologies. Upon completion of an update, built-in monitoring agents assess function through key performance metrics, error logs, and canary test cycles. Evidence of anomalies—such as loss of connectivity, critical process hangs, or performance regression—automatically triggers the rollback workflow. Rather than awaiting human intervention, devices revert to their last known state without delay, restoring full service availability.

Automated mechanisms demand thorough predeployment validation. Regular simulated failure drills—artificially inducing faults and verifying rollback performance—ensure real-world dependability. One common pitfall is overlooked dependency restoration, where underlying libraries may not revert cleanly. Countermeasures include atomic update protocols and dependency mapping tools that guarantee holistic rollback. Systems designed under these principles continuously verify performance at each update stage, closing potential gaps before they reach patient-facing services (*The Federal Risk and Management Program Dashboard*, n.d.).

Carefully orchestrated update management, integrating staged rollout, robust backup, coordinated human communication, and automated rollback, creates a foundation of trust and safety. The focus shifts naturally to life cycle completion, emphasizing secure decommissioning and data protection when medical devices reach end of service.

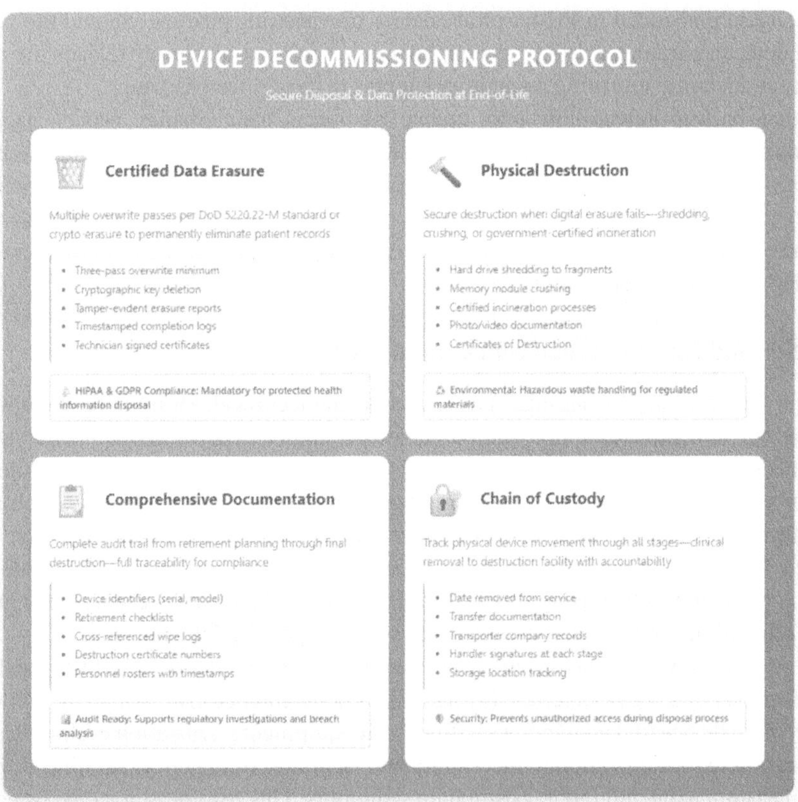

Decommissioning and Secure Disposal of Devices

Healthcare organizations must approach the decommissioning of medical devices with the same rigor used throughout the active life cycle. Protecting patient privacy and maintaining compliance depend directly on the integrity of data erasure, disposal practices, and meticulous documentation. Secure device retirement bridges the gap between daily risk management and the long-term safety and reliability expectations for electronic health technology.

Device data wiping stands as a primary defense against unauthorized data recovery. Certified data erasure protocols require practitioners to perform one or more overwrite passes—often prescribed by standards such as DoD 5220.22-M—that replace all addressable storage sectors with fixed or random patterns. This process helps ensure that residual patient records, treatment logs, or device operation histories are permanently eliminated. For newer devices supporting cryptographic storage, initiating a crypto-erasure routine by deleting and overwriting cryptographic keys can instantly render all associated encrypted data indecipherable. Teams should carefully verify the device's compatibility with such operations, referencing manufacturer documentation for built-in secure erase commands that ensure compliance with mandates like the Health Insurance Portability and Accountability Act (HIPAA) and Medical Device Regulation (MDR) (*A Guide to Data Destruction for Hospitals—ITAMG*, 2025).

Verification processes require the generation of tamper-evident erasure reports and logs. These records should itemize the storage media targeted, time stamps of each erasure, the methods utilized, and confirmation from the responsible technician. For example, after executing a three-pass overwrite conforming to DoD 5220.22-M on a retired medical imaging workstation, the IT team should receive an automated log file and, if available, a signed certificate from the device's secure erase utility. Maintaining such evidence demonstrates due diligence during audits and provides a defensible record if chain-of-custody concerns ever arise (Bream, 2025).

Legal and regulatory frameworks raise the stakes for all secure disposal activity. In the United States, HIPAA demands not just reasonable protection of patient health information, but also enforceable destruction protocols when retiring devices. The EU's General Data Protection Regulation (GDPR) stipulates explicit rights to erasure, heightening penalties for improper or incomplete destruction efforts. Improper device disposal can expose legacy patient data, resulting in multi-million-dollar fines, as well as reputational damage and mandatory breach notifications. Recent enforcement actions against clinics that failed to destroy digital storage on obsolete machines illustrate these consequences. Ensuring staff remain updated on relevant disposal statutes, industry best practices, and evolving manufacturer guidelines assists in closing compliance gaps. Many institutions arrange for annual security training, leverage regulatory update bulletins, or assign compliance officers to monitor these requirements (*A Guide to Data Destruction for Hospitals—ITAMG*, 2025).

Physical destruction is often required when digital erasure is not possible or when storage hardware cannot be reliably sanitized. Devices with nonremovable memory, unresponsive hardware, or advanced tamper-resistant features may resist traditional overwriting or cryptographic erasure. Secure destruction methods in these cases include shredding hard drives into small fragments, crushing memory modules, or using government-certified incineration processes to ensure total obliteration of data carriers. Hazardous waste handling and environmental regulations impose further responsibilities on managing debris, especially for devices containing heavy metals or other regulated substances. Partnering with certified third-party vendors, while retaining audit rights and reviewing their environmental track records, enables institutions to meet both data protection and sustainability commitments (Bream, 2025). Detailed evidence should accompany each destruction event, including photo documentation, video footage where warranted, and formal Certificates of Destruction. These records prove that both the device and the data it held no longer pose a risk to patient privacy or institutional integrity.

Comprehensive documentation underpins both regulatory compliance and operational clarity. Retirement checklists must account for all device identifiers, including model numbers, serial numbers, and asset control barcodes. Data erasure activities should provide cross-referenced wipe logs, while records of physical destruction must retain the corresponding certificates and relevant environmental compliance data. Chain of custody documentation tracks the physical movement of devices as they transition from clinical environments to storage, erasure stations, or destruction facilities. Personnel rosters and time stamps for each activity enable incident response investigations and, if necessary, forensic analysis following a breach claim. Some organizations employ audit templates that group these data points, ensuring traceability from initial retirement planning to final destruction. For example, a retirement record may include a checklist such as: device serial, date removed from service, method and date of data erasure, log file hash, destruction certificate number, transporter company, and signatures of all handlers (Bream, 2025; *A Guide to Data Destruction for Hospitals—ITAMG*, 2025).

Device decommissioning is not a final wall, but a continuation of risk-based management. Information security teams must anticipate that threat actors actively seek opportunities in device disposal and must maintain vigilance through detailed tracking, trusted vendor selection, and regular review of protocols. Awareness of emerging threats, rapid response to vulnerabilities discovered postretirement, and integration of lessons learned ensure a resilient life cycle approach as organizations turn to proactive vulnerability management in subsequent stages.

Vulnerability Management Throughout the Device Life Cycle

Secure device decommissioning and disposal represent the critical final phase in the life cycle of medical technologies, ensuring latent data and embedded software do not become exploitation vectors once devices leave service. Best practices dictate the purging of protected health information, disabling of connectivity modules, and secure destruction or recycling of device components. These measures reduce the residual attack surface and close off opportunities for attackers seeking to target devices outside active hospital oversight. Robust end-of-life protocols bolster confidence across stakeholders—healthcare providers, regulators, and patients—in the security processes safeguarding each phase from product retirement backward through the entirety of device management. This unbroken chain of vigilance carries over directly: the same principles applied at the end of device use must anchor the proactive frameworks necessary for continuously monitoring and responding to new vulnerabilities throughout the operational lifespan.

Vulnerability management in the context of medical devices comprises a structured and continuous approach to identifying, evaluating, and addressing security weaknesses or potential exploit points in device hardware and software. Core objectives for healthcare organizations include reducing the attack surface of medical systems, preventing exploitation of flaws that could threaten clinical operations or patient safety, and meeting rigorous regulatory requirements set by authorities such as the U.S. Food and Drug Administration (FDA). Specialized device considerations, such as real-time clinical use, integration with legacy networks, and the sensitivity of medical data, make it essential that the approach is both comprehensive and tailored to the unique needs of the sector (Santos, 2025).

Vulnerability disclosure programs reflect a frontline defense mechanism, inviting collaboration from security researchers, clinical staff, and even patients to surface threats that vendors or operators may have missed. These programs should define accessible, well-documented reporting channels—dedicated web portals, email addresses, or secure messaging—usable by both expert and nontechnical users. Policies ought to enshrine confidentiality, offer nonretaliation guarantees, and pledge swift, responsible communication. Such commitments, including coordinated vulnerability disclosure (CVD) timelines, ensure that newly discovered issues are quietly remediated before public exposure. For instance, a manufacturer might guarantee that reports will be acknowledged within 72 hours, and that details will only be publicized in coordination with stakeholders who can deploy patches or mitigations. Effective disclosure policies equip healthcare entities to act rapidly, patching exposed devices before adversaries can escalate attacks. They also foster trust, signaling openness to scrutiny and responsible data stewardship (Santos, 2025).

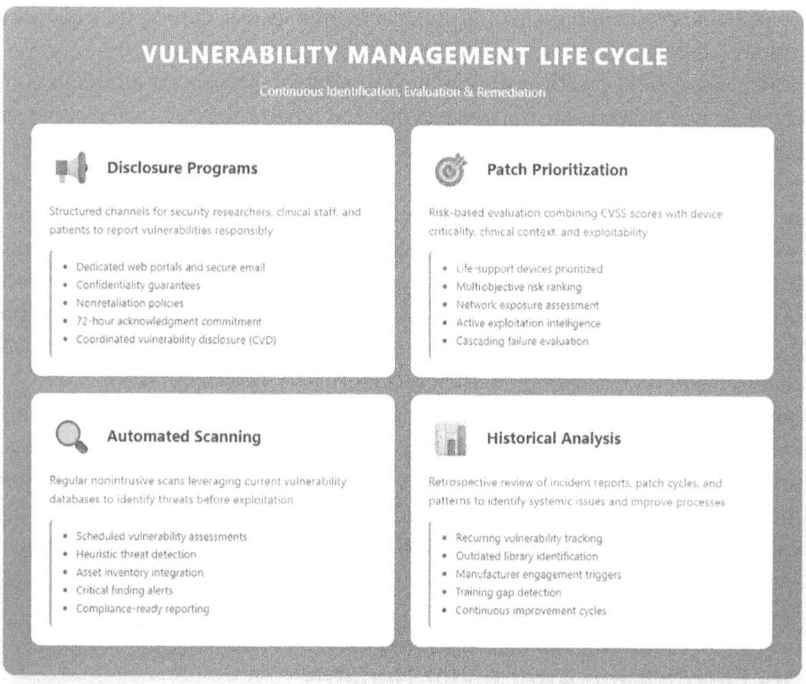

Patch prioritization frameworks deliver targeted remediation through risk-based evaluation and precise scheduling of fixes. Rather than defaulting to vendor or severity score alone, leading methodologies weigh technical factors—such as exploitability and network exposure—against clinical variables: is the device essential for life support, does the vulnerability threaten real-time safety, or could it trigger cascading failures in an interconnected care environment? Organizations adopt multiobjective approaches to rank vulnerabilities (combining, for example, CVSS [Common Vulnerability Scoring System] scores with device criticality and usage context), then structure patch deployment queues to minimize risks foremost to patient welfare. Incident response integrations allow escalation of patch deployment when active exploitation is detected in the wild or when intelligence signals imminent threat. As a practical illustration, devices connected to intensive care units or surgical workflows receive expedited attention, while updates to less critical assets follow after verification and staged rollout. Remediation plans must also detail fallback and communication procedures should patch deployment disrupt clinical operations, hence frameworks must coordinate closely with broader risk management (*A Survey on Vulnerability Prioritization: Taxonomy, Metrics, and Research Challenges*, 2025; Santos, 2025).

Automated vulnerability scanning tools are pivotal for preemptively identifying known and emerging threats across diverse medical device networks. Scheduling regular scans that leverage current vulnerability databases, heuristic detections, and asset inventory integration supports early identification of weak spots before attackers can exploit them. Automation reduces manual overhead, freeing clinical and IT staff for higher-order analysis, while embedded reporting mechanisms—configured to alert on critical findings—facilitate immediate incident escalation and follow-up. Scanning tools suited for healthcare environments must be nonintrusive, account for varied device operating systems, and provide compliance-ready reporting formats. Through integrations with asset management and patch deployment systems, organizations achieve consistent, repeatable evaluation cycles and actionable intelligence, closing information gaps that previously allowed overlooked vulnerabilities to persist (Santos, 2025).

Historical risk assessment practices underpin continuous improvement efforts in vulnerability management programs. Retrospective analysis of incident reports, previous patch cycles, and recurring vulnerability patterns enables organizations to detect underlying causes—such as outdated third-party libraries, neglected firmware revisions, or gaps in clinician training. These lessons guide refinements in scanning schedules, patch laws, and disclosure procedures, fostering learning cycles that drive ongoing risk reduction. For example, a hospital may track repeated vulnerabilities in a particular infusion pump model and proactively engage the manufacturer to address systemic flaws across the installed fleet. Regular synthesis of past outcomes arms teams to anticipate future threats and strengthens the institutional memory necessary for long-term resilience (*A Survey on Vulnerability Prioritization: Taxonomy, Metrics, and Research Challenges*, 2025).

Major regulatory changes, clinical network expansions, and the rapid adoption of new device modalities all introduce fresh compliance demands and operational uncertainties. Each development tests the adaptability and thoroughness of the vulnerability management framework, ultimately pointing toward the ongoing challenge—how organizations must coordinate technical rigor with external certification and approval requirements, ensuring uninterrupted protection and trust.

Regulatory Certification After Major Updates

Regulatory certification in the context of medical device updates and maintenance requires a continuous commitment to compliance. Certification is not static; it involves periodic reassessment as the device's software or hardware evolves. With each update, organizations must maintain up-to-date documentation that tracks and explains every significant change, ensuring that modifications do not compromise the original safety and performance commitments made to regulatory bodies. This ongoing documentation includes comprehensive update logs that detail the rationale behind each update, the exact technical modifications implemented, and their anticipated or measured impact on device operation (*Compliance Certification vs Attestation: What Is the Difference?*, 2024).

Update logs serve as the linchpin for future compliance audits and for communication with regulators. Each entry in these records should capture who authorized and performed the update, what testing preceded deployment, which systems or modules underwent change, and how the adjustments align with regulatory requirements or previously certified standards. Centralized digital record-keeping systems, such as cloud-based compliance management platforms, make these tasks efficient by providing time-stamped, version-controlled audit trails. For example, a healthcare provider that automates the tracking of firmware updates and security patches can more readily generate documentary evidence to satisfy both internal policies and external regulatory inquiries. Structured electronic logs reduce manual entry errors and can provide alerts if documentation is incomplete before or after an update, which is vital for demonstrating continual compliance (*IT Compliance Audit Guide: Key Steps & Benefits in 2025*, 2025).

Major device updates usually require a formal process of reattestation. Regulatory frameworks, such as those for ISO 13485, the U.S. FDA, or the European MDR, expect organizations to submit renewed assurances and tangible evidence that devices after modification continue to comply with approved benchmarks. Documentation typically includes updated risk assessments, verification and validation test results, revised technical files, and a clear clinical impact analysis when applicable. For example, after a cybersecurity patch that affects device connectivity, manufacturers might need to provide regression test results that confirm both the efficacy of the patch and the preservation of core patient safety features. These submissions are often required within specified timeframes, and delays or incomplete evidence can threaten the device's regulatory standing or market access (*Compliance Certification vs Attestation: What Is the Difference?*, 2024).

Timely and accurate reattestation submissions ensure that devices remain authorized for clinical use. This includes not only documentation delivered to regulators but also internal records verifying that evaluations took place prior to device release. Routine internal reviews, scheduled alongside regular software or firmware update cycles, serve as a best practice to ensure that when formal reattestation is necessary, organizations are not scrambling for historical records but can present a cohesive, verifiable package to oversight bodies.

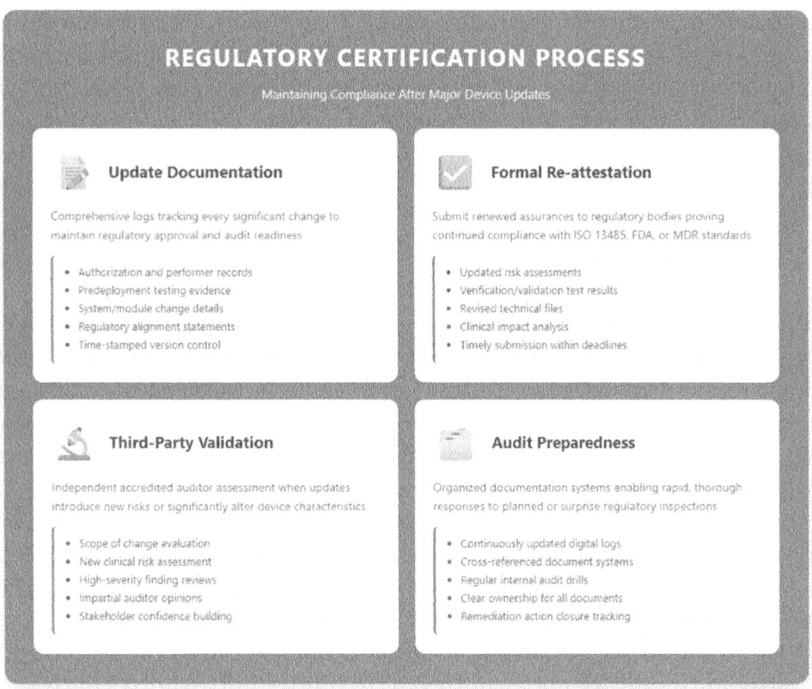

At times, regulatory authorities or internal quality processes require independent third-party validation of modifications. This is particularly important when updates introduce new risks or significantly alter the device's operating characteristics. Third-party assessments, performed by accredited bodies or auditors, provide external and impartial confirmation that the modified device continues to meet the mandated standards. Criteria for engaging external assessors usually include the scope of changes, the introduction of new clinical risks, or alignment with industry-specific triggers such as high-severity findings or critical control deficiencies previously identified in external audits (*Compliance Certification vs Attestation: What Is the Difference?*, 2024). Hypothetically, a company releasing a major software upgrade that changes diagnostic algorithms might engage an independent auditor to evaluate the update, using accepted attestation standards. The resulting auditor's opinion can then support regulatory submissions and increase stakeholder confidence.

Audit preparedness underpins the entire process of regulatory certification maintenance. Organizations committed to compliance organize their documentation for both planned and surprise inspections. Key strategies include maintaining continuously updated digital logs, employing document management systems that permit cross-referencing between design changes and related testing or safety reports, and scheduling regular internal audits to uncover and address issues before external assessments. Having clear ownership for every compliance document and keeping evidence trails easily accessible can reduce the time required to respond to regulators' inquiries. For instance, in one example from the financial sector, an effective compliance management system reduced evidence collection times from weeks to days, enabling faster and more thorough responses to auditors (*IT Compliance Audit Guide: Key Steps & Benefits in 2025*, 2025).

To ensure readiness for both scheduled and unannounced audits, teams should perform regular drills that simulate real audit conditions, review the completeness and accuracy of historical change logs, and address open gaps identified in internal reviews. Remediation actions from previous findings must be logged with specifics: responsible individuals, due dates, corrective actions, and closure evidence. These practices enable a structured response during actual audits, support uninterrupted device market access, and reinforce stakeholder trust in the company's quality and regulatory posture. Regular and proactive compliance activities ensure regulatory confidence at each stage of the device life cycle.

Bringing It All Together

Having explored comprehensive strategies for secure medical device updates, maintenance, and end-of-life management, healthcare stakeholders are now equipped to implement robust, risk-based approaches that safeguard patient safety, ensure operational continuity, and meet stringent regulatory demands. By integrating secure delivery methods, rigorous authentication, coordinated clinical communication, continuous vulnerability management, and meticulous documentation, organizations can strengthen their defenses throughout the device life cycle. Looking ahead, embracing these best practices will enable proactive identification and mitigation of emerging threats, foster regulatory compliance, and sustain trust among patients and providers alike. This foundation positions healthcare systems to adapt confidently to evolving technologies and cyber risks, ensuring resilient and secure medical device ecosystems now and into the future.

References

A Guide to Data Destruction for Hospitals—ITAMG. (2025, April 24). IT Asset Management Group. https://www.itamg.com/data-destruction-guides/a-guide-to-data-destruction-for-hospitals/.

A Survey on Vulnerability Prioritization: Taxonomy, Metrics, and Research Challenges. (2025). Arxiv.org. https://arxiv.org/html/2502.11070v1.

Bream, L. (2025, March 10). *FFIEC Data Destruction Guidelines*. Securis—Data Destruction, Secure Shredding, and Electronics Recycling. https://securis.com/news/federal-financial-institutions-examination-council-ffiec-data-destruction-guidelines/.

Compliance Certification vs Attestation: What is the difference? (2024, December 25). TrustCommunity. https://community.trustcloud.ai/docs/grc-launchpad/grc-101/compliance/compliance-certification-vs-attestation-what-is-the-difference/.

Infrastructure, N. R. C. (US) C. on M. P. and S. in H. C. A. of the N. I. (1997). *Technical Approaches to Protecting Electronic Health Information*. Www.ncbi.nlm.nih.gov; National Academies Press (US). https://www.ncbi.nlm.nih.gov/books/NBK233433/.

IT Compliance Audit Guide: Key Steps Benefits in 2025. (2025, May 20). Kraft Business Systems. https://kraftbusiness.com/blog/it-compliance-audit-guide/.

Javaid, M., Haleem, A., Singh, R. P., Suman, R. (2023). *Towards Insighting Cybersecurity for Healthcare Domains: A Comprehensive Review of Recent Practices and Trends*. Cyber Security and Applications.

Santos, O. (2025, February 3). *Evolution of Cybersecurity Vulnerability Management (2019–2024)*. Medium; AI Security Chronicles. https://aisecuritychronicles.org/evolution-of-cybersecurity-vulnerability-management-2019-2024-fec8326331dd.

Ştefan, A.-M., Rusu, N.-R., Ovreiu, E., Ciuc, M. (2024, June 1). *Empowering Healthcare: A Comprehensive Guide to Implementing a Robust Medical Information System—Components, Benefits, Objectives, Evaluation Criteria, and Seamless Deployment Strategies*. Applied System Innovation. https://doi.org/10.3390/asi7030051.

The Federal Risk And Management Program Dashboard. (n.d.). Marketplace.fedramp.gov. https://marketplace.fedramp.gov/.

19

Medical Device Security Incident Case Studies

Medical devices, once seen as isolated tools dedicated solely to patient care, have become deeply integrated into complex digital networks, exposing them to cyber threats of unprecedented scale and sophistication. Security breaches targeting these devices are no longer rare anomalies but represent a persistent and evolving challenge that can compromise patient safety, disrupt healthcare operations, and undermine trust in medical technology. Despite advances in healthcare security, many vulnerabilities remain embedded within device design, network architecture, and organizational policies—creating entry points that attackers continuously exploit. Understanding the nature of these incidents and the responses they provoke is essential for professionals tasked with safeguarding medical environments. This chapter examines critical cases that illuminate the technical flaws, human factors, and systemic weaknesses at the heart of medical device security failures, while also highlighting how these events have shaped contemporary strategies aimed at resilience and risk reduction across the healthcare sector.

Infamous Medical Device Attacks: Root Causes and Consequences

Major attacks on medical devices have exposed the underlying technical vulnerabilities in healthcare's digital infrastructure, triggering industry-wide reforms and shaping contemporary standards for medical cybersecurity. Examining significant incidents not only reveals the flaws in device design and network architecture but also underscores the dire consequences of failing to address security from the outset. Each major category of attack—implantable device hacks, ransomware-driven hospital shutdowns, breaches of cloud-integrated devices, and multidevice exploit chains—provides a unique lens through which the evolution of security threats and responses becomes evident.

Implantable medical devices, such as pacemakers and insulin pumps, have suffered from weak authentication protocols combined with rare or absent firmware updates. Notably, in 2017, security researchers demonstrated wireless hacks that could manipulate cardiac implantables by leveraging hardcoded credentials and unencrypted telemetry channels, a flaw mirrored in real-world vulnerabilities affecting manufacturers with millions of deployed devices. Attackers could capture device radio signals, replay communication to issue unauthorized commands, or inject malicious firmware when update mechanisms lacked cryptographic validation. Such technical weaknesses exposed patients to grave threats: attackers could cause device malfunctions or even modify dosage settings remotely. The public disclosure of these vulnerabilities led to regulatory advisories, urgent recall actions, and widespread panic among patients and providers. Underlying these exploits lay the absence of layered authentication and regular security patching, as well as manufacturers' limited appreciation for adversarial threat models at the time. The industry's response involved adopting robust mutual authentication, mandated over-the-air encryption, and dedicated vulnerability reporting programs, marking a shift from proprietary obscurity to coordinated disclosure and risk mitigation (He et al., 2021).

Hospital-wide ransomware attacks presented a different spectrum of risk, targeting the operational backbone of healthcare facilities rather than individual devices alone. In 2020, a major European university hospital experienced an attack that began with a phishing campaign targeting staff endpoints. Attackers then pivoted laterally through interconnected medical devices running unsupported operating systems. These legacy endpoints, lacking critical security updates, served as beachheads for the ransomware payload, which encrypted network shares and rendered clinical systems inaccessible. The entire hospital faced a forced diversion of critical cases, large-scale rescheduling, and prolonged system outages. Root causes included poorly segmented networks, pervasive use of outdated device software, and the failure to enforce least-privilege access across device communications. Affected organizations also lacked tested incident response plans tailored to networked medical technology, compounding the initial damage and exposing the sector's insufficient business continuity planning (He et al., 2021; Javaid et al., 2023). The frequency and impact of such incidents elevated cybersecurity as a board-level concern, driving hospitals to invest in network access control, micro-segmentation, comprehensive asset management, and behavioral anomaly detection specific to medical device traffic.

The move toward cloud-connected medical devices has introduced an expanded attack surface, emphasizing the importance of end-to-end digital hygiene. A significant breach occurred when attackers exploited cloud-integrated infusion pumps with inadequate encryption of transmission channels and misconfigured access permissions on their associated cloud dashboards. By targeting the weak link in device-to-cloud authentication, attackers exfiltrated both real-time patient metrics and device management data. This exploitation exposed systemic gaps in device identity verification and cloud security orchestration, impacting not just data confidentiality but also raising noncompliance issues with regional health privacy laws such as the Health Insurance Portability and Accountability Act (HIPAA) and General Data Protection Regulation (GDPR). The regulatory fallout prompted device manufacturers and health systems to embrace encrypted communication protocols, enforce multifactor authentication for cloud management platforms, and audit data access paths end-to-end (Javaid et al., 2023). The incident highlighted the necessity of continuous security validation and dynamic threat modeling throughout the life cycle of interconnected medical devices.

Multidevice attack chains demonstrate the sophistication with which adversaries exploit healthcare environments, leveraging the trust relationships between devices to escalate privileges. In one pivotal incident, researchers mapped a pathway that started with a compromise of an internet-connected imaging workstation with an exposed administrative interface. Attackers then pivoted to bedside monitoring equipment via unsecured internal protocols, eventually reaching high-privilege pharmacy dispensing machines. Each link in the chain exploited a distinct vulnerability—ranging from lack of protocol authentication and insufficient network segmentation to default credentials on embedded web servers—culminating in attackers gaining critical control without immediate detection. These chained exploits evade traditional security controls, which often focus on single-device threat models without accounting for lateral movement or privilege escalation through dependent systems (He et al., 2021). Health IT teams confronted with such incidents recognized the importance of continuous device inventory, deployment of network-wide anomaly detection, and rigorous penetration testing that simulates chained attack scenarios.

Underlying each attack, a pattern emerges: exploitations capitalize on a combination of technical oversights—untested patching processes, default configurations, and underdeveloped authentication—as well as organizational gaps, from lack of security awareness training to uncoordinated incident response. As medical device threats continue to evolve, they expose the boundaries of existing defenses and emphasize the pressing need for proactive, research-driven risk management and strong ethical foundations in vulnerability disclosure (He et al., 2021; Javaid et al., 2023).

White Hat Discoveries: Exposing Hidden Weaknesses

Across the realm of medical device security, public security research has played a pivotal role in exposing critical vulnerabilities and compelling lasting change. Ethical researchers, driven by the goal of patient safety, have systematically pushed manufacturers to address hidden flaws in devices ranging from insulin pumps to cardiac defibrillators. This partnership, at its best, combines rigorous testing, responsible disclosure, and rapid remediation, cumulatively raising the baseline for device security.

Ethical hacking—in particular, the practice of responsible, coordinated disclosure—has become a central engine for progress. When researchers identify a vulnerability and report it through structured channels, they often catalyze swift industry responses. Coordinated efforts not only ensure that vendors receive actionable details early, but also create external pressure to release timely patches. By setting transparent timelines for remediation and public disclosure, researchers and vendors have aligned incentives to prioritize patient safety. Adherence to these protocols has helped reduce legal and reputational risks for researchers, creating a foundation of mutual trust that ultimately benefits everyone involved (HackerOne, 2023).

In one prototypical case, security researchers working in a controlled lab setting discovered that a leading brand's pacemaker could be wirelessly accessed and, in theory, reprogrammed with malicious commands. By responsibly informing the manufacturer, providing technical details, and giving ample time for investigation, the researchers triggered a series of internal audits and code reviews. The result was a comprehensive firmware redesign that not only patched the immediate vulnerability but also reduced the device's attack surface by implementing strict access controls and encrypted communication protocols. These upgrades were subsequently applied to new device generations, showcasing how ethical "white-hat" activities produce broad, real-world benefits (HackerOne, 2023).

Other victories have come from collaborative industry partnerships, where vendors and independent researchers craft solutions together. Transparency around vulnerability remediation allowed hospitals and clinicians to anticipate updates, reducing patient disruption during critical patch deployments. Trust, fostered through open and respectful communication, enabled more dynamic exchanges of technical information. This environment gave rise to bug bounty programs and responsible disclosure policies that defined clear rules for reporting and recognition, driving greater engagement from the research community and more robust responses from vendors (*Vulnerability Disclosure: Risks, Significance, and Best Practices | Fortinet*, 2023).

Evident in several case studies, organizations that instituted formal responsible disclosure policies provided safe harbor for researchers and outlined concrete expectations for handling reports, such as estimated response times and regular progress updates. These practices aligned with industry best practices and supported a climate where positive feedback loops flourished. Not only did effective communication hasten the identification and resolution of vulnerabilities, but it also set a public precedent that raised the reputational profile of proactive vendors (HackerOne, 2023).

However, these advances have not been uniform. Some manufacturers responded to disclosures with denial or delay, contesting the severity or even the existence of reported vulnerabilities. In several notable incidents, disputes over intellectual property or disagreement about timing of public disclosure led to adversarial standoffs. In one case, a manufacturer's reluctance to acknowledge risk caused updates to be postponed, prolonging patient exposure and fueling public criticism. The consequences were not limited to technical failure—they included lasting damage to brand reputation and eroded trust between researchers and vendors. Lessons from such scenarios highlight the risk of ignoring researcher reports and emphasize the need for structured, respectful dialogue (HackerOne, 2023).

At the other end of the spectrum, some companies have set new standards by embracing vulnerability reports and working openly with the research community. Their swift action in releasing advisories and patches, accompanied by public recognition of researcher contributions, fostered goodwill and confidence across the healthcare sector. By moving quickly not just to secure their own products but also to urge suppliers and partners to improve, these leaders accelerated the diffusion of better security practices throughout the industry (*Vulnerability Disclosure: Risks, Significance, and Best Practices | Fortinet*, 2023).

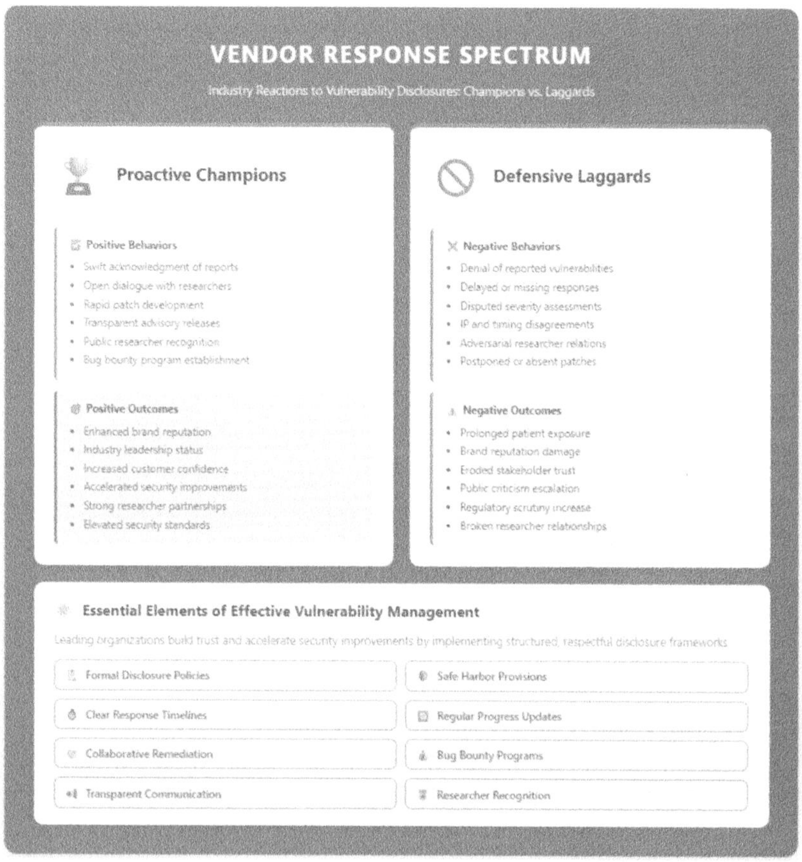

This range of industry responses underlines a broader evolution in medical device security culture. The adoption of incident response strategies that incorporate insights from public research has improved not only the speed of remediation but also the overall resilience of manufacturers to emerging threats. The involvement of ethical researchers has repeatedly proven that incident response planning is most effective when collaborative relationships and transparent communication underpin every stage. This continuous integration of public discoveries, structured partnerships, and efficient remediation has strengthened medical device defenses and paved the way for ongoing improvements that support safer outcomes for patients and healthcare providers alike (HackerOne, 2023; *Vulnerability Disclosure: Risks, Significance, and Best Practices | Fortinet*, 2023).

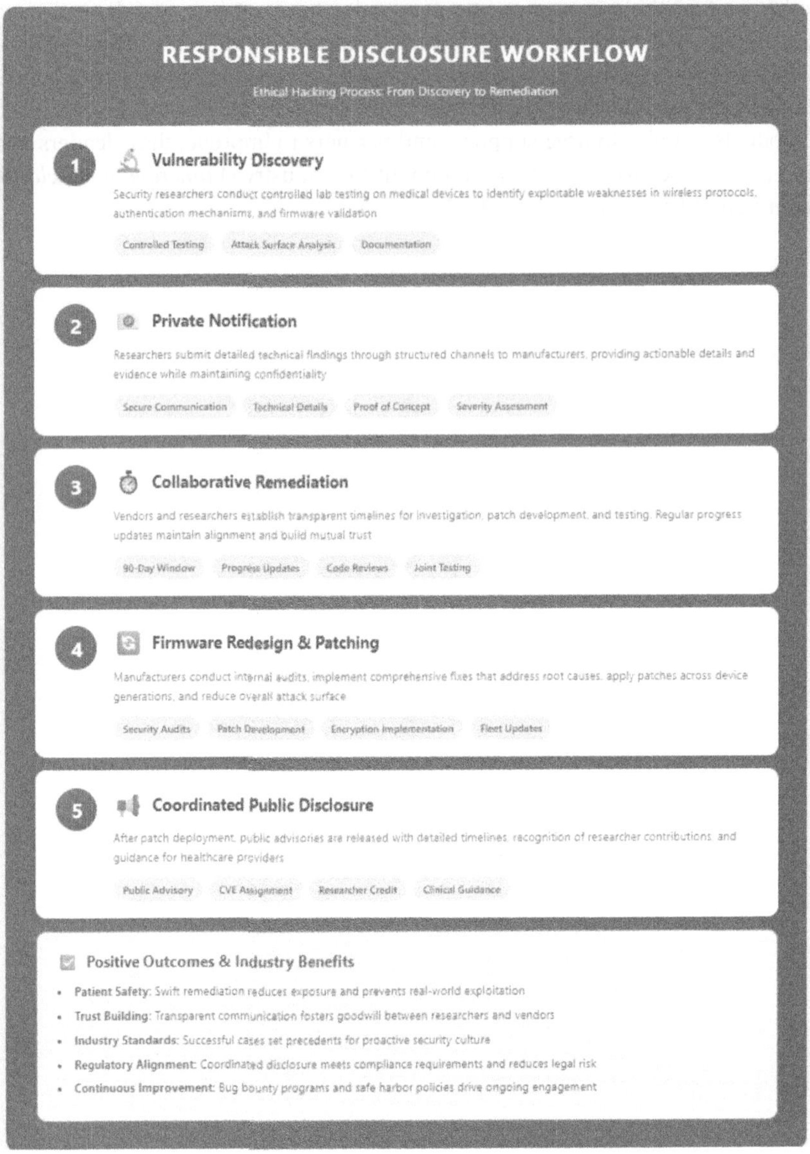

Response and Recovery: Best Practices in Action

Rapid Containment and Network Segmentation

Swift isolation of compromised medical devices remains a foundational element in modern incident response. In critical cases, immediate network segmentation stops malicious actors from moving laterally, preserving both data integrity and patient welfare. For example, during a ransomware incident at a major healthcare provider in 2017, technical teams quickly disconnected network links between imaging devices and the broader hospital information system. This decisiveness prevented the malware from spreading to infusion pumps and laboratory analyzers, averting further service outages. Implementation of "break-glass" network disconnection protocols—predetermined, rehearsed steps for physical and logical separation—enabled restoration of affected devices in a controlled sequence, validating their integrity before resuming regular operations. Such cases underscore how methodical device quarantine and stepwise rebooting contained damage, sustained core clinical capabilities, and ensured the continuous delivery of urgent care (*Incident Reporting in Healthcare: 7 Reasons to Build Resilience*, 2023).

Healthcare organizations systematically institutionalized these best practices, embedding them into incident response playbooks. Rapid isolation procedures protected patient safety not only by limiting the attack's reach but also by facilitating the prioritized recovery of devices crucial for life-sustaining therapies. Timely sequencing and integrity checks prevented inadvertent reinfection and minimized the likelihood of cascading failures, demonstrating a clear alignment with key safety and risk-reduction imperatives discussed by Mistri et al. (2023).

Incident Response Team Preparedness

Incident containment begins with well-drilled response teams integrating clinical and technical expertise. In one high-profile case, a multidisciplinary team regularly participated in simulation exercises mimicking active threats against infusion systems. Each team member's responsibilities—technical diagnosis, device shutdown, clinical impact assessment, and regulatory notification—were preassigned and reinforced through periodic drills. During an actual breach, this structure enabled parallel execution of critical tasks: IT specialists isolated the device, biomedical engineers assessed affected equipment for patient risk, and communications leads coordinated with executive leadership.

Role clarity proved vital: overlapping duties caused delays during earlier, uncoordinated incidents, while structured division minimized confusion. Team rehearsal fostered automatic, confident action, reducing reliance on ad-hoc decision-making under stress. Furthermore, these collaborative workflows enabled rapid identification of attack vectors and expedited restoration of essential devices, all while allowing clinical operations to pivot to unaffected equipment. This systematic, hands-on approach exemplified the operationalization of a safety-first culture grounded in organized, continuous learning (*Incident Reporting in Healthcare: 7 Reasons to Build Resilience*, 2023; Mistri et al., 2023).

Transparent Communication and Trust Preservation

Transparent communication during and after security incidents served as a linchpin for trust and trustworthiness. Recognizing the volatility of information flow during crises, successful organizations built strategies to distribute accurate updates at regular intervals. Secure hospital portals delivered tailored advisories to staff, explaining device status and instructing on alternative workflows. Simultaneously, dedicated patient liaisons provided direct outreach to affected individuals, sharing progress and care adjustments.

A coordinated messaging approach allowed spokespersons to brief media, preempt misinformation, and demonstrate institutional accountability. Following a series of connected device breaches in a regional hospital system, centralized communication limited rumors and misinformation among clinical staff, while visible leadership reinforced confidence in the institution's preparedness and honesty. Documenting and publicizing incident timelines, action steps, and solutions contributed to community reassurance, a practice also aligning with the need for a blame-free, transparent safety culture highlighted by Mistri et al. (2023). This openness influenced positive relationships with patients, regulators, and partners, and directly correlated with sustained trust, especially in environments where rapid spread of information could amplify negative perceptions.

Regulatory Compliance Under Duress

Fulfilling regulatory requirements, even under intense pressure, distinguished effective incident response from failed efforts. Healthcare institutions that embedded automated recordkeeping and reporting within their incident management platforms ensured that audit trails were preserved and could be reviewed by regulatory agencies postincident. This included real-time documentation of isolation actions, communications, and device recovery steps. In one large hospital, incident response software generated structured logs and notifications compliant with health data protection mandates, which were immediately shareable with authorities.

Such systems guaranteed the uninterrupted collection of evidence, supporting postincident analysis and legal defense where necessary. By adhering to established reporting timelines and maintaining open lines of communication with oversight bodies, institutions avoided fines and mitigated reputational damage. Simultaneously, these practices demonstrated a culture of rule-following driven by internal professionalism, not just fear of external enforcement (*Incident Reporting in Healthcare: 7 Reasons to Build Resilience*, 2023). Maintaining compliance amid chaos enabled organizations to stabilize operations, minimize financial risk, and continue to provide quality care—outcomes intimately linked to the core priorities of patient safety and public trust (Mistri et al., 2023).

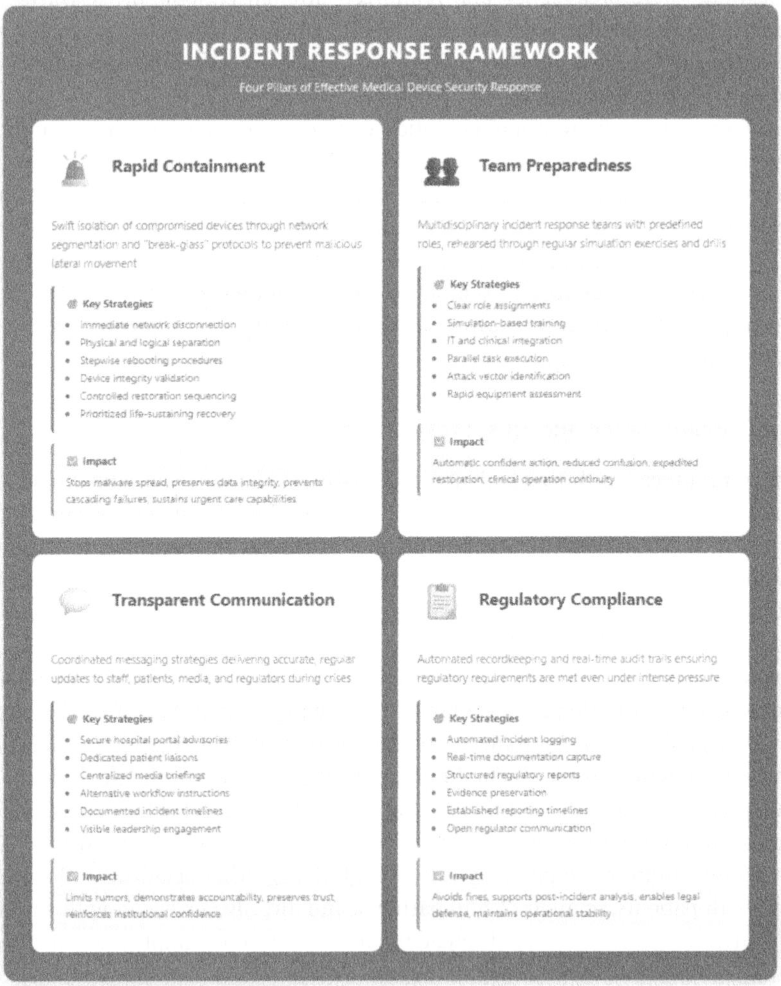

By reviewing these best practice cases, technical and clinical leaders encounter blueprint approaches for effective intervention, placing patient well-being and organizational integrity at the center of every response. When these mechanisms are missing or poorly executed, incidents escalate, causing cascading consequences that threaten care delivery and institutional viability—a reality that shapes the focus of subsequent analyses on breakdowns and failures.

Failures in Policy: Lessons from Security Missteps

Major medical device security incidents frequently originate from weak risk management practices and fragmented containment strategies, especially where flawed security policies and bureaucratic inertia create deep operational vulnerabilities. When health systems neglect the principles of systematic, process-driven cybersecurity, the resulting policy gaps directly compromise device safety and patient outcomes. These vulnerabilities often surface during patch management cycles, where institutional inefficiencies and governance lapses expose critical points of failure.

Delay in patch deployment stands out as a recurring risk, largely fueled by inflexible policies and decentralized administration. In practice, medical device fleets are often highly heterogeneous, making standardized patch application complex. Lack of clear tracking mechanisms—exacerbated by siloed IT operations and missing or outdated asset inventories—leads to inconsistent update coverage. Rollback hesitancy, where updates are reversed due to insufficiently tested procedures, allows known vulnerabilities to persist; threat actors frequently exploit this window. For example, in multiple published incidents affecting connected infusion pumps, failure to apply urgently released firmware patches due to "validation bottlenecks" resulted in attackers exploiting old vulnerabilities for weeks, sometimes gaining persistent access to device networks. These events typically traced back to ambiguous responsibility for patch ownership and weak interdisciplinary collaboration, highlighting that risk containment falters without coherent cross-departmental strategies and responsive governance protocols (Prado-Prado et al., 2020).

Access control failures further amplify the dangers of policy lapses. Medical environments often operate with permissive defaults—shared generic administrator credentials, legacy devices lacking granular role assignment, or unsegmented device networks. Over-permissive user roles—such as broad rights issued to clinical engineering contractors or temporary staff—bypass least privilege regimes. In one notable scenario, a clinical staff member used shared credentials to configure radiology devices; these were later stolen via phishing, granting attackers lateral access across several interconnected modalities. Subsequently, audit logs revealed hundreds of failed authentication attempts and multiple successful privilege escalations, all of which could have been prevented by limiting administrative rights and enforcing multifactor authentication. Erratic policy enforcement around password renewal, account deprovisioning, or network segmentation enabled breaches to escalate within device networks and delayed effective threat containment. In each case, technical shortcomings stemmed from weak governance frameworks—sometimes the result of outdated or insufficiently enforced institutional policies—rather than device-level vulnerabilities alone (Prado-Prado et al., 2020; Xian et al., 2022).

Insufficient or out-of-context staff security training is another persistent driver of medical device risk. Training programs are often generic or box-ticking exercises, focusing on basic compliance rather than contextualizing threats within real clinical or operational scenarios. Staff may be required to complete annual modules addressing password security, yet these trainings rarely cover specific adversary tactics such as phishing campaigns that target health workers or the manipulation of wireless device interfaces. When security education ignores simulation-based exercises or incident walkthroughs, employees remain unprepared for genuine attacks. In one anonymized episode, a hospital's biomedical engineering team received only policy-based security briefings rather than hands-on exercises. When a sophisticated phishing email triggered credential loss, technicians failed to recognize or respond to subsequent unusual device messages—a gap directly attributable to insufficient training. Such gaps result in slow incident detection and weak initial containment, substantially prolonging adversary dwell time on high-value device networks (Prado-Prado et al., 2020).

The deeper structural roots of these failures lie in organizational culture and management priorities. Where leadership views cybersecurity as a compliance checkbox rather than a strategic business necessity, security initiatives lose vital budget support and operational focus. Budgetary constraints lead to outmoded devices remaining in service without crucial security features or consistent patch support. Operations teams, under pressure from competing clinical agendas, may postpone or deprioritize nonurgent software updates. Bureaucratic resistance is particularly acute in highly siloed or hierarchical organizations, where process changes require escalated approvals, and responsibility for device security is diffused or ambiguous. Gaps in accountability enable the normalization of risky practices, such as the continued use of default credentials or unauthorized device modifications. Without robust performance metrics to track policy implementation or enforce consequences for nonadherence, legacy approaches persist. This maintains an environment where known and fixable vulnerabilities repeatedly drive serious incidents, from remote exploitation of diagnostic devices to sustained access via obsolete hospital information systems (Xian et al., 2022).

These interconnected failures—spanning patch management, access control, staff readiness, and organizational culture—demonstrate that institutional inertia and policy flaws do not merely delay progress but actively perpetuate systemic risk in the healthcare sector. Each incident investigated reveals not only immediate technical oversights but also the enduring influence of outdated priorities and governance models that no longer align with the evolving threat landscape. The persistent gap between policy and practice spotlights the critical need for holistic reforms that span process redesign, leadership engagement, and measurable risk accountability to enable resilient, adaptive medical device security (Prado-Prado et al., 2020; Xian et al., 2022).

Driving Improvement: Change Through Crisis

Widespread security breaches have forced healthcare organizations to reevaluate their entire approach to protecting medical devices and related infrastructure. Following major incidents, many organizations initiated comprehensive security overhauls. Immediate steps included redesigning network architecture to better segment medical device traffic from other organizational systems. Network segmentation isolated vulnerable devices, restricted lateral movement by attackers, and limited potential damage, addressing weaknesses that had been exploited during previous crises (*Ransomware Hits Healthcare through Applications*, 2025). Alongside segmentation, organizations installed continuous monitoring and anomaly alert systems that gave IT and clinical engineering teams greater visibility into activities on the network. By adopting real-time threat detection tools, hospitals could rapidly identify and respond to suspicious behavior, preventing attackers from gaining footholds that would otherwise go unnoticed for extended periods.

Investment in device management also increased. Organizations prioritized robust inventory controls to track devices throughout their life cycle, instituted strict procurement vetting to ensure security baselines were met, and implemented rigorous patch management schedules. These changes closed known vulnerabilities that adversaries had exploited and supported compliance requirements to maintain operational integrity (Junaid, 2022; *Ransomware Hits Healthcare through Applications*, 2025). Life cycle management extended to well-documented retirement processes for obsolete devices, a step that reduced reliance on unsupported equipment and minimized the risk of unmanaged attack surfaces.

The process of rewriting security policies brought further clarity and ensured enforceability. In the aftermath of incidents exposing ambiguous or incomplete guidance, policy authors worked alongside representatives from clinical, technical, and executive functions to develop actionable protocols. New policies detailed device hardening requirements, mandated frequent vulnerability assessments, and set forth minimum standards for security configurations. Staff at every level now had plain-language instructions for tasks such as password management, port deactivation, and secure device disposal. Improvements in escalation and reporting procedures made it easy for individuals to seek help or report incidents without confusion or delay, helping organizations respond to threats before they could escalate. Internal documentation standards were raised so that audit trails demonstrated ongoing compliance and provided regulators with clear evidence of due diligence (*Ransomware Hits Healthcare through Applications*, 2025).

Achieving lasting improvements depended on broader participation and ownership of security responsibilities. Leadership committed to visible demonstration of good security hygiene, modeling behaviors such as regular engagement with incident response teams and transparency in reporting new threats. Security champions were selected inside both clinical care and IT departments to serve as liaisons for policy dissemination and feedback. These individuals were equipped with ongoing education and clear mandates to drive local accountability. Organizations also launched hands-on training programs tailored to the realities of working with medical equipment. For clinical staff, this meant practical scenarios on recognizing compromised device behavior and understanding basic troubleshooting steps for potential security threats (Junaid, 2022). IT professionals received updated courses on device-specific threats, patch management, and secure deployment.

These educational efforts helped foster a shared sense of vigilance and encouraged reporting of near misses and suspicious activity. Leadership support ensured that security was seen not just as a regulatory requirement, but as an integral component of patient safety and organizational trust. Routine security drills and tabletop exercises regularly tested response readiness, ensuring that improvements were not static but embedded in daily operations.

Recognizing the limitations of tackling threats in isolation, healthcare organizations increasingly turned outward to develop collaborative partnerships. Vendor relationships were formalized with routine vulnerability disclosures, transparent security updates, and joint risk assessments before and after device deployment. Partnerships with regulatory bodies and industry groups led to co-development of new security guidelines that reflected real-world threat intelligence. For example, groups shared anonymized incident data to help create resources like improved device certification requirements and practical guides for incident containment (*Ransomware Hits Healthcare through Applications*, 2025). In some cases, data reported from breaches informed the update of national and international mandates around device connectivity and cloud services—ensuring shared lessons translated into measurable standards (Junaid, 2022).

Collective action extended to threat intelligence sharing networks, where organizations quickly distributed information about emerging attack techniques, malware signatures, and mitigation strategies. This approach enabled a more agile, community-wide response to rapidly evolving threats that individual entities could not counter alone. Industry conferences and working groups also became forums for sharing best practices and lessons learned from real incidents, further driving improvement across the ecosystem.

Every improvement took root because the reality of crisis made complacency impossible. The combination of technical upgrades, policy overhaul, staff empowerment, and external collaboration created a culture of proactive defense. Ultimately, these concerted reforms led to more resilient healthcare environments that could adapt to new threats while maintaining the safety and integrity of medical devices, thus ensuring a higher level of care and confidence for patients and providers alike (Junaid, 2022; *Ransomware Hits Healthcare through Applications*, 2025).

SECURITY TRANSFORMATION ROADMAP
Driving Comprehensive Improvement Through Crisis Response
PHASE 1
Infrastructure Overhaul
Key Actions
Network segmentation redesign
Medical device traffic isolation
Continuous monitoring deployment
Real-time anomaly alerts
Threat detection tool installation
Benefits
Limited lateral movement
Contained potential damage
Greater network visibility
Rapid threat identification
Enhanced operational integrity
PHASE 2
Device Lifecycle Management
Key Actions
Robust inventory controls
Strict procurement vetting
Rigorous patch schedules
Documented retirement processes
Security baseline enforcement
Benefits
Closed known vulnerabilities
Compliance requirements met
Reduced obsolete equipment
Minimized attack surfaces
Prevented exploitation windows
PHASE 3
Policy & Governance Reform
Key Actions
Actionable protocol development
Device hardening requirements
Minimum security configurations
Escalation procedure clarity
Audit trail documentation
Benefits
Clear staff guidance
Reduced reporting confusion
Faster threat response
Regulatory compliance evidence
Enforced due diligence
PHASE 4
Staff Empowerment & Culture
Key Actions
Leadership security modeling
Security champion selection
Hands-on scenario training
Regular security drills
Near-miss reporting encouragement
Benefits
Shared vigilance culture
Increased threat reporting
Improved response readiness
Patient safety integration
Organizational trust building
PHASE 5
Collaborative Partnerships
Key Actions
Vendor vulnerability disclosure
Joint risk assessments
Regulatory guideline co-development
Threat intelligence sharing
Industry best practice forums
Benefits
Community-wide defense
Agile threat response
Updated certification standards
Practical containment guides
Resilient healthcare ecosystem

Lessons Learned

The examination of notable medical device security incidents has illuminated critical vulnerabilities and systemic weaknesses that, when addressed, can significantly enhance healthcare cybersecurity. Understanding the root causes—from technical design flaws to policy failures—and learning from both setbacks and ethical research practices equip professionals across healthcare sectors with actionable insights. Moving forward, integrating rigorous security measures, fostering transparent collaboration between manufacturers and researchers, and embedding robust response protocols into organizational culture will be essential. By prioritizing comprehensive risk management, continuous education, and cross-industry cooperation, healthcare organizations can build resilient systems that not only defend against evolving threats but also safeguard patient safety and trust in an increasingly connected medical landscape.

References

HackerOne. (2023). *Why You Need Responsible Disclosure and How to Get Started.* Www.hackerone.com. https://www.hackerone.com/knowledge-center/why-you-need-responsible-disclosure-and-how-get-started.

He, Y., Aliyu, A., Evans, M., Luo, C. (2021). *Healthcare Cyber Security Challenges and Solutions Under the Climate of COVID19: A Scoping Review.* Journal of Medical Internet Research; ncbi. https://doi.org/10.2196/21747.

Incident Reporting in Healthcare: 7 Reasons to Build Resilience. (2023, August 31). SafeQual. https://www.safequal.net/building-resilience-7-key-reasons-to-enhance-incident-reporting-in-healthcare/.

Junaid, S. B. (2022, October 3). Recent Advancements in Emerging Technologies for Healthcare Management Systems: A Survey. Healthcare. https://pmc.ncbi.nlm.nih.gov/articles/PMC9601636/.

Javaid, M., Haleem, A., Singh, R. P., Suman, R. (2023*).* Towards *I*nsighting Cybersecurity for Healthcare *D*omains: A comprehensive *R*eview of *R*ecent *P*ractices and *T*rends. Cyber Security and Applications.

Mistri, I. U., Badge, A., Shahu, S. (2023). *Enhancing Patient Safety Culture in Hospitals.* Cureus. https://doi.org/10.7759/cureus.51159.

Prado-Prado, J. C., García-Arca, J., Fernández-González, A. J., Mosteiro-Añón, M. (2020, July 1). *Increasing Competitiveness Through the Implementation of Lean Management in Healthcare.* International Journal of Environmental Research and Public Health. https://doi.org/10.3390/ijerph17144981.

Ransomware Hits Healthcare Through Applications. (2025). Bankinfosecurity.com. https://www.bankinfosecurity.com/blogs/ransomware-hits-healthcare-through-applications-p-3913.

Vulnerability Disclosure: Risks, Significance, and Best Practices | Fortinet. (2023). Fortinet. https://www.fortinet.com/uk/resources/cyberglossary/vulnerability-disclosure.

Xian, M., Zhao, C., Zhou, Y. (2022, July 6). *From Bureaucratic Coordination to a Data-Driven Model: Transformation and Capacity Building of Community-Based Prevention and Control of Public Health Events.* International Journal of Environmental Research and Public Health. https://doi.org/10.3390/ijerph19148238.

20

Future Directions and Ultimate Mandate: Securing the Connected Patient

A single overlooked vulnerability in a connected medical device can have consequences that ripple far beyond technical inconvenience. Imagine an infusion pump delivering critical medication paused abruptly by a ransomware attack, or a remote cardiac monitor transmitting corrupted data, leaving clinicians blind to a patient's true condition. These incidents are no longer hypothetical scenarios but growing realities that challenge the very foundations of patient safety. As healthcare becomes increasingly reliant on interconnected technologies, the stakes of cybersecurity rise exponentially, demanding more than reactive measures—they require a fundamental rethinking of responsibility, education, policy, and innovation. The path forward calls for unwavering commitment from every stakeholder involved, where security is not a negotiable feature but an ethical imperative embedded in design, training, governance, and culture. This chapter engages with these demands, framing the critical actions necessary to secure the future of connected patient care amid an evolving landscape of threats and technological progress.

The Ethical Imperative: Why Perfection in Device Cybersecurity Is Not Optional

The stakes of securing connected medical devices could not be higher. In every moment of device operation—whether inside the intensive care unit, implanted within a vulnerable patient, or monitoring vitals remotely—the balance between cybersecurity and operational convenience is fundamentally a matter of patient safety vs. expediency. Decisions that sacrifice robust cybersecurity for efficiency can expose glaring vulnerabilities. There have been incidents where minor lapses, such as the use of default passwords or hastily bypassed authentication to expedite workflow, have resulted in unauthorized access to life-sustaining equipment, threatening catastrophic consequences for patients (Mensah et al., 2024). These are not hypothetical risks: even a single unauthorized command or subtle data manipulation can directly interfere with treatment delivery, lead to misdiagnosis, or even contribute to fatalities. Thus, policies and engineering priorities must always place uncompromising security above all practical justifications for temporary convenience.

In today's healthcare environment, the curve of digital threats targeting medical devices continues to steepen, with attackers exploiting both legacy systems and the latest networked technologies. Ransomware, once primarily a threat to databases and administrative networks, now regularly takes aim at medical devices, such as infusion pumps that control critical medication dosing. Instances have been reported where attackers encrypted device firmware or hijacked remote updates, paralyzing operations and putting patients at risk unless exorbitant ransoms were paid (Clarke & Martin, 2023). The proliferation of connected devices, spanning everything from network-enabled ventilators to wearable cardiac monitors, has also expanded the attack surface, introducing vulnerabilities in remote access protocols, firmware supply chains, and wireless communication standards. When even a single endpoint goes unpatched or poorly secured, an entire hospital network—and the well-being of every patient connected to it—can be placed in jeopardy (Mensah et al., 2024).

High-profile breaches have shown that the consequences extend far beyond operational disruption. A compromised infusion pump or disabled monitoring station directly impacts patient care, with delayed treatment, delivery of incorrect medication doses, or system outages leading to immediate risk to life. Financially, the fallout from such incidents can include regulatory fines, legal liabilities, and the enormous cost of system restoration (Clarke & Martin, 2023). These expenses, however, pale in comparison to the human costs. Consider the scenario in which a ransomware attack disrupts vital patient monitoring systems during surgery or corrupts real-time electronic health record (EHR) data that clinicians rely upon for critical care decisions. In such moments, minutes of inaccessibility or unreliable data can prove fatal. Furthermore, breaches undermine the foundational trust that patients place in the healthcare system—once lost, this trust is hard to regain and may inflict permanent brand and reputation damage on both the provider and the manufacturer (Mensah et al., 2024).

This reality defines an uncompromising ethical mandate for all stakeholders. For manufacturers, the responsibility begins with the design and development of devices, demanding security-centric engineering that proactively addresses vulnerabilities and anticipates attack vectors. Embedded encryption, secure boot mechanisms, strict authentication protocols, and rigorous vulnerability testing form the baseline. Health professionals who deploy and maintain these systems bear the duty to guard against lapses, ensuring that only authorized personnel have access and that every patch or update is applied rigorously (Mensah et al., 2024). IT leaders must enforce network segmentation, conduct continuous monitoring for anomalies, and foster an organizational culture in which reporting and remediating security gaps is not merely encouraged but expected (Clarke & Martin, 2023).

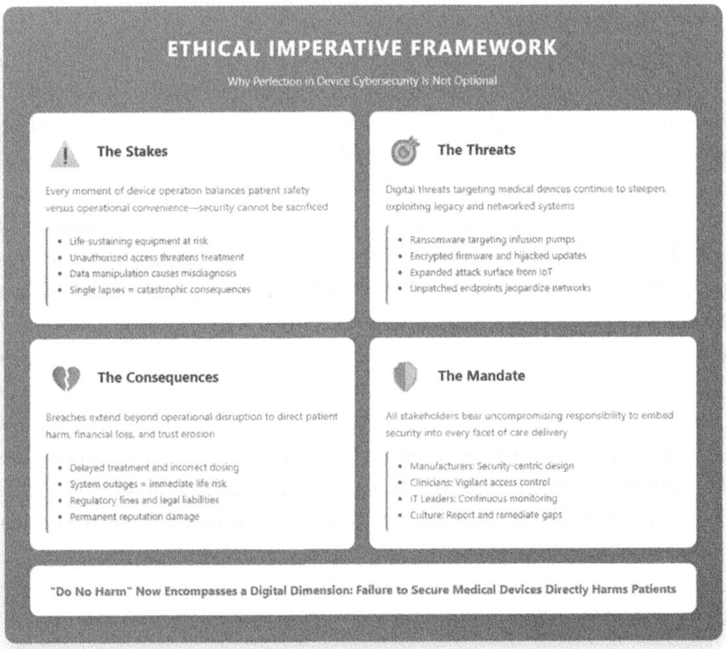

Ethically, when healthcare practitioners or technical managers compromise on security—even unintentionally—they also compromise their commitment to professional integrity, patient welfare, and public trust. The Hippocratic imperative to "do no harm" now encompasses a digital dimension: failure to secure medical devices can directly lead to patient harm. For example, circumventing proper device configuration to reduce setup times or sharing login credentials under workload pressure can create vulnerabilities that attackers exploit, with consequences reaching far beyond individual patients. Conversely, a culture of accountability and vigilance—where every update, access control, and compliance standard is followed with precision—actively protects patient safety and the collective reputation of healthcare organizations (Mensah et al., 2024).

To rise to this mandate, the current workforce and incoming professionals must possess both conceptual and practical mastery over device cybersecurity. Achieving this requires a shift in priorities, where security is viewed as a core ethical responsibility on par with clinical competence and patient care. This perspective must infuse every facet of clinical, technical, and engineering decision-making, recognizing that lives depend not just on advanced medical knowledge or innovative device functionality, but on the strength of the digital safeguards that protect these capabilities at every level (Clarke & Martin, 2023).

Integrating Cybersecurity into Medical Education

At the foundation of healthcare and engineering education, uncompromising cybersecurity must be established as a core tenet, tightly interwoven with students' understanding of technical excellence and ethical duty. Curriculum reform is paramount; cybersecurity topics can no longer exist as elective interests or optional modules, but instead should be systematic requirements infusing every level of instruction. Foundational competencies should not only include robust training in the engineering and operational aspects of device protection but also emphasize risk assessment, threat analysis, and the implementation of practical protocols unique to clinical and healthcare environments (Lewis et al., 2024). Graduates entering the increasingly digital, interconnected landscape need technical fluency in secure software development, network defense principles, and vulnerability management—alongside a sophisticated awareness of real-world threats facing patient-facing technologies, their users, and those charged with ensuring safety.

Embedding such comprehensive preparation demands a cross-disciplinary approach. It is critical that core education programs dissolve traditional silos, merging expertise from information security, biomedicine, medical ethics, regulatory compliance, and systems engineering. Course structures must enable medical and engineering students to jointly approach pivotal issues, equipping all with a working knowledge of both device functionality and the potential attack vectors exploiting technical and human weaknesses. For example, a curriculum might require medical and engineering cohorts to jointly analyze a hypothetical vulnerability in the firmware of an infusion pump, collaboratively designing improved authentication controls and incident response protocols. Capstone projects could unite students in modeling adversary scenarios against a hospital imaging device, drawing on both clinical workflows and cybersecurity defensive patterns (Lewis et al., 2024). Through such exercises, future professionals cultivate vital communication skills, a sense of shared responsibility, and proficiency in threat modeling from multiple disciplinary vantage points.

However, classroom instruction and theoretical study alone are insufficient to foster the intuition and situational readiness necessary for defending against advanced threats. To nurture true competency, immersive practical experiences must become standard in medical and engineering education. Security-focused laboratories, simulation environments, and scenario-based exercises should be incorporated as mandatory components rather than optional supplements. These might include tabletop exercises dramatizing the progression of a ransomware incident affecting medication delivery devices within a simulated hospital environment. In such scenarios, student teams assume different professional roles—clinician, IT security specialist, and administrator—grappling with device lockdowns, data integrity crises, and patient care disruptions. Penetration testing labs are equally essential; learners identify and exploit vulnerabilities on a segmented replica of a healthcare network, then implement and validate patches while maintaining operational continuity. Continual exposure to complex, unpredictable exercises trains students in the mental flexibility, critical analysis, and operational discipline so essential for reliable cyber defense in clinical settings (Khader et al., 2021; Lewis et al., 2024).

Graduates' transition from undergraduate study to residency or early-career technical practice marks a further critical juncture for ingraining security-first thinking. Clinical rotations and engineering residencies should embed direct engagement with device security challenges encountered in live healthcare settings. Structured programs may require residents to assess ongoing device security measures, participate in multidisciplinary incident response efforts, and conduct peer reviews evaluating cyber risk in actual device usage. A medical resident might work in partnership with a clinical IT team to investigate an anomalous device behavior alert and oversee the deployment of threat mitigation protocols in real time, while an engineering trainee develops secure integration solutions for legacy monitoring systems with updated authentication mechanisms. These immersive experiences not only enrich practical understanding but also accelerate the transition toward a workforce equipped with hands-on, context-aware judgment about security—competence that purely classroom-based learning can rarely deliver (Lewis et al., 2024).

Professional credentialing and accreditation bodies play a pivotal role in reinforcing this culture of excellence. Evolving accreditation standards must explicitly mandate demonstrable cybersecurity competency as a precondition for all licenses and certifications relevant to device design, development, implementation, and clinical use. Rigorous risk assessment sections, device configuration proficiency tests, and cybersecurity scenario evaluations should be included in board exams for medical device engineers and clinical practitioners alike (Khader et al., 2021; Lewis et al., 2024). Regulatory authorities have both opportunity and obligation to enforce industry-wide adoption of these elevated standards, ensuring that every new entrant to the profession meets contemporary security expectations and ethical imperatives. As institutional actors uphold these requirements, the responsibility to regularly update and refine curricula remains—demanding close alignment between educational content, professional standards, and the rapidly evolving spectrum of threats targeting connected medical technologies. Organizational leadership thus becomes the steward of an ongoing process, sustaining an adaptive, resilient approach to security education that anchors patient safety and professional integrity at its core.

Proactive Policy, Advocacy, and Regulation for a Connected Future

Building a resilient healthcare infrastructure depends fundamentally on education and ensuring competence in cybersecurity. Professional competency lays the groundwork for creating an environment where best practices are part of the organizational culture, not just periodic compliance activities. For instance, the U.S. Department of Health and Human Services (HHS) and the American Medical Association (AMA) emphasize making cybersecurity resources accessible to all levels of healthcare personnel, reinforcing the expectation that awareness and preparedness are shared responsibilities across clinical and administrative domains (*Physician Cybersecurity*, n.d.). This baseline knowledge empowers practitioners to anticipate risks and underscores the urgency for policy frameworks that go beyond immediate threats to anticipate evolving cyber challenges.

Forward-looking policy requires more than static rules; it demands dynamic standards capable of recognizing, adapting to, and even predicting emerging vulnerabilities. International standards, such as ISO/IEC 27001 for information security and IEC 62304 for software life cycle processes, establish a robust, flexible foundation that can evolve as threats shift (Reddy, 2024). These frameworks are not strictly prescriptive—they encourage organizations to continuously reevaluate security postures in response to updated knowledge about risk factors. A proactive approach to standardization enables healthcare systems to address vulnerabilities before they manifest as exploits, closing windows of exposure and fostering a posture of preparedness rather than recovery. Requirements for regular security risk assessments, such as those mandated by the Health Insurance Portability and Accountability Act (HIPAA) and supported through the Office for Civil Rights' (OCR) Security Risk Assessment (SRA) Tool, institutionalize this forward-thinking perspective (*Physician Cybersecurity*, n.d.).

Industry advocacy groups and cross-sector coalitions offer another layer of critical support for advancing cybersecurity in patient care. Such groups often move ahead of regulatory mandates, encouraging adoption of best practices that adapt more rapidly to the evolving threat landscape. The Healthcare Sector Coordinating Council (HSCC) is one example, producing training series and resources that translate complex technical topics into accessible guidance for frontline clinicians. These initiatives create shared responsibility within the sector, establishing a collaborative culture that prioritizes continuous improvement, transparency, and accountability (*Physician Cybersecurity*, n.d.). Collaborative platforms not only facilitate the rapid dissemination of innovative solutions, such as threat intelligence sharing networks and collective incident response playbooks, but also enable consensus around emerging security challenges. The collective impact achieved through these voluntary alliances can be seen in improved adoption of encryption norms, timely patch management, and coordinated vulnerability disclosures, all of which have contributed to decreasing the frequency and impact of breaches.

Expanding the circle of cybersecurity awareness is a policy imperative that extends well beyond technical professionals. Strategic public awareness campaigns play a pivotal role in activating patients, families, and nontechnical caregivers as stakeholders in the security conversation. For example, government-led video series and educational toolkits distributed by health agencies promote foundational "cyber-hygiene" and convey the implications of poor practices for patient safety (*Physician Cybersecurity*, n.d.). When the general public understands the risks and becomes alert to social engineering threats or the misuse of medical devices, this heightened vigilance drives greater demand for robust protections. These initiatives often result in observable changes—from increased patient engagement in device setup security to healthcare staff following multifactor authentication protocols or encrypting personally owned devices.

Legislation and regulatory mandates remain a cornerstone for shaping secure connected care, but their highest value emerges through preemptive action. Security by design—ensuring every new device and platform incorporates cybersecurity considerations from inception—becomes an enforceable expectation through clarified regulatory frameworks. Ongoing requirements for risk assessment and incident reporting are no longer discretionary but are codified, making continuous vigilance routine (Reddy, 2024). Regulatory interventions such as comprehensive breach notification rules and detailed device approval processes have triggered sector-wide improvements, reducing the scale and frequency of incidents. Early regulatory engagement, as seen with mandatory SRA tools and implementation guides released by HHS, also lowers postincident remediation costs and simplifies enforcement by establishing unambiguous chains of responsibility (*Physician Cybersecurity*, n.d.).

Meaningful enhancements in sector-wide resilience often follow from these policy interventions. The harmonization of global standards for software-driven medical devices, for instance, allows cross-border sharing of security best practices and speeds the detection and mitigation of vulnerabilities in globally distributed products (Reddy, 2024). Ongoing alignment with standards organizations also enables stakeholders to address risks such as data leaks or AI model manipulation before they escalate into patient safety crises, delivering long-term societal benefit.

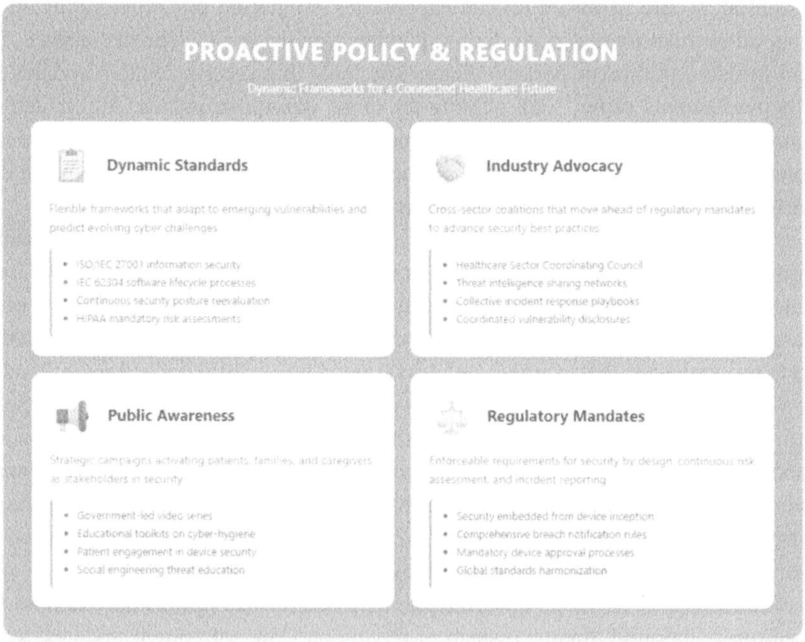

From this perspective, proactive policy, industry advocacy, public engagement, and regulatory foresight interconnect to form an enduring defense against cyber threats. These efforts establish the groundwork for a new phase of innovation—where continuous adaptation and creative problem-solving are necessary for securing tomorrow's interconnected and data-rich healthcare settings. With the right structures in place, healthcare organizations are positioned to embrace not only today's best practices but also tomorrow's transformative technologies.

Continuous Innovation: Security that Evolves with Technology

Continuous innovation forms the backbone of uncompromising cybersecurity for connected medical devices, building upon the essential groundwork of informed policy and proactive governance. As healthcare technology evolves, organizations must move beyond anticipatory rule-making and align with nimble, adaptive security practices that respond to real-time shifts in medical and cyber risk. The ever-increasing complexity and interdependence of networked medical devices—from implantable cardiac monitors to networked intravenous pumps—require a mindset that keeps pace not only with technological advances but with the rapidly changing threat landscape.

Healthcare organizations strengthen cyber defense by adopting Life Cycle Security Engineering, which ensures that security is no longer a discrete phase but a sustained undertaking embedded into every step of the device journey. Security by default begins as early as the architectural design of a new device. Developers integrate defense mechanisms, such as secure boot, encrypted communications, and role-based access controls. These proactive elements set a strong foundation. Attention then shifts to development and testing, where automated code analysis tools identify vulnerabilities before devices reach clinical trials or procurement evaluations. Built-in updatability becomes essential: secure firmware update capabilities, supported by cryptographically verified code signing, allow manufacturers to deploy patches rapidly as new threats surface. Hospitals benefit from infusion pumps or portable diagnostic tools that receive regular updates based on vulnerability assessments rather than waiting for large, disruptive recalls. Automated anomaly detection embedded within such devices enables near-instant identification of abnormal usage patterns or potential intrusions, empowering healthcare professionals to act before harm can occur (*The Future of Compliance: Shift to Proactive Security | CSA*, 2025).

Security as a product feature moves the conversation away from "minimum viable compliance" to a philosophy where protection enhances brand value, patient confidence, and regulatory engagement. Leading manufacturers treat security as a hallmark of product quality by publicizing their vulnerability disclosure programs with formal mechanisms for clinical users and security researchers to report defects. For example, vendors supporting coordinated vulnerability disclosure build trust with hospitals that conduct rigorous due diligence before procurement. Some companies continuously publish threat models and share recent security audit summaries with both potential and current customers. This transparency gives assurance to hospital IT departments and clinical engineers, who must balance usability, interoperability, and regulatory compliance. Where users witness clear, ongoing communication around risks and countermeasures, adoption rates and confidence in connected medical devices increase significantly.

Rapid iteration and agile development strategies help healthcare organizations and manufacturers swiftly counter emerging threats. Instead of relying on slow, infrequent updates, teams issue small, frequent releases, applying patches for newly discovered vulnerabilities—such as Bluetooth flaws in wireless patient monitors—within days instead of months (*What Is DevOps? Principles, Benefits Tools*, 2025). Integrating Agile and DevOps methodologies with clinical stakeholder feedback ensures fast detection and remediation of unforeseen interaction risks. For example, when clinicians notice that a new software update might inadvertently impede workflow or add friction to life-critical tasks, their input informs the next cycle of improvements while keeping security controls robust. Continuous code scanning, behavioral telemetry, and automated deployment pipelines streamline this process, reducing manual overhead and preventing human error (What Is DevOps? Principles, Benefits Tools, 2025).

Open source and transparency movements drive major advances in medical device security. Widely used cryptographic and authentication libraries that are subject to peer review expose vulnerabilities that might otherwise go unnoticed in proprietary modules. By tapping into global communities, device-makers gain early warnings about risks, such as cryptographic flaws or implementation weaknesses, that isolated internal audit teams may not catch. Security experts form working groups, sometimes coordinated by regulators or international standards bodies, to focus on the unique needs of clinical environments. Some vendors push transparency further by releasing selected security configurations, application programming interfaces (APIs), and audit logs, inviting independent verifications of new releases (*The Future of Compliance: Shift to Proactive Security | CSA*, 2025). Such practices foster a form of collaborative assurance that reinforces institutional trust and supports safe device rollout across global markets.

To operationalize these innovations, healthcare organizations establish cross-functional teams that unite security engineers, software developers, clinical stakeholders, and IT operations personnel. Security becomes a shared responsibility, embedded in training programs and daily operational routines. Organizations deploy automated monitoring and response systems while granting engineering teams the autonomy to apply policy-as-code and infrastructure-as-code for rapid remediation, lowering attack windows in production environments. Adaptive threat modeling, not static checklist compliance, forms the core of ongoing assessment, ensuring that every software iteration, configuration change, or new integration is accompanied by an updated risk perspective. A culture fostering open communication, disciplined incident response, and routine skills development ensures that vigilance and leadership continuously reinforce the technical protections that define uncompromising cybersecurity in modern healthcare.

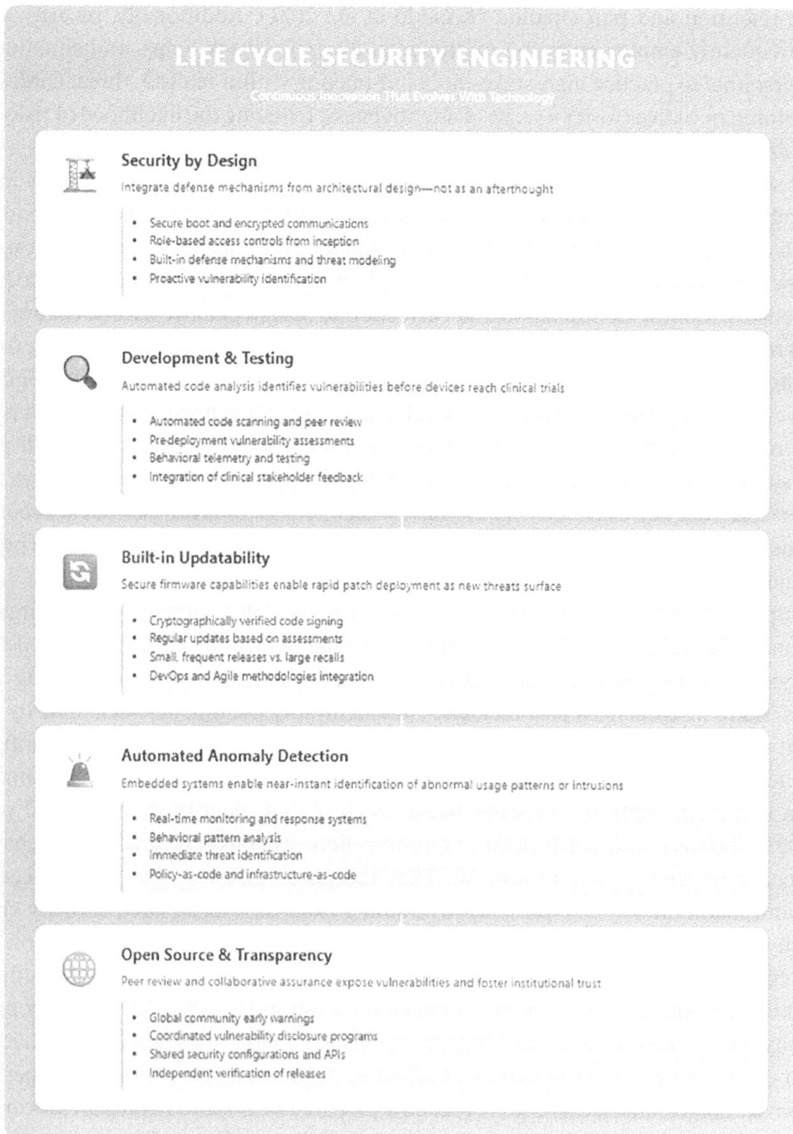

Personal and Organizational Actions for a Secure Tomorrow

Continuous education and training are the pillars upon which resilient cybersecurity for connected medical ecosystems is built. The threat landscape for connected patient care evolves rapidly, demanding that healthcare professionals—from device engineers to clinicians—update their knowledge continuously. Structured professional development routines center on delivering brief, recurring learning experiences that align with emerging threats and evolving technologies. Monthly seminars, targeted micro-learning modules, and scenario-based simulations represent concrete examples of dynamic training approaches. Some organizations use gamified learning platforms, offering interactive tasks and real-world attack simulations that encourage stronger knowledge retention and participation (Khando et al., 2021). Additionally, security awareness programs frequently employ engaging multimedia, direct feedback loops, and situational drills, allowing personnel to practice high-stakes decision-making within realistic threat contexts. These targeted training initiatives foster a sense of attentiveness, reducing the likelihood of risky behavior and equipping all stakeholders to respond effectively to security events.

A robust security culture within healthcare organizations transforms best practices into unwavering mandates. The enculturation of a zero-tolerance approach to security lapses requires organizations to articulate security as a fundamental value in every operation and decision. Codified policies set expectations for uncompromising compliance, while clearly delineated accountability structures ensure prompt identification and correction of unsafe practices. Effective organizations implement regular audits, internal phishing campaigns, and role-based access reviews; these activities hold individuals directly accountable for their actions and incentivize constant vigilance. For instance, a clinical department that fails a simulated phishing test must participate in additional, immediate retraining followed by repeated assessment. By making repercussions explicit, such as limiting system privileges or triggering security incident reviews for repeat infractions, organizations underscore that lapses in security are never tolerated. Without these robust accountability mechanisms, well-meaning but neglected processes expose connected patient care to preventable breaches and patient harm (Willie, 2023).

Leadership engagement fundamentally shapes the security climate of an organization. Security-conscious leaders embody accountability, transparency, and decisiveness, modeling these behaviors visibly across all levels of operations. Effective leaders invest in learning alongside their teams, share insights from past incidents, and demystify the link between security protocols and patient safety. When engineering leads regularly review and refine device configurations with their teams, or clinical supervisors start daily briefings with cybersecurity reminders and recognition of vigilant staff, the message becomes clear that security is shared and prioritized (Willie, 2023). Leaders also drive open communication, establishing predictable channels for reporting concerns without fear of reprisal. They recognize security wins, provide constructive feedback, and maintain a visible presence in training exercises. Designated security champions within each team work closely with leadership, acting as peer mentors and information conduits, helping to cultivate a culture where awareness, rapid response, and adaptation are routine.

For organizations and individuals to maintain momentum and ensure accountability, progress in cybersecurity must be measurable and transparent. Quantifiable metrics provide objective evidence of program effectiveness. Training participation rates, posttraining quiz scores, and performance metrics from simulated attack exercises reveal how prepared staff are to manage real-world threats (Khando et al., 2021). Incident response metrics—such as average time to detection, speed of escalation, and thoroughness of root cause analysis—expose strengths and gaps in operational readiness. Adherence to policy, including documented compliance with software update schedules and device vulnerability scans, is routinely tracked and reported in real-time dashboards available to leadership. Tracking metrics around vulnerability remediation cycles and the effectiveness of access controls provides actionable data for continuous improvement (Willie, 2023). Organizations that publicize results, both internally and at times with external partners, build trust by demonstrating commitment and transparency.

Organizations have gone further by incentivizing sustained security performance. Periodic recognition for teams that make measurable advances, such as reducing response times or achieving zero lapses within a quarter, helps sustain engagement. Openly sharing how data trends inform resource allocation and policy shifts reassures all participants that strategic decisions are grounded in facts, not inertia. Regular review of these results allows recalibration of training focus, investment in new technologies, and adjustment of accountability measures if specific objectives are not met. Data-driven decision-making strengthens buy-in, as professionals see the direct impact of their actions on organizational security posture and, ultimately, the safety of connected patients.

Through practical training programs, a relentless zero-tolerance culture, visible leadership, and transparent performance measurement, personal and organizational responsibility drives daily progress in securing connected patient care. These elements transform technical innovation and strategy into sustained, actionable behaviors, forming the ultimate mandate for uncompromising security in healthcare's digital future (Khando et al., 2021; Willie, 2023).

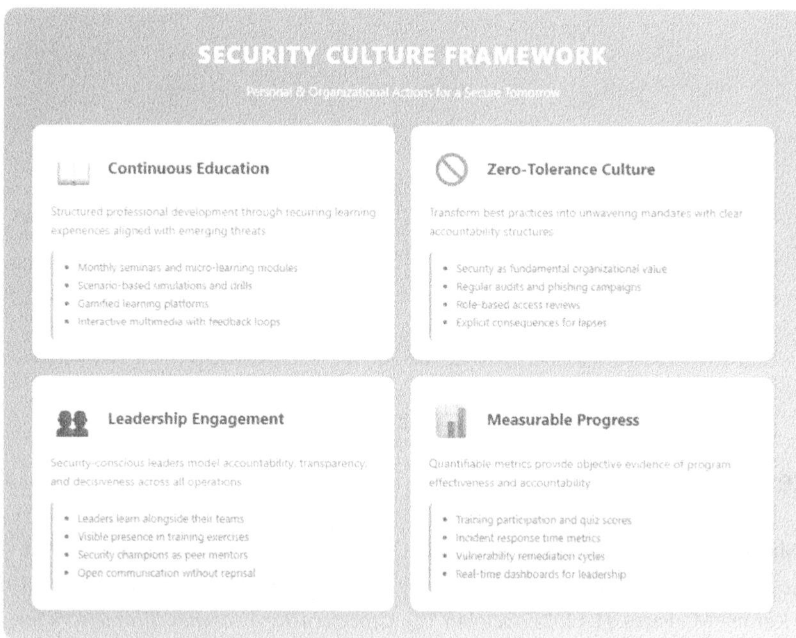

Bringing It All Together

Now that we have outlined the critical need for flawless cybersecurity in connected medical devices, the path forward demands unwavering commitment from all stakeholders to embed security into every stage of device design, education, policy, and practice. By integrating comprehensive cybersecurity training within healthcare and engineering curricula, fostering proactive policies and collaborative advocacy, and continuously advancing technical innovations tailored to evolving threats, we can build resilient systems that safeguard patient safety without compromise. Equally important is cultivating a culture of accountability and leadership engagement that translates strategy into consistent action across organizations. Embracing this multifaceted approach positions the healthcare sector not only to respond effectively to present challenges but also to anticipate and neutralize future risks, ensuring secure, trustworthy connected care for years to come.

References

Clarke, M., Martin, K. (2023, August 25). *Managing Cybersecurity Risk in Healthcare Settings.* Healthcare Management Forum; SAGE Publishing. https://doi.org/10.1177/08404704231195804.

Khader, M., Karam, M., Fares, H. (2021, October 12). *Cybersecurity Awareness Framework for Academia*. Information. https://doi.org/10.3390/info12100417.

Khando, K., Gao, S., Islam, S. M., Salman, A. (2021). *Enhancing Employees Information Security Awareness in Private and Public Organisations: A Systematic Literature Review*. Computers Security; Sciencedirect. https://doi.org/10.1016/j.cose.2021.102267.

Lewis, K. O., Popov, V., Syeda Sadia Fatima. (2024, January 23). *From Static Web to Metaverse: Reinventing Medical Education in the Post-Pandemic Era*. Annals of Medicine; Informa. https://doi.org/10.1080/07853890.2024.2305694.

Mensah, N. K., Adzakpah, G., Kissi, J., Taylor-Abdulai, H., Johnson, S. B., Agbeshie, P. A., Opoku, C., Abakah, J., Osei, E., Agyekum, A. Y., Boadu, R. O. (2024, January 1). *Health Professionals' Ethical, Security, and Patient Safety Concerns Using Digital Health Technologies: A Mixed Method Research Study*. Health Services Insights; SAGE Publishing. https://doi.org/10.1177/11786329241303379.

Physician Cybersecurity. (n.d.). American Medical Association. https://www.ama-assn.org/practice-management/sustainability/physician-cybersecurity.

Reddy, S. (2024, December 24). *Global Harmonisation of AI-Enabled Software as a Medical Device Regulation: Addressing Challenges and Unifying Standards.* Mayo Clinic Proceedings: Digital Health; Elsevier. https://doi.org/10.1016/j.mcpdig.2024.100191.

The Future of Compliance: Shift to Proactive Security | CSA. (2025). Cloudsecurityalliance.org. https://cloudsecurityalliance.org/blog/2025/02/04/the-future-of-compliance-is-here-automation-intelligence-and-a-shift-to-proactive-security.

What is DevOps? Principles, Benefits Tools. (2025, July 22). SentinelOne. https://www.sentinelone.com/cybersecurity-101/cybersecurity/what-is-devops/.

Willie, M. M. (2023, June 8). *The Role of Organizational Culture in Cybersecurity: Building a Security-First Culture.* ResearchGate; unknown. https://www.researchgate.net/publication/371399113_The_Role_of_Organizational_Culture_in_Cybersecurity_Building_a_Security-First_Culture.

21

Conclusion

"When the pager in the hospital's emergency room suddenly stopped working, a team scrambled to keep critical care uninterrupted. Behind the scenes, a connected infusion pump had been compromised, causing cascading failures across the network. The incident was not just a technical malfunction but a moment that tested the very foundation of trust between patients and caregivers. This scenario is far from unique; it reflects the complex challenges inherent in securing the digital systems that increasingly govern modern healthcare. As medical devices evolve, so too do the risks they face, underscoring the urgency and weight of protecting these technologies—not only to safeguard data but to preserve lives."

Reaffirming the Mission and Synthesizing the Security Landscape

Protecting patient lives stands as the absolute priority in healthcare. In modern clinical environments, digital devices no longer act as isolated tools at the periphery of care; instead, they work at the heart of diagnosis, monitoring, and treatment. The ethics and professionalism embedded in medical practice now extend to managing technical risk, as nearly every new medical device introduced into the hospital ecosystem comes equipped with internet connectivity, artificial intelligence, and remote management features. As this digital transformation advances, securing medical devices is no longer simply a matter of compliance. It becomes a vital duty grounded in the fundamental aim of preserving patient safety and public trust (Accorian, 2025; Senior, 2025).

Early generations of medical devices operated in isolation, limited in capability and scope. Today's medical equipment is integrated within vast networks linking patient records, physician workstations, and cloud-based diagnostic platforms. This interconnectedness multiplies points of vulnerability. A single, poorly secured device—such as an insulin pump, a pacemaker, or an infusion system—can become a foothold for attackers seeking to infiltrate hospital networks. An escalation in ransomware attacks, such as the incident that crippled the UK's National Health Service by locking critical records and putting patient care at risk, reveals just how deeply cybersecurity is tied to operational continuity and patient well-being (Accorian, 2025). Attackers have also targeted wireless implants and smart monitoring systems, with outcomes ranging from the theft of protected health information to the possibility of life-altering or even fatal device manipulation.

Key threat actors targeting healthcare include financially motivated cybercriminals deploying ransomware, nation-state hackers focused on intelligence gathering, and hacktivists with ideological aims. Healthcare's reliance on legacy systems and vast data repositories only increases the temptation for attackers. Common attack vectors involve phishing, supply-chain compromises, unpatched vulnerabilities, and exploitation of weak authentication or misconfigured cloud interfaces. Concrete examples include adversaries manipulating AI diagnostic models to produce inaccurate results or leveraging exposure in device firmware to push malicious updates remotely (Accorian, 2025).

Regulatory oversight shapes the standards for medical device security. Frameworks such as the U.S. Food and Drug Administration (FDA)'s cybersecurity guidance, EU Medical Device Regulation (MDR), and the National Institute of Standards and Technology (NIST) Cybersecurity Framework require end-to-end risk management and continuous monitoring. These regulations do not just insist on initial compliance, they drive a dynamic process of ongoing assessment, patching, and adaptation to emerging threats. The FDA, for instance, expects manufacturers to perform risk assessments before devices are introduced to the market and to issue timely software updates in response to newly discovered vulnerabilities (Senior, 2025). Likewise, the General Data Protection Regulation (GDPR)'s strict requirements for personal data protection apply to health organizations worldwide and mandate immediate breach notification and responsible handling of sensitive data (Senior, 2025).

Effective security begins with thorough risk assessment. Threat modeling identifies possible attack pathways—such as unauthorized remote access or data exfiltration. Vulnerability analysis pinpoints technical weaknesses, like unpatched code or insufficient encryption. Security risk management develops mitigation plans tailored to the device and its role in patient care. These methods weave seamlessly into best practices for development and deployment. For example, implementing security-by-design principles ensures that authentication, encryption, and audit logging are built into devices from the outset. Automated testing during software development helps detect weaknesses before a product enters clinical settings. Once deployed, hospitals may employ continuous real-time monitoring to detect suspicious device behavior or attempted breaches.

Practical scenarios illustrate these approaches. A device manufacturer might require developers to follow secure coding guidelines and run automated penetration tests, ensuring that vulnerabilities do not reach patients. Meanwhile, hospital IT managers could configure network segmentation to isolate devices by risk level, reducing the impact of compromise. Applying secure, encrypted communications between devices and monitoring systems further protects against data interception.

Incident response strategies must match the speed and complexity of modern threats. Coordinating among clinical staff, IT personnel, and external partners becomes essential. When Universal Health Services suffered a major ransomware attack, more than 400 clinics went offline; a rapid response was needed to contain the threat while protecting critical care and informing patients (Accorian, 2025). Effective plans establish clear roles, ensure swift detection, and mandate open communication to minimize service disruption and risk to patients. Regular drills and real-world simulations test and refine these organizational responses.

A secure healthcare ecosystem emerges only when these components work together. The evolution of devices, expanding threat awareness, compliance, risk assessment, sound engineering, and robust incident response jointly enable a resilient posture. Precision in technical execution must always be matched by an unwavering commitment to ethical and professional standards. Professionals safeguarding medical devices shape the future of patient safety through both their expertise and their moral responsibility. Recognizing this intersection helps every stakeholder serve as a guardian against the evolving dangers of a connected healthcare world (Accorian, 2025; Senior, 2025).

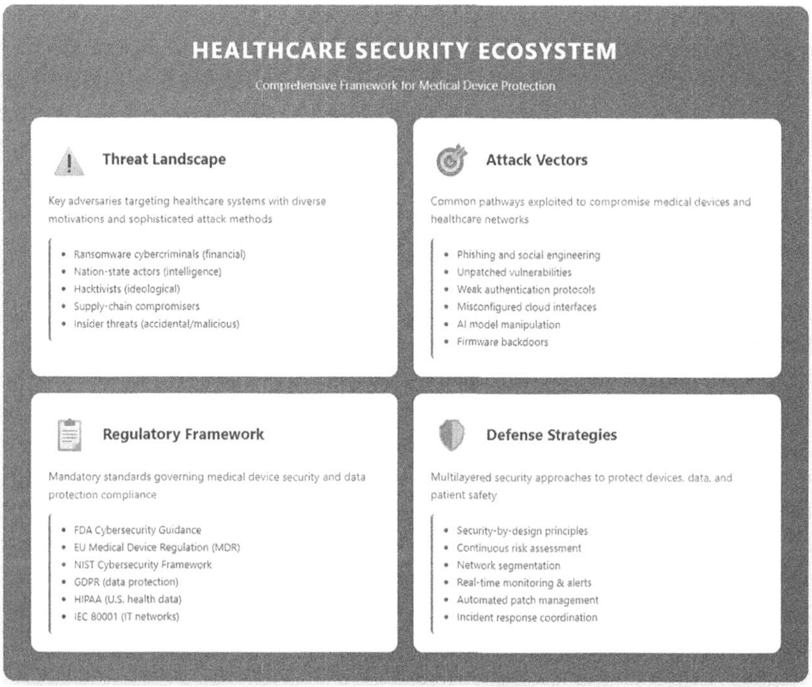

Transformative Insights on Ethical Accountability

The pursuit of patient safety forms the heart of ethical responsibility in healthcare cybersecurity. Every professional in this landscape, whether engaged in engineering, compliance, management, or research, bears a moral duty to mitigate risks that threaten not just data, but human well-being. The digitalization of patient information and widespread integration of networked devices mean that a compromised database or malfunctioning device can jeopardize diagnoses, delay treatments, and, in acute cases, threaten lives (Aldosari, 2025; Javaid et al., 2023). A single breach ripples out, affecting both the patient's experience of care and the public's faith in the healthcare system.

Viewing cybersecurity solely through the lens of technical requirements belies its greater significance as an ethical and professional mandate. The fundamental tenets of professional integrity and patient trust depend on the responsible stewardship of sensitive information and critical service continuity. Trust forms the cornerstone of care; patients entrust their most intimate stories, and failures in cybersecurity can erode that relationship, leaving patients vulnerable not only to identity theft and fraud but to the denial or compromise of medically necessary services (Aldosari, 2025). It follows that the professional ethic in this context is defined not only by technical mastery but also by the earnest prioritization of human safety, privacy, and dignity.

The belief that concealing system details or vulnerabilities—security through obscurity—can ensure safety is a fallacy with especially high stakes in healthcare. Several well-documented ransomware attacks on hospitals have exposed this error. When attackers encrypted hospital records and demanded payment, some care facilities were forced to divert critical care or postpone chemotherapy and may have experienced fatal delays in emergency response. Incidents involving compromised network-connected medical devices have also resulted in altered device behavior, disrupting drug deliveries and interfering with patient monitoring sessions (Javaid et al., 2023). These events underscore the need for transparency, robust security assessments, and third-party scrutiny. Open vulnerability disclosure programs, together with industry-wide collaboration on emerging threats, offer ethical countermeasures to the secretive treatment of security flaws. Professionals in this sphere cannot afford insularity; they must foster open communication, welcome third-party validation, and integrate industry best practices throughout the process chain.

A proactive commitment to security by design anchors the ethical posture required of every specialist in healthcare technology. Embedding cybersecurity at each stage—from the initial concept to postmarket surveillance—represents both a legal expectation and a moral standard. Regulatory agencies such as the FDA require demonstration of cybersecurity plans as part of their premarket submission process, reflecting the expectation of safety by design. Employing robust authentication for clinical staff, integrating hardware-based root-of-trust modules in critical devices, and planning for continuous patch management are not simply technical recommendations; each action serves as a concrete realization of the duty to minimize harm. For instance, organizations that routinely isolate network segments and routinely update medical equipment reduce the scope and impact of any incident, thereby translating abstract ethical imperatives into tangible benefits for patients and staff alike (Aldosari, 2025).

Legal compliance through frameworks such as HIPAA, GDPR, IEC 80001, and U.S. FDA guidance forms only the basic threshold for professional conduct. While adherence to these statutes is mandatory, legal minimums are not substitutes for moral discernment and forward-thinking advocacy. An ethical cybersecurity leader recognizes that meeting existing standards is necessary but insufficient; true integrity emerges when professionals anticipate new threats, adapt policy and process ahead of regulatory updates, and champion interventions that are reactive and preventive. Teams who engage employees in trainings to recognize phishing attempts, protect against social engineering, and institute strong breach notification and response plans exemplify compliance that advances patient interests, rather than simply checking regulatory boxes (Aldosari, 2025; Javaid et al., 2023).

By internalizing cybersecurity best practices as part of professional identity, healthcare technologists and administrators can catalyze a positive organizational culture. Cultivating an environment where security education is continuous, vulnerabilities are openly reported, and incidents are addressed transparently models ethical leadership. One hospital's early adoption of public vulnerability disclosure with coordinated incident response led not only to reduced breach frequency but to an enhanced sense of trust among employees, patients, and external partners. Such stories reinforce the role of practitioners not only as stewards of compliance but as architects of the trust and resilience underpinning modern healthcare.

This orientation becomes the foundation for true professional readiness: the ability to act with both technical acuity and unwavering ethical clarity, fortifying healthcare against harm while modeling integrity at every level (Aldosari, 2025; Javaid et al., 2023).

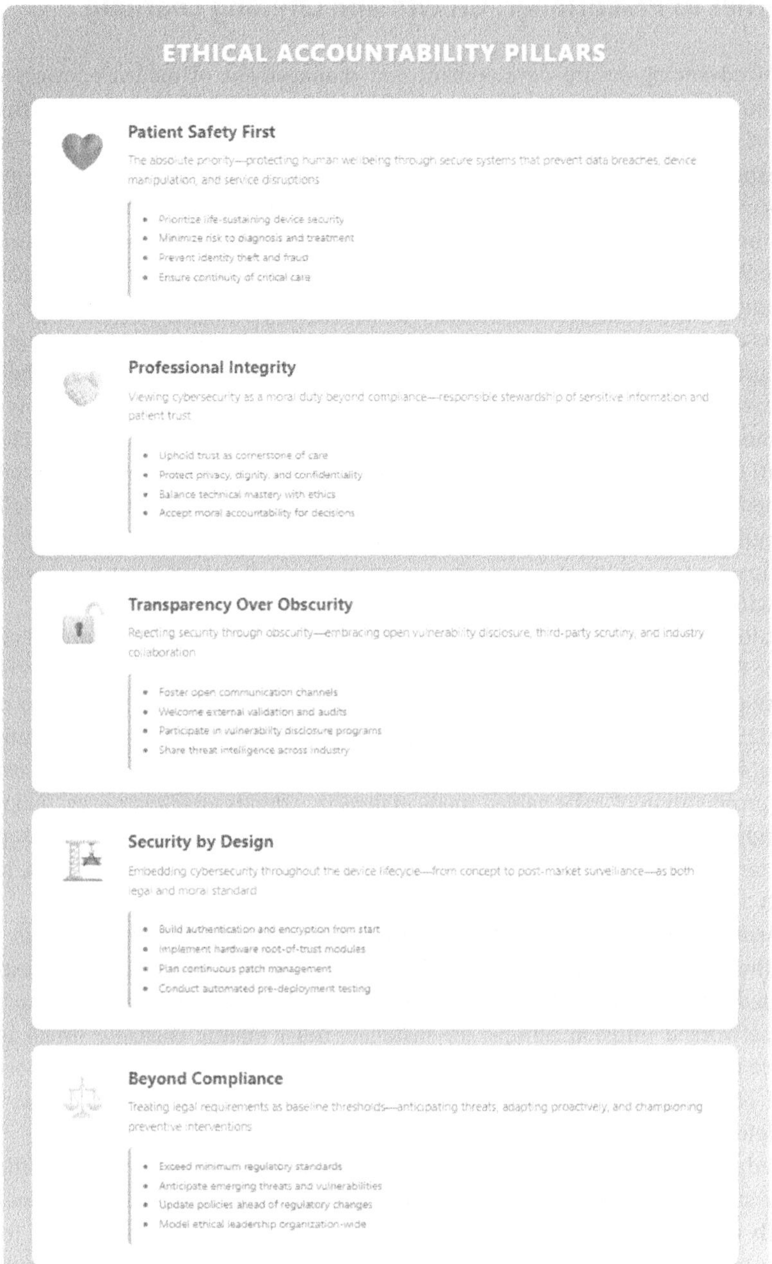
ETHICAL ACCOUNTABILITY PILLARS
Patient Safety First
The absolute priority—protecting human wellbeing through secure systems that prevent data breaches, device manipulation, and service disruptions
Prioritize life-sustaining device security
Minimize risk to diagnosis and treatment
Prevent identity theft and fraud
Ensure continuity of critical care
Professional Integrity
Viewing cybersecurity as a moral duty beyond compliance—responsible stewardship of sensitive information and patient trust
Uphold trust as cornerstone of care
Protect privacy, dignity, and confidentiality
Balance technical mastery with ethics
Accept moral accountability for decisions
Transparency Over Obscurity
Rejecting security through obscurity—embracing open vulnerability disclosure, third-party scrutiny, and industry collaboration
Foster open communication channels
Welcome external validation and audits
Participate in vulnerability disclosure programs
Share threat intelligence across industry
Security by Design
Embedding cybersecurity throughout the device lifecycle—from concept to post-market surveillance—as both legal and moral standard
Build authentication and encryption from start
Implement hardware root-of-trust modules
Plan continuous patch management
Conduct automated pre-deployment testing
Beyond Compliance
Treating legal requirements as baseline thresholds—anticipating threats, adapting proactively, and championing preventive interventions
Exceed minimum regulatory standards
Anticipate emerging threats and vulnerabilities
Update policies ahead of regulatory changes
Model ethical leadership organization-wide

From Learner to Practitioner: Growth and Lifelong Learning

The work of advancing secure development and management of medical devices derives its meaning from an unwavering commitment to patient safety and ethical stewardship. Rigorous exploration of systems, regulatory frameworks, and technical countermeasures transforms theory into lived expertise. Practitioners who have undertaken this journey have grown not only as technical specialists but as accountable leaders, equipped to weigh risk and responsibility at every point in the healthcare device life cycle. Every lesson in this field—whether legal, technological, or human-centered—has served as a catalyst for professional growth through a landscape fraught with stakes and continuous evolution.

Healthcare cybersecurity now demands not just initial competence but continuous mastery. Ethical and professional obligations extend far beyond a baseline understanding, obliging experts to pursue learning as a perpetual condition of their role. AI-driven clinical systems and interconnected medical devices expand both the promise and the risk of modern practice. In this environment, the principles of transparency, human oversight, and system accountability matter as much as technical control, forming the bedrock of secure and ethical innovation (*HIPAA Is Not Enough: Why AI and Emerging Tech Demand a New Risk Standard*, 2025). Oversight of algorithms, thorough impact assessments, and proactive governance are all part of a routine that must evolve in lockstep with regulatory movements and emerging technical exposures.

The regulatory landscape remains fluid, compelling practitioners to keep informed of new standards like the EU AI Act, tightening requirements around risk management, data governance, and cybersecurity protocols (*HIPAA Is Not Enough: Why AI and Emerging Tech Demand a New Risk Standard*, 2025). This dynamism increases the importance of seeking up-to-date guidance, participating in ongoing training programs, and adapting policies as new threats emerge. For example, the continuous proliferation of IoT-enabled medical devices extends attack surfaces, making regular education and technical training essential in protecting sensitive health systems from sophisticated threats like targeted ransomware campaigns or lateral movement of malware across legacy endpoints (He et al., 2021).

Active learning fortifies cybersecurity expertise. Technical staff and administrators, by staying attuned to the latest threat research and compliance guidance, can be agile in responding to unprecedented attack vectors. A cybersecurity professional recognizing anomalous command activity in networked infusion pumps, for instance, must rapidly correlate threat intelligence feeds, enforce improved segmentation policies, and communicate new response protocols to both clinical and IT teams. This vigilance not only thwarts potential breaches but also transcends individual defense, nurturing a culture of shared responsibility.

Ongoing adaptation establishes trust with end users and patients. Compliance officers might encounter a sudden regulatory update requiring multifactor authentication and audit logging on telehealth monitoring devices. Their preparedness depends on maintaining up-to-date documentation, mapping regulation onto real organizational workflows, and validating system changes through rigorous testing as well as front-line user feedback. These actions foster not only compliance but resilience against regulatory risk and operational disruption, elevating the organizational standard of care (*HIPAA Is Not Enough: Why AI and Emerging Tech Demand a New Risk Standard*, 2025).

Engineering teams, meanwhile, benefit from integrating postmarket surveillance mechanisms directly into device design and life cycle management. This enables early detection of emerging software vulnerabilities, continuous patch validation, and the discovery of new threat signatures. By systematizing incident response and supporting root cause analysis, organizations minimize operational downtime and supply-chain risk (He et al., 2021). Technical measures, such as encrypted telemetry and automated endpoint analytics, bolster the security posture when coupled with coordinated response plans, board-level risk communication, and alignment with enterprise risk management frameworks.

Professional growth in this discipline is inseparable from active contributions to the broader healthcare cybersecurity ecosystem. Participating in industry workshops, pursuing advanced certifications, and collaborating with clinical leaders amplify not only the practitioner's expertise but also the sector's collective intelligence and adaptability. Engaged professionals foster secure-by-design principles at the innovative core of their products and processes, embedding continuous improvement in device safety, patient trust, and operational integrity (He et al., 2021).

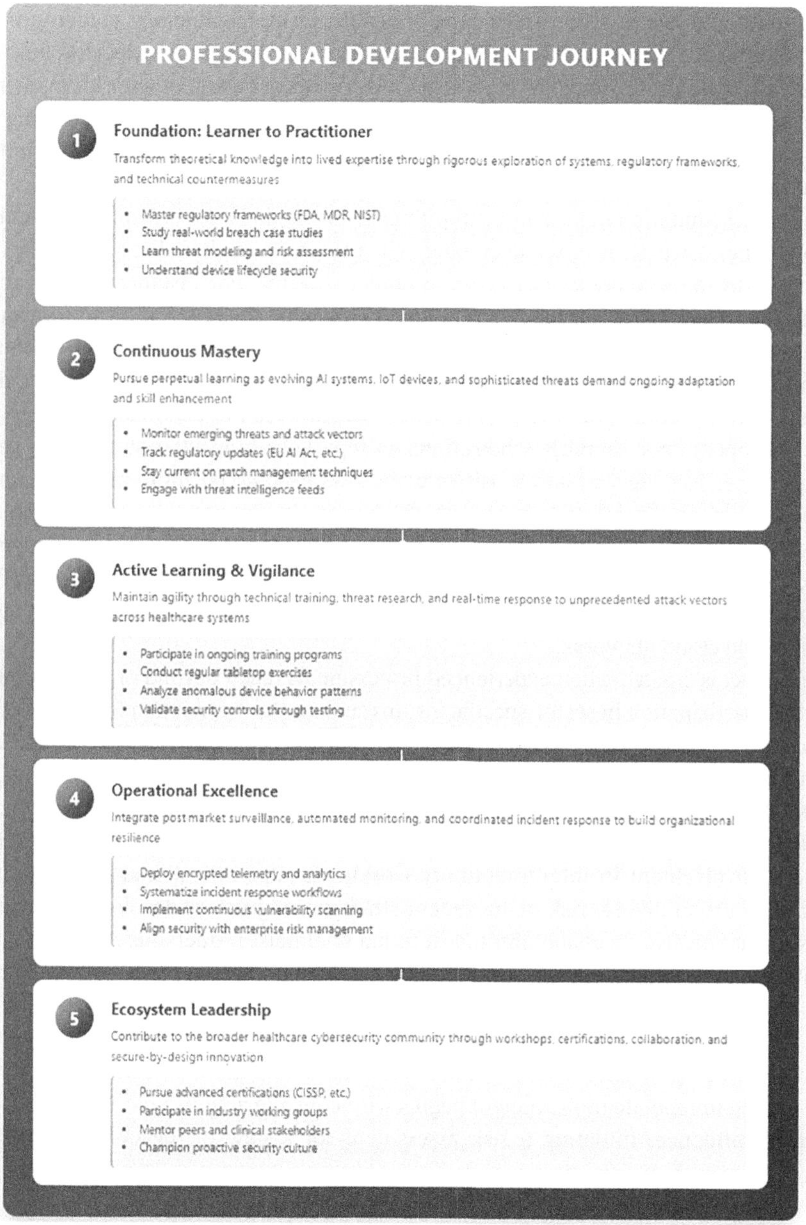

Sustained vigilance, reinforced by peer-reviewed research and industry best practices, empowers practitioners to move from isolated problem-solving toward collaborative solution-building. Harnessing their skills, they help organizations transcend reactive defense, embedding resilience, anticipation of evolving threats, and effective governance into the very fabric of healthcare delivery. Building upon these new capabilities, practitioners contribute to an ecosystem where the safety and advancement of healthcare remain ever in focus, cultivating a future that values both innovation and the enduring ethics of care.

Active Contribution, Advocacy, and Practical Next Steps

As leaders and practitioners at the intersection of healthcare and technology, readers are uniquely positioned to advance medical device cybersecurity. The knowledge and expertise built through experience, sustained professional development, and critical engagement with the evolving threat landscape are themselves powerful assets. By sharing technical insights, advocating for best practices, and mentoring peers, each professional strengthens the collective capability to secure healthcare environments and protect patient lives.

Knowledge dissemination holds a forceful influence in shaping both organizational culture and tangible outcomes. Security is most sustainable when treated as a shared responsibility across all roles, rather than the sole domain of IT or compliance teams. For instance, organizations that conducted comprehensive, cross-functional debriefs following the WannaCry ransomware attack demonstrated remarkable improvements during later incidents, achieving faster response times and minimizing operational disruptions (Elendu et al., 2024). Open discussions about incident findings, transparent reporting, and interdepartmental learning lead to actionable changes in policy and day-to-day operations. Similarly, when clinicians report phishing attempts to IT and participate in simulation exercises, devices and sensitive workflows become far more resilient to persistent threats. By equipping clinical end users with the knowledge to recognize and promptly report anomalies, employees become a first line of defense, as seen in organizations that successfully combine technical controls with effective training (Ribeiro et al., 2025). These behavioral shifts lay the groundwork for cultures of vigilance, which remain essential as medical devices integrate further with hospital and cloud networks.

Broader impact is created when experienced professionals reach beyond organizational boundaries. Regular participation in sector-specific forums such as the Association for the Advancement of Medical Instrumentation (AAMI) Medical Device Security Working Group or the IEEE Technical Committee on Security & Privacy can stimulate innovation and help establish consensus-driven guidelines. Healthcare Information Sharing and Analysis Centers (H-ISAC) provide actionable threat intelligence, real-time alerts, and case studies relevant to regional and international stakeholders. Involvement in interdisciplinary working groups or national consortia facilitates first-hand exposure to diverse risk management strategies, regulatory developments, and operational lessons not always accessible through internal channels. Leadership engagement in these communities enables organizations to benchmark security programs, contribute to data-sharing initiatives, and remain current on mitigation techniques tailored to sector-specific challenges (Ribeiro et al., 2025). These activities not only sharpen technical skills and regulatory awareness but also reinforce the principle that information-sharing is foundational to collective cyber defense.

A clear path to immediate improvement begins with concrete actions tailored to one's expertise and scope of influence. Initiating a risk assessment for a new or legacy connected medical device involves assembling stakeholders to review device connectivity, evaluate software vulnerabilities, assess vendor support models, and document any unmanaged communications. Integrating security by design requires working with engineering and design partners from premarket phases, validating that basic controls—such as authentication, audit logging, and encrypted data transmission—are implemented before device prototypes reach validation testing. Cross-functional Tabletop Exercises, orchestrated with clinical, IT, and biomedical engineering colleagues, help teams rehearse coordinated responses to device security incidents, clarifying escalation paths and identifying potential breakdowns before threats occur.

Each action faces familiar barriers, including resistance to changing workflows, budgetary constraints, and uncertainty about emerging requirements. Overcoming these obstacles often relies on clear communication of security's role in supporting operational resilience and patient safety, not just compliance. Demonstrating the benefit of proactive controls—such as reduced downtime, prevention of costly breaches, and enhanced reputation—encourages greater buy-in from leadership and end users alike (Elendu et al., 2024; Ribeiro et al., 2025).

For continued professional growth, readers should reference industry frameworks and tools for practical implementation and benchmarking. The NIST SP 800-53 catalog provides comprehensive controls applicable to networked medical devices, while NIST SP 800-82 offers detailed guidance specifically for industrial control systems in clinical contexts. ISO/IEC 80001 outlines risk management for medical IT networks, and the Health Information Trust Alliance (HITRUST) Common Security Framework provides a harmonized approach for regulatory compliance. Specialty certifications such as CISSP (with a healthcare focus) and the AAMI Certified Professional in Medical Device Security demonstrate advanced competence and enable structured, ongoing learning. Leveraging these resources empowers professionals to further embed security into clinical innovation and device life cycle management, adapting to rapidly evolving regulatory environments.

Progress in medical device cybersecurity thrives on professionals committed to translating technical knowledge into active leadership, mentorship, and collaboration. The examples shared reflect the profound difference stakeholders at every level can make by advocating for smart, practical improvements and by fueling a culture of mutual accountability and knowledge exchange. As the field advances, this collective, purposeful engagement builds the strongest foundation for patient safety and sector-wide advancement, underscoring the vital role of every contributor in the broader movement for healthcare security (Elendu et al., 2024; Ribeiro et al., 2025).

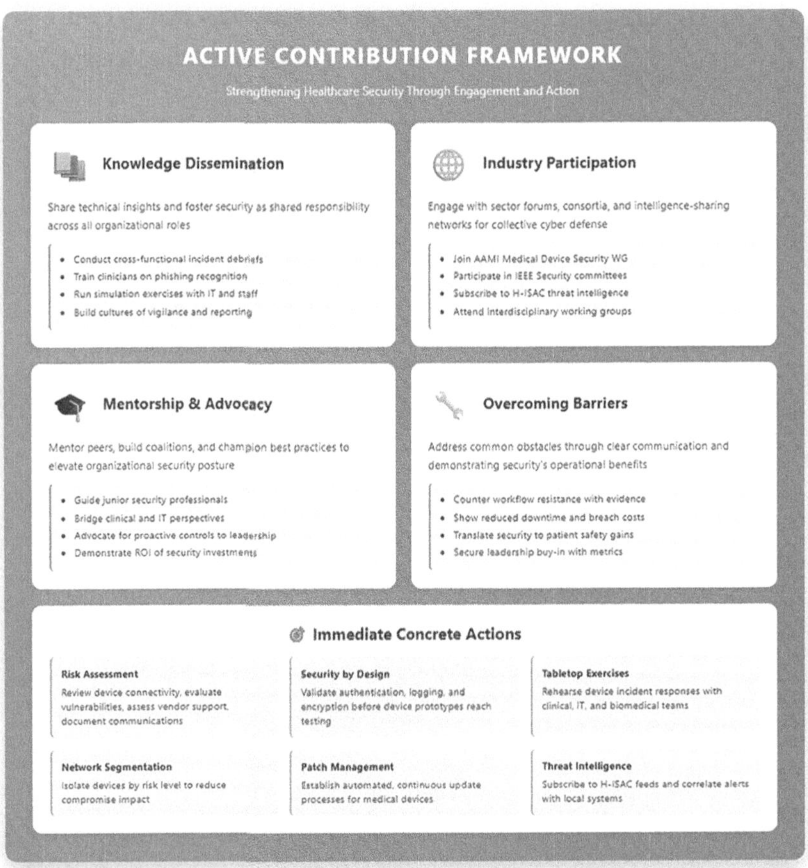

Gratitude, Connection, and a Motivational Call to Action

Your expertise in securing the healthcare landscape constitutes more than a technical achievement; it is a vital expression of ethical stewardship and a demonstration of unshakable commitment to patient well-being. The collective progress in medical device security reflects the sum of your diligence, adaptability, and readiness to uphold the foundational promise of medicine—first, do no harm. The evolving digital environment brings with it growing risks to patient data, device integrity, and clinical operations, making your sustained vigilance indispensable in protecting confidentiality and patient trust (Filkins et al., 2016; Mensah et al., 2024).

Your engagement in proactive risk management, especially amid rising threats presented by interconnected systems and the rapid development of digital health technologies, positions you at the center of advancing patient safety. Each day, you navigate and overcome multidisciplinary obstacles spanning technology, regulation, and ethics. For example, when you lead your team to implement secure design and engineering protocols in developing networked infusion pumps or diagnostic equipment, you exemplify how collaborative innovation addresses real vulnerabilities. By driving the integration of cybersecurity standards such as encryption and robust access controls into device life cycles, you enable clinical effectiveness while ensuring compliance with data protection regulations, echoing the overarching mission that guides your field (Mensah et al., 2024).

Championing secure health technology also requires forging strong alliances across roles—whether you are a biomedical engineer working alongside software developers, an IT leader building bridges with clinical stakeholders, or a regulatory specialist interpreting evolving standards. Your ability to establish and nurture these partnerships directly enhances the resilience of device ecosystems. When you facilitate interdisciplinary tabletop exercises to test incident response protocols or ensure regular security-focused training for staff, you foster a shared sense of purpose. Such actions transform isolated compliance measures into a culture of collective responsibility in which each contributor internalizes their vital role in upholding privacy, integrity, and system availability (Filkins et al., 2016; Mensah et al., 2024).

Recognition of your integrity extends to your response in crises. Imagine the scenario of a major hospital network facing an active ransomware event. Your readiness to deploy mitigation processes, communicate rapidly across teams, and restore affected systems could mean the difference between a brief disturbance and the catastrophic erosion of patient trust or safety. Successful identification and remediation of a device vulnerability safeguard not only operational continuity but also the personal dignity and confidentiality of those who rely on healthcare systems. Each new update you promote, each access policy you refine, serves as a bulwark against external and internal threats—ranging from sophisticated cybercriminals to inadvertent lapses by authorized personnel (Filkins et al., 2016; Mensah et al., 2024).

Your role extends beyond immediate defense measures to include the deliberate cultivation of cyber-awareness across your organization. Mentoring colleagues on proper data handling, raising awareness of phishing schemes, and securing the responsible use of mobile or cloud-enabled medical devices all deepen organizational readiness. You contribute to building a workforce where cyber literacy and ethical conduct become ingrained, not only within the technical sphere but throughout every aspect of clinical and research practice. When you share incident lessons through internal briefings, participate in external threat-sharing alliances, or contribute to industry best practice guidelines, you reinforce a feedback loop of continual improvement and collaboration (Filkins et al., 2016).

You are not alone in this pursuit. The broader professional community stands committed to supporting your growth, facilitating connections with peers, and empowering you to confront new threats. Communities of practice, industry working groups, and regulatory partnerships provide living networks for knowledge exchange, advocacy, and mutual accountability. These structures enable you to stay agile, anticipate emerging risks, and embrace innovations that keep pace with adversarial tactics in the digital domain.

The immense privilege and responsibility entrusted to you as a guardian of digital health empower you to shape the very future of patient care. Visualize yourself safeguarding a pediatric intensive care unit's wireless monitoring systems or engineering a next-generation defibrillator that resists evolving attack vectors. Your expertise and decisions define both the experience and the safety of patients whose voices remain unheard in technical debates yet whose lives depend on your work. Your actions merge innovation with humility, embodying the social contract at the core of healthcare—a trust that binds professionals, patients, and society.

Thank you for your perseverance, courage, and vision. May this chapter serve as both recognition of the journey thus far and a call to continued leadership. Stand proud as a steward of health and technology; your commitment sustains the resilience and promise of connected care, forging a future where security and compassion remain inseparable (Filkins et al., 2016; Mensah et al., 2024).

Final Thoughts

As we integrate the knowledge and principles outlined in this chapter, it is clear that securing medical devices demands continuous vigilance, ethical commitment, and collaborative expertise across disciplines. The evolving digital healthcare landscape presents complex risks, but by embracing proactive risk management, security-by-design approaches, and transparent communication, professionals can safeguard patient safety and uphold public trust. Moving forward, healthcare cybersecurity specialists, engineers, compliance officers, and IT managers must remain agile learners and advocates, adapting to emerging threats and regulatory changes while fostering a culture of shared responsibility. Together, these concerted efforts will strengthen the resilience of interconnected medical systems, ensuring that innovation advances hand in hand with the enduring mission to protect human life and dignity in healthcare's digital future.

References

Accorian. (2025, May 23). *Securing Complying with Medical Device Security (Med Dev Sec)—Accorian.* Accorian. https://www.accorian.com/securing-complying-with-medical-device-security-med-dev-sec.

Aldosari, B. (2025, May 6). *Cybersecurity in Healthcare: New Threat to Patient Safety.* Cureus; Springer Science and Business Media LLC. https://doi.org/10.7759/cureus.83614.

Elendu, C., Omeludike, E. K., Oloyede, P. O., Obidigbo, B. T., Omeludike, J. C. (2024, September 27). *Legal Implications for Clinicians in Cybersecurity Incidents: A Review.* Medicine; Wolters Kluwer. https://doi.org/10.1097/md.0000000000039887.

Filkins, B. L., Kim, J. Y., Roberts, B., Armstrong, W., Miller, M. A., Hultner, M. L., Castillo, A. P., Jean-Christophe Ducom, Topol, E. J., Steinhubl, S. R. (2016, March 15). *Privacy and Security in the Era of Digital Health: What Should Translational Researchers Know and Do About It?* American Journal of Translational Research. https://pmc.ncbi.nlm.nih.gov/articles/PMC4859641/.

He, Y., Aliyu, A., Evans, M., Luo, C. (2021). *Healthcare Cyber Security Challenges and Solutions Under the Climate of Covid19: A Scoping Review.* Journal of Medical Internet Research; ncbi. https://doi.org/10.2196/21747.

HIPAA is Not Enough: Why AI and Emerging Tech Demand a New Risk Standard. (2025, July 31). Censinet.com; Censinet. https://www.censinet.com/perspectives/hipaa-is-not-enough-why-ai-and-emerging-tech-demand-a-new-risk-standard.

Javaid, M., Haleem, A., Singh, R. P., Suman, R. (2023). *Towards Insighting Cybersecurity for Healthcare Domains: A Comprehensive Review of Recent Practices and Trends*. Cyber Security and Applications.

Mensah, N. K., Adzakpah, G., Kissi, J., Taylor-Abdulai, H., Johnson, S. B., Agbeshie, P. A., Opoku, C., Abakah, J., Osei, E., Agyekum, A. Y., Boadu, R. O. (2024, January 1). *Health Professionals' Ethical, Security, and Patient Safety Concerns Using Digital Health Technologies: A Mixed Method Research Study*. Health Services Insights; SAGE Publishing. https://doi.org/10.1177/11786329241303379.

Ribeiro, A., Casserly, S., Al-Oteiby, A. M., Scharra, M., Workentin, B., Koehler, A., Ribeiro, Sheila, A. (2025, February 23). *Prioritizing Organizational Cybersecurity Governance, Boosting Operational Resilience Across OT, ICS Environments*. Industrial Cyber. https://industrialcyber.co/features/prioritizing-organizational-cybersecurity-governance-boosting-operational-resilience-across-ot-ics-environments/.

Senior. (2025, June 27). *What are Regulatory Requirements in Healthcare?* Security Compass. https://www.securitycompass.com/blog/regulatory-requirements-in-healthcare/.

Index

d

f

i

n

o

p

q

r

Printed and bound by CPI Group (UK) Ltd, Croydon, CR0 4YY
16/08/2026
14942718-0003